Aerogels I
Preparation, Properties and Applications

Edited by

Inamuddin[1], Tauseef Ahmad Rangreez[2], Mohd Imran Ahamed[3] and Rajender Boddula[4]

[1]Department of Applied Chemistry, Faculty of Engineering and Technology, Aligarh Muslim University, Aligarh-202 002, India

[2]Department of Chemistry, National Institute of Technology, Srinagar, Jammu and Kashmir 190006, India

[3]Department of Chemistry, Faculty of Science, Aligarh Muslim University, Aligarh-202 002, India

[4]CAS Key Laboratory of Nanosystem and Hierarchical Fabrication, National Center for Nanoscience and Technology, Beijing 100190, PR China

Published as part of the book series
Materials Research Foundations
Volume 84 (2020)
ISSN 2471-8890 (Print)
ISSN 2471-8904 (Online)

Print ISBN 978-1-64490-098-7
eBook ISBN 978-1-64490-099-4

Distributed worldwide by

Materials Research Forum LLC
105 Springdale Lane
Millersville, PA 17551
USA
https://www.mrforum.com

Manufactured in the United States of America
10 9 8 7 6 5 4 3 2 1

Table of Contents

Preface

Aerogels are extremely porous, the lightest solid materials with huge surface areas and nanoscale pore sizes mainly in the micro- and mesoporous regimes. They are made of polymers with a solvent to form a gel, and then removing the liquid from the wet gels and replacing it with air without collapsing the 3D network structure. These unique characteristics make them promising materials that can be used in many applications including energy storage, thermal storage, catalysis, water splitting, environmental remediation, and among others. Therefore, to gather viewpoints and opportunities of aerogels are needed in the materials research community. This book exclusively focuses on the properties and applications of aerogels. Chapters examine the synthetic methodologies, characterization tools, various types of organic, inorganic and hybrid aerogels, composites, and its applications in energy and environmental science. The chapters are written by leading experts in this field. The book targets the need of scientists, faculty, and postgraduate students working with aerogels and their potential applications and should be of interest to readers working in the areas of chemistry, physics, polymer science, nanotechnology, and material science. This book includes the following eleven chapters.

Chapter 1 demonstrates the importance of nanocellulose aerogels. Several production technologies are detailed, as well as the properties of the different types of aerogels. This chapter shows the use of these materials as absorbents, gas filters, packaging materials, energy storage, electrical devices, thermal insulations, and fire-retardants, as well as their pharmaceutical and biomedical applications.

Chapter 2 overviews various types of porous aerogels including silicate and non-silicate aerogels, organic/natural, composite/hybrid aerogels, polymeric, carbon-based (mostly CNTs and graphene-based) and biogels. The unique properties displayed by these aerogels together with their use in environmental, biomedical, catalytic, and other advanced applications have been discussed.

Chapter 3 describes the principles for creating hybrid silica-based aerogel and an overview of the strategies and recent reports on the literature. The objective is to discuss the advantages and limitations of the techniques that use polymers, biomolecules, and graphene, in addition to silica, to create hybrid aerogels.

Chapter 4 presents a comprehensive study of the synthesis methodology, physico-chemical properties, and the application of the silica aerogel. It also gives an overview of the sol-gel synthesis for silica aerogels including an introduction to hydrophobic silica aerogels. The applications of silica aerogels are discussed in full detail.

Chapter 5 describes carbon aerogel, their classification in detail, their different types of properties. The classification is made based on their flexibility, dimension, doping with the different metal atom. Furthermore, their application in a diversified field has been well described. Overall, this chapter is a general description of carbon aerogel.

Chapter 6 discusses the different types of magnetic aerogels such as cellulose, magnetic graphene, carbon, magnetic silica, and magnetic pectin-based aerogels along with their applications.

Chapter 7 discusses the various properties of aerogels. They are used as sound, heat, and electric insulators. Depending on the solid backbone, they can be used in space materials manufacturing or marine industry, even they can present biocompatibility and can be used for biomedical applications and as drug delivery systems.

Chapter 8 focuses on customizing techniques like modulating the pore structure, modifying the surface, coating of the surface, and post-treatment. Additionally, the commercial uses of aerogels and their products, an unbroken view of industrial aerogel suppliers are discussed. The chapter also discusses plausible substitute sources for raw materials as well as precursors.

Chapter 9 discusses future perspectives for aerogel applications. Several applications such as functional foods, thickeners, stabilizers, and scaffolding in tissue repair are reported. Additionally, polymeric aerogels as impact-absorbing materials, catalyst supports, and aerospace components are presented. Lastly, the perspectives for carbon aerogel and inorganic aerogels are discussed.

Chapter 10 discusses the potential applications of aerogels, a list of patents in the different areas of aerogels such as medical, organic synthesis, and diverse applications of aerogels. This chapter also focuses on the recently granted patents on the aerogels.

Chapter 11 discusses the synthesis of aerogel by the sol-gel process. The chapter focuses on current development, classification, and different properties and preparation methods of distinct types of aerogels. The application of aerogels in the future in the various fields is also reviewed.

Key features:
1. Gives a detailed account of aerogels properties and applications.
2. Covering the basic concepts, methodologies, properties, and problems associated with aerogels.
3. Fits the background of various science and engineering disciplines.
4. Provides cutting-edge advances for aerogel technology.

Inamuddin[1], Tauseef Ahmad Rangreez[2], Mohd Imran Ahamed[3] and Rajender Boddula[4]

[1]Department of Applied Chemistry, Faculty of Engineering and Technology, Aligarh Muslim University, Aligarh-202 002, India

[2]Department of Chemistry, National Institute of Technology, Srinagar, Jammu and Kashmir 190006, India

[3]Department of Chemistry, Faculty of Science, Aligarh Muslim University, Aligarh-202 002, India

[4]CAS Key Laboratory of Nanosystem and Hierarchical Fabrication, National Center for Nanoscience and Technology, Beijing 100190, PR China

Aerogels I: Preparation, Properties and Applications
Materials Research Foundations **84** (2020) 1-33

Materials Research Forum LLC
https://doi.org/10.21741/9781644900994-1

Chapter 1

Nanocellulose Aerogels

Elaine Crisitna Lengowski[1]*, Pedro Henrique Gonzalez de Cademartori[2], André Luiz Missio[3], Rodrigo Coldebella[4], Eraldo Antonio Bonfatti Júnior[5]

[1] Faculty of Forestry Engineering, Graduate Program in Forestry and Environmental Sciences (PPGCFA), Federal University of Mato Grosso (UFMT), Fernando Corrêa da Costa St, 2367 - Boa Esperança, Cuiabá, MT 78068-600, Brazil

[2] Department of Forest Engineering and Technology (DETF), Graduate Program in Forest Engineering (PPGEF) / Graduate Program in Engineering and Materials Science (PIPE),Federal University of Paraná (UFPR), Pref. LothárioMeissner Av., 632, Jardim Botânico, Curitiba, PR 80.210-170, Brazil

[3] Graduate Program in Materials Science and Engineering, Technology Development Center, Federal University of Pelotas, Pelotas, Brazil

[4] Graduate Program in Forest Engineering, Federal University of Santa Maria (UFSM), Roraima Av., 1000, University City, Santa Maria, RS 97105-900, Brazil

[5] Department of Forest Engineering and Technology (DETF), Graduate Program in Forest Engineering (PPGEF), Federal University of Paraná (UFPR), Pref. LothárioMeissner Av., 632, Jardim Botânico, Curitiba, PR 80.210-170, Brazil

Email: elainelengowski@gmail.com

Abstract

Nanocellulose is a biodegradable material, which comes from natural sources and has properties such as good chemical compatibility, low density, high mechanical properties and low thermal conductivity, characteristics that make it a potential material for several applications. Among the potential forms and applications is the production of nanocellulose aerogel. This chapter summarizes the main ways of obtaining nanocellulose aerogel and its properties. The main applications, like materials absorbents, carbon porous materials, gas filters and membranes, packaging materials, biomedical and pharmaceutical, electrical devices, energy storage systems, thermal insulation and fire-retardant materials are also discussed.

Keywords

Cellulose, Nanotechnology, Biorefinery, Natural Composites, Advanced Materials

Contents

1. Introduction

The use of materials from natural sources to replace synthetic polymers is the focus of researches [1]. As the largest component of plant biomass, cellulose is the most abundant organic polymer on the planet, its annual production is estimated at 7.5×10^{10} tons per year [2,3], as a result, cellulose gains increasing and considerable attention in the development of new materials. Each cellulose molecule is composed of microfibrils, which are 1-2 micrometers long and 3-4 nm in diameter [2]. Microfibril is composed of highly ordered regions (crystalline) and disordered regions, (amorphous) [4].

The processing of cellulose by different routes and methods results in a class of materials called nanocellulose. The acid and enzymatic hydrolysismethods preserve the crystalline region and obtain the crystalline nanocellulose (CNC) whereas the mechanical methods or the combination of chemical or enzymatic methods with the mechanical methods lead to the formation of the nanofibrillated nanocellulose (CNF) [5,6]. Due to its small size, nanocellulose has interesting characteristics, such as low density, high strength, and good reactivity to modify the surface chemistry and flexibility with chemical inertness [7,8].

Nanocellulose has been intensively researched in the last decades for several applications, such as food packaging [9,10], food industry [11–14], encapsulation and controlled drug release [15,16], reinforcement additive [5] and currently research is growing on nanogellulose aerogels [7,17].

The nanocellulose aerogel is a highly porous solid of ultra-low density with nanometric pore sizes formed by replacement of liquid in a gel with gas [18]. Aerogels are produced by drying a suspension in gel, usually hydrogel, by replacing the water with an organic solvent, forming an organogel. This organogel is dried by critical drying conditions or freeze drying process, preserving the three-dimensional porous structure. The density of the aerogels is between 0.001-0.2 g cm^{-3}, pores of dimensions of 2-50 nm and a porosity above 90% [17].

Due to the great potential of this material for the development of new products, this chapter contemplates the main forms of production, characteristics and research already developed using the nanocellulose aerogel.

2.   Production processes of nanocellulose aerogels

Three steps are mandatory to prepare both cellulose nanocrystal (CNC) and/or cellulose nanofiber (CNF) aerogels: sol-gel transition (gelation), aging and drying [19]. These CNC and CNF-based aerogels can be produced by combining organic and inorganic materials in their structure, such as alginate [20,21], collagen [22,23], polyvinyl alcohol [24,25], surfactants [26,27], nylon [28], chitosan [29], TiO$_2$, chitosan, polyvinyl acetate (PVA), carbon nanotubes (CNT) and silver (Ag) nanoparticles [30]. In the production of nanocellulose aerogels, one of the main challenges is to preserve the porous structure after the solvent (organic solvents or water) removal, since this can result in significant damages of cracking, collapse and warping of the material [7].

The first step of nanocellulose aerogels production; sol-gel transition, refers to the formation of a gel with an interconnected polymer network. Here, a high-stable gel is formed due to the large number of hydroxyl (OH) groups present in the cellulose chemical structure and their physical crosslinking; attributed to the hydrogen bonds [19], considering water as media. However, this sol-gel synthesis can be performed with organic solvents, i.e., without water, for example: replacing hydroxyl groups of cellulose by acetyl groups, and increasing solubility in acetone [31]. The shape of the final aerogel depends on the operational conditions adopted during the gel formation [32]. The second step; aging, represents the time required for crosslinking and curing to form a high-stable gel. The drying step is very important and responsible to reduce the surface tension into the pores during the formation of the aerogel network [19]. The most common drying steps to prepare nanocellulose aerogels are freeze drying and critical point drying (or supercritical drying) [7]. Consequently, the aerogel properties will vary according to the drying method. Supercritical drying will preserve the three-dimensional network and create very small pores, while freeze drying will result in an aerogel with macropores and low specific surface area [33–35]. These drying steps can contain additional steps of solvent exchange (diffusion controlled process) before and if required by drying, or post-processing chemical modifications like chemical vapor deposition (CVD), silanization, calcination, pyrolysis and organic/inorganic coatings [30,32].

Freeze casting and vacuum filtration are other drying techniques applied to prepare nanocellulose aerogels. Both techniques are relatively simple, but they present limitations related to the properties needed for an aerogel. For example: freeze casting is an interesting way to align the nanocellulose fibrils in the direction of ice crystal growth, but their aggregation decreases substantially the application of CNF aerogel [36]. This step is complicated because the particles are subject to elongational flow and the ice crystals grow at different speeds in length in relation to thickness [36]. Vacuum filtration followed by solvent exchange and drying at room temperature result in an aerogel with high density and undesirable surface area [37]. The two most common and applied drying techniques – freeze and supercritical drying are better described below.

Freeze drying process: Freeze drying is a simple method to prepare aerogels based on the freezing step followed by the sublimation step of material suspension [35], where nanocellulose water suspensions have received a great deal of attention as raw material due to their unique properties. Usually, the operational conditions of freeze-drying process are based on the use of vacuum (pressure <100 mbar) and temperature between -70 and -20 °C [32], considering water as the fluid inside the network pores of the gel [38]. Lyophilization process is very delicate, since it can result in undesirable characteristics in the sample or causes CNF agglomeration due to changes in the water removal process. Drying hydrogels is a challenge, as it is almost impossible to avoid partial collapse of the structure during solvent removal [36,39–41]. Freezing step refers to a controlled decrease of temperature in a specific environment, such as refrigerators and liquid nitrogen. The rate of freezing influences the structure of the gel. Rapid freezing can preserve the original structure, but slow freezing can segregate the solvent and dispersed phase [7,42]. This step can result in growing ice crystals, damaging the morphology of the material obtained in the sol-gel transition (gelation) [43]. Growth of ice crystals can be avoided by using spray-freeze-drying [44] or using tert-butanol instead of water as solvent [45]. On the other hand, sublimation step principle avoids the liquid-vapor interface, making possible to create nanocellulose aerogels with desirable properties like high surface area, low density and high porosity. This step is crucial for adequate processing of aerogels because it allows the replacement of frozen liquid solvent for air into the pores [46]. However, many important characteristics of nanocellulose aerogels may affect the efficiency of sublimation, such as quantity of cellulose, temperature of processing, dimension and physical aspect [47]. Thus, the open porous materials created by freeze-drying process usually have pore size around several micrometers, which are a replica of the crystals formed in the solvent extraction by sublimation during the lyophilization [48]. This can lead to referring these materials in the literature as cryogels instead aerogels [38]. As previously mentioned, freeze drying conditions drive the properties of nanocellulose aerogels, and they have been explored under different ways (Table 1) [33,35,44,47,49–69]. Also, the most critical limitations of freeze drying is the high energy consumption, long time required for aerogels' preparation (tens of hours), batch processes and the possible formation of microcrystals [32,43].

Table 1   Summary of recent strategies adopted to prepare nanocellulose aerogels (Table 1) [33,35,44,47,49–69].

CNF aerogels						
Material	Filler	Processing			Post-modification	Ref.
		Pre-modification	Freezing	Drying		
Nanofibrillated cellulose from *Pinus roxburghii*	-	Mechanical stirring / Ultrasonication / Centrifugation / Precooling	Liquid nitrogen	Vacuum freeze dryer	-	[64]
TEMPO-CNF from spruce wood pulp		Osmotic concentration using Dextran	Pouring + Deep-freezer Spraying + Deep-freezer	Freeze-dryer	-	[44]
MTMS-CNF from eucalypt pulp	-	-	Liquid nitrogen	Vacuum freeze dryer	-	[47]
TEMPO-CNF from poplar wood and CNF from cotton fibers	-	-	Refrigerator	Freeze-dryer	-	[35]
TEMPO-CNF	-	Osmotic concentration using Dextran	Deep-freezer	Freeze-dryer	-	[33]
BTCA-CNF	-	-	Freeze casting	Freeze-dryer + vacuum oven	Impregnation with cellulose acetate or acetylated CNC	[65]

MTMS-CNF	-	-	Freezer	Lyophilizer	Silinization by CVD	[66]
HDTMS-CNF		-	Liquid nitrogen	Freeze dryer	Chemical reaction with HDTMS	[67]
TEMPO-CNF from softwood pulp	-	Sonication / Magnetic stirring	Freezer	Freeze dryer	-	[68]
CNF from *Eucalyptus* wood pulp	-	Mechanical stirring	-	Freeze dryer	-	[69]
TOCN/GO softwood bleached kraft pulp	-	Sonication / Magnetic stirring / solvent exchange	Refrigerator	Freeze dryer	-	[49]
TEMPO-CNF/CD-x	-	Mechanical stirring	-	Freeze dryer	Condensation reaction between CNF-based skeleton and amino groups of CD.	[50]
CNF-SB from hardwood kraft pulp	-	Mechanical stirring / Heating	Freezer	Freeze dryer	-	[51]
PEDOT/PSS/ CNF	-	Mechanical stirring	Freezer	Freeze dryer	EG heated at 150°C under vacuum for 30 min	[52]
PMSQ-CNF from Radiata pine wood powder	-	Mechanical stirring / Heating / Solvent exchange	-	CO_2 supercritical drying		[53]
Hydrophobic CNF	-	Mechanical stirring	Liquid nitrogen	Freeze dryer	-	[54]
Nematic*i*-CNF	-	Solvent exchange	-	Supercritical fluid extraction	PMMA coating	[55]
PVA/CNF Moso bamboo	-	Mechanical stirring / Solvent exchange	Freezer	Freeze dryer	Silylation	[56]

CNF-PANI	-	Mechanical stirring / Supramolecular assembly	Liquid nitrogen / Freezer	Freeze dryer	-	[57]
CNC aerogels						
CNC CNC-POEGMA	-	-	Freezer	Lyophilizer	-	[58]
CNC from wood pulp	-	-	Freezer	Lyophilizer	Polyamide-epichlorohydrin (Kymene)	[59]
CHO-CNC NHNH$_2$-CNC	PPy-NF PPy-CNT MnO2-NP		Freezer	Freeze-dryer	-	[127]
TEMPO-CNC CHO–CNC NH$_2$NH–CNC	-	Solvent exchange	Freezer	Critical point dryer Pressurized gas expansion vessel	-	[61]
CNF from *Eucalyptus* wood pulp	-	Mechanical stirring	-	Freeze dryer	-	[69]
MCC-PDA	-	Mechanical stirring	Liquid nitrogen	Freeze dryer	-	[62]
CNC POEGMA	-	Pressure-aided freeze casting	Freezer	Lyophilizer	-	[63]

CNF = cellulose nanofibrils; CNC = cellulose nanocrystals; TEMPO = 2,2,6,6-Tetramethylpiperidin-1-yl)oxyl; POEGMA = poly(oligo(ethylene glycol) methyl ether methacrylate; BTCA = 1,2,3,4-butanetetracarboxylic acid; MTMS = Trimethoxymethylsilane; HDTMS = hexadecyltrimethoxylan;CVD = chemical vapor deposition; CD = carbon dots; TOCN/GO = (TEMPO)-oxidized cellulose nanofibril/graphene oxide; SB = sodium bicarbonate; PEDOT/PSS = polysilane and poly(3,4-ethylene dioxythiophene)/poly(styrene sulfonate); EG =ethylene glycol; PMSQ = polymethylsilsesquioxane; SiO2 = silicon dioxide; PMMA = Poly(methyl methacrylate). Nematic *i*-CNF = TEMPO-Oxidized individualized cellulose nanofibrils; PANI = polyaniline; MCC = microcrystalline cellulose; pda = polydopamine; poegma = poly(oligoethylene glycol methacrylate).

Supercritical drying process: This method uses supercritical fluids – usually supercritical carbon dioxide (scCO$_2$) - as an alternative to traditional drying process, in order to maintain the extraordinary properties like high porosity and texture of wet gel in a dry form [70], as well as creating small pores in the aerogel structure [43] and no cracks and capillary stresses [38]. Supercritical drying using carbon dioxide (CO$_2$) can be applied to prepare different types of aerogels, and it is a safe and environmentally friendly process [32]. The CO$_2$ is widely used as supercritical fluid due to many positive aspects, especially its lower critical point (temperature of 31.3 °C and pressure of 72.9 atm), non-inflammability and non-toxicity [7,38]. This drying process is based on the extraction of the organic solvent in a single-phase mixing step using compressed CO_2 with no liquid–gas interfaces. This extraction allows the complete substitution of the organic solvent for CO_2 into the gel structure, and subsequent release of CO_2 by slow depressurization [71]. The rate of depressurization can affect the aerogel structure dried with CO_2, since a high rate induces deformation because of the rapid expansion of CO_2. This explains the preference for slow depressurization to prepare nanocellulose aerogels [7,72]. The efficiency of supercritical drying of wet gels containing cellulose-based materials can be improved by immersion in intermediate solvents miscible in CO_2, such as ethanol and acetone, to avoid the liquid-vapor interface and, consequently, damages on the aerogel structure [43,73]. This solvent exchange is necessary because CO_2 are immiscible in water [7], which is the main media used to prepare CNF and CNC suspensions.

3.   Properties of nanocellulose aerogels

The current interest in nanocellulose aerogels is based on their great properties, especially biodegradability, renewability, high porosity, low density and good mechanical strength [74]. Microscopically, aerogels are composed of fine networks of grouped nanoparticles. These materials generally have unique properties, including high strength, low density and high surface area, when the volume of solids is related to the volume of foam [75]. However, most aerogels are mechanically brittle, and their physical properties are correlated with their density [34] which, in turn, is related to the shrinkage that the foam undergoes during the manufacturing process [76]. Trying to eliminate these undesirable characteristics, or even to promote some specific characteristic, such as mechanical reinforcement, foams have often been formed with the union of two or more polymers [20]. The fact that the foams have a well-structured morphology and porosity can be an advantage in several applications, for example, in modifying their surface to repel water [77] and in biodegradable damping foams [78], replacing expanded polystyrene.

In general, aerogels have porous solids with up to 99.8% air and density lower than 4 mg/cm^3 [79]. The use of nanocellulose as matrix or filler to prepare aerogels improves many characteristics of the final products, mainly thermal stability [34,80], surface area [53,81], density [53,81,82], porosity [35,81,83], surface reactivity [82,83] and mechanical properties [35,37,79]. Other important physical aspects desired for aerogels are related to thermal, electrical and acoustic properties. Nanocellulose aerogels usually have low thermal conductivity, considerable ability to absorb sound and low dielectric loss [7]. For example: Chen et al. [35] found low thermal conductivity (<0.016 $Wm^{-1}K^{-1}$) and high

Materials Research Forum LLC
https://doi.org/10.21741/9781644900994-1

sound absorption ability at high frequencies (57.1 and 54.1 % at 4000 Hz) for unmodified and modified CNF aerogels. This low thermal conductivity can be explained by low density (close to 0.005 gcm^{-3}) and high porosity of the CNF aerogels. The authors also suggested this low thermal conductivity as a good indicator for thermal insulation applications of CNF aerogels [35]. Table 2 [44,64,66,67,73,81,84–89] summarizes some properties found in investigations with nanocellulose aerogels, focusing on physical aspects of density, porosity and surface area.

Even both CNF and CNC are from renewable materials, like wood and natural fibers, the aerogels prepared with each type of nanocellulose have specific characteristics that drive their applications (Table 2[44,64,66,67,73,81,84–89]). For example: CNF network has high chain entanglement which results in a highly stable aerogel. On the contrary, CNC aerogel has high porosity and surface area, but are more fragile than traditional aerogels, like silica-based aerogels [30].

Variation of stiffness, morphology, wettability, roughness, porosity, surface area and density can be adjusted by changing the loading levels of nanocellulose (CNF and CNC) before the drying step [45,77,81]. The difference in the cellular structure of foams influences their mechanical behavior [90]. Porosity, pore size and internal structure are the main factors that directly affect the performance of these materials[23]. For example: Rafieian et al. [67] found a significant decrease of porosity(99.5 – 99.1%) with increase the CNF loading (1.2 - 0.6%) for unmodified aerogels. On the contrary, drying method does not influence substantially the density and porosity of bio aerogels, since both most common drying techniques achieve low densities and high porosities [44].

Nanocellulose aerogels naturally have the capacity to absorb liquids in large quantities, especially because of their high surface area and porosity [91]. Density also influences the water absorption capacity of CNF aerogels, especially for values lower than 4 kgm^{-3}. The calculated absorbency increased with density of aerogels. The values are 25, 46, 83, 91, 91 and 94% for aerogels with density of 1.7, 2.7, 4.0, 5.2, 6.3 and 8.1 kg m^{-3}, respectively [88]. The authors explained this behavior; as the aerogels with higher density have smaller pores, thicker pore walls and, consequently, more capacity to hold absorbed water into their structure.

Table 2 *Summary of physical properties found in recent investigations with nanocellulose aerogels [44,64,66,67,73,81,84–89].*

Material	Processing	Surface area (m^2g^{-1})	Density (kgm^{-3})	Porosity (%)	Postprocessing	Reference
CNF aerogels						
CNF	Freeze drying	15.2 - 92.3	8.1 - 20.3	99.4 - 98.6		[64]
CNF	Freeze drying Supercritical drying	20 - 350	-	97.3 - 99.7		[73]
TEMPO-CNF	Freeze drying	80 - 100	12 – 33	98 – 99		[44]
Carboxy-methylated CNF	Freeze drying	11 – 42	4 – 14	99.1 – 99.8	Silanization by CVD	[75]
CNF	Vacuum drying	-	20	98.6	Fluorination by CVD	[87]
CNF	Freeze drying	-	1.7 – 8.3	99.5 – 99.9	Silanization by CVD	[88]
CNF	Freeze drying	3.8	46	97.08	Silanization by CVD	[66]
CNF	Freeze drying	261 - 299	8 – 17.5	98.8 – 99.5	Chemical reaction with HDTMS	[67]
CNC aerogels						
COOH-CNC	Supercritical drying	43 - 429	5.7 - 220	86.1 – 99.6	Atomic layer deposition of Al_2O_3	[89]
CNC	Freeze drying	216 - 605	78 - 155	91 - 95	-	[81]
Sulfated CNC Phosphate CNC	Supercritical drying	130 - 190	10 – 19	98.8 – 99.3	-	[84]
CNC	-	91.47 - 93.89	-	-	-	[85]

CNF = cellulose nanofibrils; CNC = cellulose nanocrystals; TEMPO = 2,2,6,6-Tetramethylpiperidin-1-yl)oxyl; cvd = chemical vapor deposition; MTMS = methytrimethoxysilane; HDTMS = hexadecyltrimethoxylan.

Regarding surface area of CNC aerogels, low values are attributed to CNC aggregation, while high surface area is influenced by non-smooth surface and monodispersion of CNC [81,92]. Heath et al. [81] affirmed that they found some of the highest surface area of CNC aerogels reported in the literature, reaching values close to 605 $m^2 g^{-1}$, which is true if comparing values of nanocellulose aerogels obtained in some previous studies like 91-93 $m^2 g^{-1}$ [85], 130-190 $m^2 g^{-1}$ [84], 3.8 $m^2 g^{-1}$ [66] and 15-92 $m^2 g^{-1}$ [64]. Low surface areas is not desired for aerogels, and usually it can be related to the presence of macropores (large size pores) in their structure, probably because the use of lyophilization and slow freezing to prepare the aerogels [66]. The authors explained that slow freezing results in macropores due to the fiber movement, which does not occur in rapid freezing. The choice of drying method is significant and affects the surface area of nanocellulose aerogels, since the use of critical point method result in higher surface areas of aerogels compared to the freeze drying method [64].

Density of nanocellulose aerogels is typically low, but the range of density vary substantially as a function of raw materials and operational conditions. Previous studies found extraordinary low densities between 1 to 10 kg m^{-3} [44,75,88], which is equal or lower than aerogels manufactured with extraordinary materials, such as graphene (2 kg m^{-3}) [93] and carbon nanotubes (CNT) (4 kg m^{-3}) [94]. Heath et al. [81] found CNC-aerogels with density between 78 to 155 kg m^{-3}, where the porosity decreased in line with increasing density. The authors affirmed that no shrinkage of hydrogel during the solvent exchange and drying steps allow to predict the theoretical density and porosity of the aerogel. Corroborating this statement, Shoseyov et al. [95] mentioned that the solvent removal forms a continuous and layered porous structure, resulting in a spongy structure, and its compactness depends on the initial concentration of solids in the solution. Also, Heath et al. [81] suggested the absence of shrinkage due to the high modulus of CNC, which avoids the collapse of sections between hydrogen bond. This behavior occurs for unmodified CNF aerogels, as found by Rafieian et al. [67], in which bulk densities (11-17.5 kg m^{-3}) increased with increase the CNF loadings (0.6 – 1.2%).

Comparison of mechanical properties obtained for nanocellulose aerogels is difficult because of the variation of materials and test methods [30]. However, some aspects can be highlighted. Young's modulus increases with increasing density of CNF aerogels; ultralow density and high porosity negatively affect maximum compressive stress and Young's modulus [67]; chemical crosslinking of CNC results in great differences in cyclic mechanical tests (wet and dry conditions), as observed by Osorio et al. [84] for S-CNC (sulfated CNC) aerogels and P-CNC (phosphate CNC) aerogels applied as bone tissue scaffolds. S-CNC aerogels presented higher compressive strength at 90% than P-CNC aerogels. On contrary, S-CNC and P-CNC aerogels presented the same flexibility and shape recovery. This flexibility can facilitate the bone regeneration. Besides higher number of crosslinking groups, higher mechanical properties of S-CNC aerogels positively influenced the lesser degradation rates and higher long-term stability of scaffold *in vivo* [84]. Likewise, this efficient shape recovery after compression higher than 99% and immersion in water also was observed by Chen et al. [35] for TEMPO-CNF aerogels. The authors explained this behavior because of the robust interconnected network and stable porous structure observed in the nanocellulose aerogels.

Chemical modification of nanocellulose is an interesting way to facilitate some steps during the preparation of aerogels, and to produce aerogels with outstanding properties for added-value applications, and to avoid problems during applications due to their possible redispersion in water and tendency to disassemble [30].

Surface reactivity is an important aspect explored in the preparation of nanocellulose aerogels. The increase of surface hydrophobicity of nanocellulose aerogels is easily achieved by silanization, as described by Zanini et al. [47], Oliveira et al. [66], Nguyen et al. [96] and Zhang et al. [91]. A higher contact angle and its time-dependent stability, i.e., a more hydrophobic and stable surface, can be positive for applications related to oil/water separation processes [47]. According to the same authors, the silanized CNF aerogel-oil interaction occurs mainly due to van der Waals and intramolecular interaction forces, in which the chemically modified aerogel surface requires lesser energy to absorb the oil.

The potential of CNF aerogels as an amphiphilic material was investigated by Jiang and Hsieh [88]. They observed a natural amphiphilic characteristic through the absorption of 210 and 375 g g^{-1} of water and chloroform, respectively. But the post processing of CNF aerogels with CVD by silanization using triethoxyl(octyl) silane resulted in a more hydrophobic and oleophilic aerogel, with 139–356 times absorption of non-polar hydrocarbons, polar aprotic solvents and oils [88]. Likewise, fluorination of nanocellulose aerogels is another interesting technique to achieve super hydrophobicity and superoleophobicity with notable applications as gas permeable, buoyant and dirt-repellent coatings for sensors [87]. Superoleophobicity of CNF aerogels also was observed by Aulin et al. [77]. In this investigation, they related the superoleophobicity to the synergistic effect of low surface energy of PFOTS coating, surface roughness and re-entrant topography of CNF aerogels [77]. Furthermore, chemical modification can improve mechanical properties of nanocellulose aerogels, as previously described by Jiang et al. [88] for salinized-CNF aerogels and as suggested by the findings of Huang et al. in 2013 [97].

4.   Applications of nanocellulose aerogels

As a new 3D material with high porosity and ultralow density [98], nanocellulose aerogels provides many applications like material absorbents, gas filters and membranes, packaging materials, energy storage systems and electrical devices, thermal insulation and materials with fire-retardant properties, pharmaceutical and biomedical applications. This section presents the applications of this material in several industrial fields.

4.1   Materials absorbents

Aerogels are materials made of the liquid part replacement of a gel with a gas [99,100]. In general aerogels are known for three specific characteristics: high porosity, low density and high specific surface area, these properties are desirable in materials for absorbent uses [101]. When made of nanocellulose, in addition to the aforementioned properties, the material will have high strength, liquid-crystalline properties, biodegradability, environmental friendly, general biocompatibility and ease to surface modification [3,4,100,102–104].

Absorbent materials are routinely used for separating oil/water emulsification and in oil spills [105]. Oil/water emulsification is a very common effluent in the food and oil industry and oil spills are very harmful environmental disasters, not only for marine life but also in the economic production involved in this environment [106]. The use of porous and absorbent material is the most effective and inexpensive way to separate the oil/water mixture [100], through physical adsorption the separation of these solutions is facilitated, which increases the interest of researchers and companies in the development of absorbent materials [107,108], many of them having nanocellulose as chemical base.

Mostly the nanocellulose aerogels research is based in CNF, the high length (many micrometers long) and the presence of crystalline and amorphous regions make the nanofibers very flexible and easy interlacing [5,45,87,99], and are indicated for oil absorbents, filters and antibacterial substrates [99].

Through lyophilization followed by functionalization with titanium dioxide from CNF hydrogels, Korhonen et al. [109] produced a hydrophobic and oleophilic material. With porosity greater than 98% and density between 20 to 30 mg m^{-3}, this CNF-aerogel, highly water-repellent, was able to absorb and retain oil. The chemical vapor deposition (CVD) is the most widely used hydrophobization technique to produce CNF-aerogels. With this method Cervin et al. [75] produced several CNF-aerogels, with a range of porosity from 99.1 to 99.8% and a ultra-light density of 4–14 mg cm^{-3}. These highly hydrophobic aerogels showed a 150° water contact angle, selective to n-hexadecane were able to absorb up to 45 times their own weights in oil.

In addition to the use of CNF to produce absorbent aerogels, there are successful experiences with bacterial nanocellulose and cellulose nanocrystal, as in the case of the research developed by Zhang et al. [91] that produced BC-aerogels with low density (6.77 mg cm^{-3}), high specific surface area (169.1 m^2 g^{-1}) and high porosity (99.6%), capable of absorbing a wide range of oils and organic solvents. In the case of CNC-aerogel Yang and Cranston [99] prepared CNC-aerogels based on hydrazone crosslinking of hydrazide and aldehyde-functionalized CNCs and achieved ultralight density (5.6 mg cm^{-3}) and high porosity (99.6%). With high mechanical resistance, these CNC-aerogels were able to quickly absorb water, dimethyl sulfoxide (DMSO), ethanol and dodecane.

4.2 Gas filters and membranes

For the cleaning of surgical environments, medical and pharmaceutical industry, food industry and research laboratories, two types of filters have stood out for their high efficiency in air filtration: High-efficiency particulate air (HEPA) and ultralow penetration air (ULPA). Made up of a mesh of fibers arranged at random, HEPA filters are generally used in nanoparticle containment rooms [110], removing at least 99.97% of suspended particles of 0.3 microns in diameter, the minimum resistance of the filter to the air flow, or pressure drop, is normally specified at around 300 Pa at its nominal flow [111]. An ULPA filter can remove from the air at least 99.999% of any airborne particles with a size of 100 nanometres (0.1 μm) or larger [111]. Cellulose-based aerogels find exceptional applications in air filters [43], such as the case of pioneering research developed by Nemoto et al. [112], who showed that the reinforcement of CNF in HEPA or ULPA filters improving the air filter quality.

Due to the rapid increase in industrial carbon dioxide emissions in recent years [113], the development of effective and low-cost gas separation technologies has gained importance [114]. In view of this Gebald et al. [115] developed an absorbent material made of amine-based CNF, capable of absorbing up to 1.4 mmol g^{-1} of CO_2, and, for the same purpose, Sehaqui et al. [116] combined oxidized nanofibrillated cellulose (NFC) and a high molar mass polyethylenimine (PEI), and fabricated a direct air capture membrane with 2.22 mmol g^{-1} CO_2 capture capacity.

4.3 Packaging materials

The ability of a material to absorb energy and plastically deform without fracturing is called toughness [117], in the international system of units (SI) toughness is measured in units of joule per cubic meter (J m^{-3}) and it is calculated using area underneath the stress–strain curve [117,118]. Significant toughness is required for the development of packaging [119], as in the case of materials which has good capacity to absorb impact, like expanded polystyrene (EPS) [120], polypropylene (PP) [121], cellulose fibers [122] and wood [123].

In the research carried out by Donius et al. [124] an anisotropic nanocellulose montmorillonite aerogels prepared by freeze casting able to absorb a lot of energy per unit volume (46.2 kJ m^{-3}), this is an indicator for potential application in packaging. Cellulose-derived materials have the advantages of being renewable and biodegradable [5], compared to other petroleum-based materials such as some difficult-to-recycle polymers used in packaging.

In addition to good mechanical resistance, the application of cellulosic biomaterials in the production of food packaging is favored due to biodegradability, non-toxicity to humans and the possibility of adding preservative or bioactive substances to produce active food packaging like antimicrobial, antioxidant and aromatic food packaging [125].

4.4   Energy storage systems and electrical devices

The flexibility, low density and high strength and the flexibility of nanocellulose-based aerogels are important their use in energy storage systems and electrical devices [7], but for these proposes good electrical conductivity is also required [126]. The challenge for this application is to increase the electrical conductivity or capacitance of nanocellulose aerogels, either by chemical modification or by adding materials with good electrical conductivity, without impairing the aforementioned properties and, mainly, maintaining the continuous network of connections within the aerogels.

Capacitance or electrical capacity is the scalar quantity that measures the energy storage capacity in electrical equipment and devices, its unit is given in farad, represented by the letter "F"[126], the incorporation of polypyrrole nanofibers (PPyNF), polypyrrole-coated carbon nanotubes (PPy-CNT) and manganese dioxide nanoparticles (MnO_2-NP) gave to CNC-aerogel an excellent capacitance retention at high charge–discharge rates for supercapacitor devices with low internal resistance[127]. In an experiment involving NFC-based aerogels and rapid layer-by-layer assembly method (LbL), Hamedi et al. [128] developed a very high charge-storage capacity material, with calculated capacitance of 419 ± 17 F g^{-1}.

As examples of improvement in the electrical conductivity of nanocellulose aerogels we have the work of Pääkköö et al. [129] that by incorporating highly porous polyaniline (PANI) in CNF foams created a material with conductivity of $7x10^{-3}$ S cm^{-1}. On the other hand, the synthesis of CNF/polypyrrole/silver nanoparticles hybrid foams resulted in low electrical conductivity ($0.5x10^{-2}$ S cm^{-1}) [130], showing that even the incorporation of a good electrical conductor alone may not improve the conductivity of nanocellulose-based aerogels, as the electrical conductor can harm the formation of internal networks.

4.5   Thermal insulation and fire-retardant materials

The thermal conductivity of an aerogel is due to the conduction of the solid network, conduction of the gas phase and radiation inside or through the pores [131], the contribution of the solid part decreases as the material is more porous. The gas phase, on the other hand, contributes to the elastic collision between the gas molecules, thus reducing the size of the pores, suppressing both gaseous thermal transport and radiation [131].

Among the potential applications of nanocellulose aerogels is thermal insulation [132]. In addition to naturally low thermal conductivity, the nanocellulose aerogel cellular structure significantly reduces air circulation, further reducing conductivity. Its transparency also inhibits the transmission of radioactivity from infrared radiation. Some works report that the thermal conductivity of nanocellulose aerogels is below 25 mW m^{-1}K^{-1}, which is like air conductivity. The lowest conductivity in relation to precursor material-cellulose (40-50 mW m^{-1} K^{-1}) is justified by the small size of the particles and pores that are formed during processing [7,44,133].

Additives such as graphene oxide and sepiolite in the formation of nanocellulose aerogels decrease the thermal conductivity in 3 mWm^{-1} K^{-1}in relation to films without additives [134]. Multiscale porous fibers made by cellulose acetate and Poly (acrylic acid) sheath and CNF aerogel in the core showed good thermal insulation capacity in the range of -20 to 150 °C [132]. Another method of improving thermal properties is by reducing the pore size through drying methods such as spray freeze-dried instead of conventional freeze drying [44].

He et al. [135] developed a strategy to improve the absorption, mechanical properties and hinder flammability through the incorporation of aluminum hydroxide nanoparticles into cellulose aerogels via an in-situ sol-gel process followed by freeze-drying. The formation of composite aerogels with cellulose and silica significantly improves the flame retardant performance, achieving self-extinguishing of the post-ignition flame [136]. The use of magnesium hydroxide nanoparticles synthesized with nanocellulose gel followed by freeze-drying produces aerogels with improved flame properties [137]. Other inorganic additives such as clay and graphene have the ability to improve flammability [134].

Despite being considered an excellent thermal insulator, nanocellulose aerogels are easy to ignite, which ends up limiting their application to different uses. Because of this, many researches focus on the use of flame retardants, especially for technological applications [138].

4.6　Pharmaceutical and biomedical applications

A controlled drug delivery system allows the introduction of a drug into the human body, controlling rate, time and location of its release within the organism[139,140]. Nanocellulose has the potential for pharmaceutical field, mainly in the controlled drug release and in improving the solubility of poor water solubility drugs [141,142]. In case of nanocellulose aerogels and foams the drug delivery rate has relation with porosity, so it can be administered by changing the porous structure [143,144]. Also, there are modifications that can improve the drug load capacity, as example the polyethylenimine-grafted CNF (PEI-CNF) foams, that can contain 20 times more drug then CNF foams without modification [145].

A very prominent biomedical application is the regenerative medicine, in view of this several materials have been introduced as potential 3D cell culture scaffolds, these include protein extracts, peptide amphiphiles, synthetic and naturals polymers [146]. 3D cell culture scaffolds consist of an interconnected porous network that favors cellular infiltration [147], the relation between cells and the scaffold is determined by the scaffold material characteristics. In this case the materials used need intrinsic biocompatibility and proper chemistry to induce molecular biorecognition from cells [148]. Bacterial nanocellulose has biocompatibility and the durability, than makes it a promising material for 3D cell culture scaffolds [149], and also the CNF aerogel shows potential, in the research developed by Liu et al. [13] in a 3D TEMPO CNF aerogel scaffolds wherein the cell mortality was less than 5%.

For pharmaceutical or biomedical applications, the use of nanocellulose aerogels needs to be evaluated by in vitro dissolution tests and by toxicological tests, because the aerogels is obtained by chemical and physical processes, and it is necessary to verify the non-attendance of substances that can harm the human organs [150].

5.　Final considerations

Nanocellulose provides interesting properties to produce aerogels with high resistance, low density, high surface area and high porosity, in addition to being biodegradable and renewable.

Aerogels produced with nanocelluloses can present several variations in rigidity, morphology, wettability, roughness, porosity, surface area and density, being linked to the levels of nanocellulose load (CNF and CNC) before the drying step, reaction time in the aging step, the use of additives and surface chemical modification.

The most critical aspect in the production of nanocellulose aerogels is the aggregation of the nanocellulose, a fact that influences the density and surface area of the aerogels and their mechanical properties. Another factor of great relevance is the choice of the drying method, since the use of the critical point method results in larger surface areas of aerogel compared to the freeze-drying process.

The studies carried out show that the aerogels physical properties and variable surface chemistry are useful to increase its absorbing capacity of bacteria and viruses, electrochemical energy storage, air filtration.

In addition, the high porosity provides its use in the field of insulating and fire-retardant materials, electrical and energy storage devices, CO_2 capture and antibacterial applications.

Studies are still needed to improve the properties of aerogels and understand the changes that occur in the drying step for the various potential applications of this material.

References

[1]   S. Wang, A. Lu, L. Zhang, Recent advances in regenerated cellulose materials, Prog. Polym. Sci. 53 (2016) 169–206. https://doi.org/10.1016/j.progpolymsci.2015.07.003

[2]   N. Lavoine, I. Desloges, A. Dufresne, J. Bras, Microfibrillated cellulose - Its barrier properties and applications in cellulosic materials: A review, Carbohydr. Polym. 90 (2012) 735–764. https://doi.org/10.1016/j.carbpol.2012.05.026

[3]   Y. Habibi, L.A. Lucia, O.J. Rojas, Cellulose nanocrystals: Chemistry, self-assembly, and applications, Chem. Rev. 110 (2010) 3479–3500. https://doi.org/10.1021/cr900339w

[4]   M.A.S.A. Samir, F. Alloin, A. Dufresne, Review of recent research into cellulosic whiskers, their properties and their application in nanocomposite field, Biomolecules (2005). https://doi.org/10.1021/BM0493685

[5]   E.C. Lengowski, E.A. Bonfatti Júnior, M.M.N. Kumode, M.E. Carneiro, K.G. Satyanarayana, Nanocellulose in the paper making, in: Inamuddin, S. Thomas, R.K. Mishra, A.M. Asiri (Eds.), Sustainable Polymer Composites and Nanocomposites, Springer International Publishing, Cham, 2019: pp. 1027–1066. https://doi.org/10.1007/978-3-030-05399-4_36

[6]   J. Rojas, M. Bedoya, Y. Ciro, Current trends in the production of cellulose nanoparticles and nanocomposites for biomedical applications, in: Cellul. - Fundam. Asp. Curr. Trends, InTech, 2015. https://doi.org/10.5772/61334

[7]   N. Lavoine, L. Bergström, Nanocellulose-based foams and aerogels: Processing, properties, and applications, J. Mater. Chem. A. 5 (2017) 16105–16117. https://doi.org/10.1039/c7ta02807e

[8]   T. Abitbol, A. Rivkin, Y. Cao, Y. Nevo, E. Abraham, T. Ben-Shalom, S. Lapidot, O. Shoseyov, Nanocellulose, a tiny fiber with huge applications, Curr. Opin. Biotechnol. 39 (2016) 76–88. https://doi.org/10.1016/j.copbio.2016.01.002

[9]   L.R. Amparo, M.J. Rovira, M.M. Sanz, L.G. Gómez-Mascaraque, Nanomaterials for food applications, n.d

[10]   M.A. Hubbe, A. Ferre, P. Tyagi;, Y. Yin, C. Salas, L. Pal, O.J. Rojas, Nanocellulose in thin films, coatings, and plies for packaging applications: A Review Hubbe | BioResources, BioResources. 12 (2017) 2143–2233. https://ojs.cnr.ncsu.edu/index.php/BioRes/article/view/BioRes_12_1_2143_Hubbe_Re view_Nanocellulose_Thin_Films_Coatings_Plies (accessed January 30, 2020)

[11]   L. Liu, F. Kong, Influence of nanocellulose on in vitro digestion of whey protein isolate, Carbohydr. Polym. 210 (2019) 399–411. https://doi.org/10.1016/j.carbpol.2019.01.071

[12]   L. Liu, W.L. Kerr, F. Kong, D.R. Dee, M. Lin, Influence of nano-fibrillated cellulose (NFC) on starch digestion and glucose absorption, Carbohydr. Polym. 196 (2018) 146–153. https://doi.org/10.1016/j.carbpol.2018.04.116

[13]   Z. Liu, M. Zhang, B. Bhandari, Y. Wang, 3D printing: Printing precision and application in food sector, Trends Food Sci. Technol. 69 (2017) 83–94. https://doi.org/10.1016/j.tifs.2017.08.018

[14]   A. Derossi, R. Caporizzi, D. Azzollini, C. Severini, Application of 3D printing for customized food. A case on the development of a fruit-based snack for children, J. Food Eng. 220 (2018) 65–75. https://doi.org/10.1016/j.jfoodeng.2017.05.015

[15]   H. Du, W. Liu, M. Zhang, C. Si, X. Zhang, B. Li, Cellulose nanocrystals and cellulose nanofibrils based hydrogels for biomedical applications, Carbohydr. Polym. 209 (2019) 130–144. https://doi.org/10.1016/j.carbpol.2019.01.020

[16]   T.H. Tan, H.V. Lee, W.A. Yehya Dabdawb, S.B.B.O.A.A. Hamid, A review of nanocellulose in the drug-delivery system, in: Materials for biomedical engineering:

Nanomaterials based drug delivery, Elsevier, 2019: pp. 131–164.
https://doi.org/10.1016/b978-0-12-816913-1.00005-2

[17]   A. Zaman, F. Huang, M. Jiang, W. Wei, Z. Zhou, Preparation, properties, and applications of natural cellulosic aerogels: A review, Energy Built Environ. (2019). https://doi.org/10.1016/j.enbenv.2019.09.002

[18]   H. Sehaqui, M. Salajková, Q. Zhou, L.A. Berglund, Mechanical performance tailoring of tough ultra-high porosity foams prepared from cellulose i nanofiber suspensions, Soft Matter. 6 (2010) 1824–1832. https://doi.org/10.1039/b927505c

[19]   E. Barrios, D. Fox, Y.Y. Li Sip, R. Catarata, J.E. Calderon, N. Azim, S. Afrin, Z. Zhang, L. Zhai, Nanomaterials in advanced, high-performance aerogel composites: A review, Polymers (Basel). 11 (2019) 726. https://doi.org/10.3390/polym11040726

[20]   H.F. Ko, C. Sfeir, P.N. Kumta, Novel synthesis strategies for natural polymer and composite biomaterials as potential scaffolds for tissue engineering, Philos. Trans. R. Soc. A Math. Phys. Eng. Sci. 368 (2010) 1981–1997. https://doi.org/10.1098/rsta.2010.0009

[21]   N. Lin, C. Bruzzese, A. Dufresne, TEMPO-oxidized nanocellulose participating as crosslinking aid for alginate-based sponges, ACS Appl. Mater. Interfaces. 4 (2012) 4948–4959. https://doi.org/10.1021/am301325r

[22]   T. Lu, Q. Li, W. Chen, H. Yu, Composite aerogels based on dialdehyde nanocellulose and collagen for potential applications as wound dressing and tissue engineering scaffold, Compos. Sci. Technol. 94 (2014) 132–138. https://doi.org/https://doi.org/10.1016/j.compscitech.2014.01.020

[23]   A.B. Castro-Ceseña, T.A. Camacho-Villegas, P.H. Lugo-Fabres, E.E. Novitskaya, J. McKittrick, A. Licea-Navarro, Effect of starch on the mechanical and in vitro properties of collagen-hydroxyapatite sponges for applications in dentistry, Carbohydr. Polym. 148 (2016) 78–85.
https://doi.org/https://doi.org/10.1016/j.carbpol.2016.04.056

[24]   A. Tang, S. Zhao, J. Song, Structure control and characterization of 3D porous scaffold based on cellulose-nanofibers for tissue engineering, Cailiao Yanjiu Xuebao/Chinese J. Mater. Res. 28 (2014) 721–729

[25]   A. Liu, L. Medina, L.A. Berglund, High-strength nanocomposite aerogels of ternary composition: poly(vinyl alcohol), clay, and cellulose nanofibrils, ACS Appl. Mater. Interfaces. 9 (2017) 6453–6461. https://doi.org/10.1021/acsami.6b15561

Materials Research Forum LLC
https://doi.org/10.21741/9781644900994-1

[26]   R. Li, J. Du, Y. Zheng, Y. Wen, X. Zhang, W. Yang, A. Lue, L. Zhang, Ultra-lightweight cellulose foam material: preparation and properties, Cellulose. 24 (2017) 1417–1426. https://doi.org/10.1007/s10570-017-1196-y

[27]   B.L. Tardy, S. Yokota, M. Ago, W. Xiang, T. Kondo, R. Bordes, O.J. Rojas, Nanocellulose–surfactant interactions, Curr. Opin. Colloid Interface Sci. 29 (2017) 57–67. https://doi.org/https://doi.org/10.1016/j.cocis.2017.02.004

[28]   H. Yousefian, D. Rodrigue, Morphological, physical and mechanical properties of nanocrystalline cellulose filled Nylon 6 foams, J. Cell. Plast. 53 (2016) 253–271. https://doi.org/10.1177/0021955X16651241

[29]   Y. Wang, K. Uetani, S. Liu, X. Zhang, Y. Wang, P. Lu, T. Wei, Z. Fan, J. Shen, H. Yu, S. Li, Q. Zhang, Q. Li, J. Fan, N. Yang, Q. Wang, Y. Liu, J. Cao, J. Li, W. Chen, Multifunctional bionanocomposite foams with a chitosan matrix reinforced by nanofibrillated cellulose, ChemNanoMat. 3 (2017) 98–108. https://doi.org/10.1002/cnma.201600266

[30]   K.J. De France, T. Hoare, E.D. Cranston, Review of hydrogels and aerogels containing nanocellulose, Chem. Mater. 29 (2017) 4609–4631. https://doi.org/10.1021/acs.chemmater.7b00531

[31]   C. Tan, B.M. Fung, J.K. Newman, C. Vu, Organic aerogels with very high impact strength, Adv. Mater. 13 (2001) 644–646. https://doi.org/10.1002/1521-4095(200105)13:9<644::AID-ADMA644>3.0.CO;2-#

[32]   I. Smirnova, P. Gurikov, Aerogel production: Current status, research directions, and future opportunities, J. Supercrit. Fluids. 134 (2018) 228–233. https://doi.org/https://doi.org/10.1016/j.supflu.2017.12.037

[33]   C. Jiménez-Saelices, B. Seantier, B. Cathala, Y. Grohens, Effect of freeze-drying parameters on the microstructure and thermal insulating properties of nanofibrillated cellulose aerogels, J. Sol-Gel Sci. Technol. 84 (2017) 475–485. https://doi.org/10.1007/s10971-017-4451-7

[34]   Y. Kobayashi, T. Saito, A. Isogai, Aerogels with 3D ordered nanofiber skeletons of liquid-crystalline nanocellulose derivatives as tough and transparent insulators, Angew. Chemie Int. Ed. 53 (2014) 10394–10397. https://doi.org/10.1002/anie.201405123

[35]   W. Chen, Q. Li, Y. Wang, X. Yi, J. Zeng, H. Yu, Y. Liu, J. Li, Comparative study of aerogels obtained from differently prepared nanocellulose fibers, ChemSusChem. 7 (2014) 154–161. https://doi.org/10.1002/cssc.201300950

[36]   P. Munier, K. Gordeyeva, L. Bergström, A.B. Fall, Directional freezing of
nanocellulose dispersions aligns the rod-like particles and produces low-density and
robust particle networks, Biomacromolecules. 17 (2016) 1875–1881.
https://doi.org/10.1021/acs.biomac.6b00304

[37]   M.S. Toivonen, A. Kaskela, O.J. Rojas, E.I. Kauppinen, O. Ikkala, Ambient-dried
cellulose nanofibril aerogel membranes with high tensile strength and their use for
aerosol collection and templates for transparent, flexible devices, Adv. Funct. Mater.
25 (2015) 6618–6626. https://doi.org/10.1002/adfm.201502566

[38]   K. Ganesan, T. Budtova, L. Ratke, P. Gurikov, V. Baudron, I. Preibisch, P.
Niemeyer, I. Smirnova, B. Milow, Review on the Production of Polysaccharide
Aerogel Particles., Mater. (Basel, Switzerland). 11 (2018).
https://doi.org/10.3390/ma11112144

[39]   Y. Peng, D.J. Gardner, Y. Han, Drying cellulose nanofibrils: in search of a suitable
method, Cellulose. 19 (2012) 91–102. https://doi.org/10.1007/s10570-011-9630-z

[40]   P. Srinivasa, A. Kulachenko, C. Aulin, Experimental characterisation of
nanofibrillated cellulose foams, Cellulose. 22 (2015) 3739–3753.
https://doi.org/10.1007/s10570-015-0753-5

[41]   K.S. Gordeyeva, A.B. Fall, S. Hall, B. Wicklein, L. Bergström, Stabilizing
nanocellulose-nonionic surfactant composite foams by delayed Ca-induced gelation, J.
Colloid Interface Sci. 472 (2016) 44–51.
https://doi.org/https://doi.org/10.1016/j.jcis.2016.03.031

[42]   S. Deville, Freeze-Casting of porous biomaterials: structure, properties and
opportunities, Materials (Basel). 3 (2010) 1913–1927

[43]   A. Zaman, F. Huang, M. Jiang, W. Wei, Z. Zhou, Preparation, Properties, and
applications of natural cellulosic aerogels: A review, Energy Built Environ. 1 (2020)
60–76. https://doi.org/https://doi.org/10.1016/j.enbenv.2019.09.002

[44]   C. Jiménez-Saelices, B. Seantier, B. Cathala, Y. Grohens, Spray freeze-dried
nanofibrillated cellulose aerogels with thermal superinsulating properties, Carbohydr.
Polym. 157 (2017) 105–113. https://doi.org/10.1016/j.carbpol.2016.09.068

[45]   H. Sehaqui, Q. Zhou, O. Ikkala, L.A. Berglund, Strong and tough cellulose
nanopaper with high specific surface area and porosity, Biomacromolecules. 12 (2011)
3638–3644. https://doi.org/10.1021/bm2008907

[46]   G. Zu, T. Shimizu, K. Kanamori, Y. Zhu, A. Maeno, H. Kaji, J. Shen, K.
Nakanishi, Transparent, Superflexible Doubly Cross-Linked

Polyvinylpolymethylsiloxane Aerogel Superinsulators via Ambient Pressure Drying, ACS Nano. 12 (2018) 521–532. https://doi.org/10.1021/acsnano.7b07117

[47]   M. Zanini, A. Lavoratti, L.K. Lazzari, D. Galiotto, M. Pagnocelli, C. Baldasso, A.J. Zattera, Producing aerogels from silanized cellulose nanofiber suspension, Cellulose. 24 (2017) 769–779. https://doi.org/10.1007/s10570-016-1142-4

[48]   T. Köhnke, A. Lin, T. Elder, H. Theliander, A.J. Ragauskas, Nanoreinforced xylan–cellulose composite foams by freeze-casting, Green Chem. 14 (2012) 1864–1869. https://doi.org/10.1039/C2GC35413F

[49]   Q. Yang, J. Yang, Z. Gao, B. Li, C. Xiong, Carbonized cellulose nanofibril/graphene oxide composite aerogels for high-performance supercapacitors, ACS Appl. Energy Mater. (2019). https://doi.org/10.1021/acsaem.9b02195

[50]   B. Wu, G. Zhu, A. Dufresne, N. Lin, Fluorescent aerogels based on chemical crosslinking between nanocellulose and carbon dots for optical sensor, ACS Appl. Mater. Interfaces. 11 (2019) 16048–16058. https://doi.org/10.1021/acsami.9b02754

[51]   M. Farooq, M.H. Sipponen, A. Seppälä, M. Österberg, Eco-friendly flame-retardant cellulose nanofibril aerogels by incorporating sodium bicarbonate, ACS Appl. Mater. Interfaces. 10 (2018) 27407–27415. https://doi.org/10.1021/acsami.8b04376

[52]   J. Zhou, Y.-L. Hsieh, Conductive polymer protonated nanocellulose aerogels for tunable and linearly responsive strain sensors, ACS Appl. Mater. Interfaces. 10 (2018) 27902–27910. https://doi.org/10.1021/acsami.8b10239

[53]   G. Hayase, K. Kanamori, K. Abe, H. Yano, A. Maeno, H. Kaji, K. Nakanishi, Polymethylsilsesquioxane–cellulose nanofiber biocomposite aerogels with high thermal insulation, bendability, and superhydrophobicity, ACS Appl. Mater. Interfaces. 6 (2014) 9466–9471. https://doi.org/10.1021/am501822y

[54]   S. Zhou, T. You, X. Zhang, F. Xu, Superhydrophobic cellulose nanofiber-assembled aerogels for highly efficient water-in-oil emulsions separation, ACS Appl. Nano Mater. 1 (2018) 2095–2103. https://doi.org/10.1021/acsanm.8b00079

[55]   S.F. Plappert, S. Quraishi, J.-M. Nedelec, J. Konnerth, H. Rennhofer, H.C. Lichtenegger, F.W. Liebner, Conformal ultrathin coating by scCO$_2$-mediated pmma deposition: a facile approach to add moisture resistance to lightweight ordered nanocellulose aerogels, Chem. Mater. 30 (2018) 2322–2330. https://doi.org/10.1021/acs.chemmater.7b05226

[56]   X. Zhang, H. Wang, Z. Cai, N. Yan, M. Liu, Y. Yu, Highly compressible and hydrophobic anisotropic aerogels for selective oil/organic solvent absorption, ACS Sustain. Chem. Eng. 7 (2019) 332–340. https://doi.org/10.1021/acssuschemeng.8b03554

[57]   D.C. Wang, H.Y. Yu, D. Qi, M. Ramasamy, J. Yao, F. Tang, K. (Michael) C. Tam, Q. Ni, Supramolecular self-assembly of 3D conductive cellulose nanofiber aerogels for flexible supercapacitors and ultrasensitive sensors, ACS Appl. Mater. Interfaces. 11 (2019) 24435–24446. https://doi.org/10.1021/acsami.9b06527

[58]   T. Or, S. Saem, A. Esteve, D.A. Osorio, K.J. De France, J. Vapaavuori, T. Hoare, A. Cerf, E.D. Cranston, J.M. Moran-Mirabal, Patterned cellulose nanocrystal aerogel films with tunable dimensions and morphologies as ultra-porous scaffolds for cell culture, ACS Appl. Nano Mater. 2 (2019) 4169–4179. https://doi.org/10.1021/acsanm.9b00640

[59]   V.C.F. Li, C.K. Dunn, Z. Zhang, Y. Deng, H.J. Qi, Direct Ink Write (DIW) 3D Printed cellulose nanocrystal aerogel structures, Sci. Rep. 7 (2017) 8018. https://doi.org/10.1038/s41598-017-07771-y

[60]   X. Yang, K. Shi, I. Zhitomirsky, E.D. Cranston, Cellulose nanocrystal aerogels as universal 3D lightweight substrates for supercapacitor materials, Adv. Mater. 27 (2015) 6104–6109. https://doi.org/10.1002/adma.201502284

[61]   D.A. Osorio, B. Seifried, P. Moquin, K. Grandfield, E.D. Cranston, Morphology of cross-linked cellulose nanocrystal aerogels: cryo-templating versus pressurized gas expansion processing, J. Mater. Sci. 53 (2018) 9842–9860. https://doi.org/10.1007/s10853-018-2235-2

[62]   X. Wei, T. Huang, J. Nie, J. Yang, X. Qi, Z. Zhou, Y. Wang, Bio-inspired functionalization of microcrystalline cellulose aerogel with high adsorption performance toward dyes, Carbohydr. Polym. 198 (2018) 546–555. https://doi.org/https://doi.org/10.1016/j.carbpol.2018.06.112

[63]   T. Or, K. Miettunen, E.D. Cranston, J.M. Moran-Mirabal, J. Vapaavuori, Cellulose nanocrystal aerogels as electrolyte scaffolds for glass and plastic dye-sensitized solar cells, ACS Appl. Energy Mater. 2 (2019) 5635–5642. https://doi.org/10.1021/acsaem.9b00795

[64]   P. Gupta, B. Singh, A.K. Agrawal, P.K. Maji, Low density and high strength nanofibrillated cellulose aerogel for thermal insulation application, Mater. Des. 158 (2018) 224–236. https://doi.org/https://doi.org/10.1016/j.matdes.2018.08.031

[65] J. Wei, S. Geng, J. Hedlund, K. Oksman, Lightweight, flexible, and multifunctional anisotropic nanocellulose-based aerogels for CO2 adsorption, Cellulose. (2020). https://doi.org/10.1007/s10570-019-02935-7

[66] P.B. de Oliveira, M. Godinho, A.J. Zattera, Oils sorption on hydrophobic nanocellulose aerogel obtained from the wood furniture industry waste, Cellulose. 25 (2018) 3105–3119. https://doi.org/10.1007/s10570-018-1781-8

[67] F. Rafieian, M. Hosseini, M. Jonoobi, Q. Yu, Development of hydrophobic nanocellulose-based aerogel via chemical vapor deposition for oil separation for water treatment, Cellulose. 25 (2018) 4695–4710. https://doi.org/10.1007/s10570-018-1867-3

[68] J. Li, K. Zuo, W. Wu, Z. Xu, Y. Yi, Y. Jing, H. Dai, G. Fang, Shape memory aerogels from nanocellulose and polyethyleneimine as a novel adsorbent for removal of Cu(II) and Pb(II), Carbohydr. Polym. 196 (2018) 376–384. https://doi.org/https://doi.org/10.1016/j.carbpol.2018.05.015

[69] T. Zhang, Y. Zhang, X. Wang, S. Liu, Y. Yao, Characterization of the nano-cellulose aerogel from mixing CNF and CNC with different ratio, Mater. Lett. 229 (2018) 103–106. https://doi.org/https://doi.org/10.1016/j.matlet.2018.06.101

[70] C.A. García-González, M.C. Camino-Rey, M. Alnaief, C. Zetzl, I. Smirnova, Supercritical drying of aerogels using CO_2: Effect of extraction time on the end material textural properties, J. Supercrit. Fluids. 66 (2012) 297–306. https://doi.org/https://doi.org/10.1016/j.supflu.2012.02.026

[71] J. Quiño, M. Ruehl, T. Klima, F. Ruiz, S. Will, A. Braeuer, Supercritical drying of aerogel: In situ analysis of concentration profiles inside the gel and derivation of the effective binary diffusion coefficient using Raman spectroscopy, J. Supercrit. Fluids. 108 (2016) 1–12. https://doi.org/https://doi.org/10.1016/j.supflu.2015.10.011

[72] L.M. Sanz-Moral, M. Rueda, R. Mato, Á. Martín, View cell investigation of silica aerogels during supercritical drying: Analysis of size variation and mass transfer mechanisms, J. Supercrit. Fluids. 92 (2014) 24–30. https://doi.org/https://doi.org/10.1016/j.supflu.2014.05.004

[73] K. Sakai, Y. Kobayashi, T. Saito, A. Isogai, Partitioned airs at microscale and nanoscale: thermal diffusivity in ultrahigh porosity solids of nanocellulose, Sci. Rep. 6 (2016) 20434. https://doi.org/10.1038/srep20434

[74] H. Liu, B. Geng, Y. Chen, H. Wang, Review on the aerogel-type oil sorbents derived from nanocellulose, ACS Sustain. Chem. Eng. 5 (2017) 49–66. https://doi.org/10.1021/acssuschemeng.6b02301

[75] N.T. Cervin, C. Aulin, P.T. Larsson, L. Wågberg, Ultra porous nanocellulose aerogels as separation medium for mixtures of oil/water liquids, Cellulose. 19 (2012) 401–410. https://doi.org/10.1007/s10570-011-9629-5

[76] S. Mueller, J. Sapkota, A. Nicharat, T. Zimmermann, P. Tingaut, C. Weder, E.J. Foster, Influence of the nanofiber dimensions on the properties of nanocellulose/poly(vinyl alcohol) aerogels, J. Appl. Polym. Sci. 132 (2015). https://doi.org/10.1002/app.41740

[77] C. Aulin, J. Netrval, L. Wågberg, T. Lindström, Aerogels from nanofibrillated cellulose with tunable oleophobicity, Soft Matter. 6 (2010) 3298–3305. https://doi.org/10.1039/C001939A

[78] T. Lindström, C. Aulin, Market and technical challenges and opportunities in the area of innovative new materials and composites based on nanocellulosics, Scand. J. For. Res. 29 (2014) 345–351. https://doi.org/10.1080/02827581.2014.928365

[79] H. Kargarzadeh, J. Huang, N. Lin, I. Ahmad, M. Mariano, A. Dufresne, S. Thomas, A. Gałęski, Recent developments in nanocellulose-based biodegradable polymers, thermoplastic polymers, and porous nanocomposites, Prog. Polym. Sci. 87 (2018) 197–227. https://doi.org/https://doi.org/10.1016/j.progpolymsci.2018.07.008

[80] B.N. Nguyen, E. Cudjoe, A. Douglas, D. Scheiman, L. McCorkle, M.A.B. Meador, S.J. Rowan, Polyimide cellulose nanocrystal composite aerogels, Macromolecules. 49 (2016) 1692–1703. https://doi.org/10.1021/acs.macromol.5b01573

[81] L. Heath, W. Thielemans, Cellulose nanowhisker aerogels, Green Chem. 12 (2010) 1448–1453. https://doi.org/10.1039/C0GC00035C

[82] M. Cai, S. Shafi, Y. Zhao, Preparation of compressible silica aerogel reinforced by bacterial cellulose using tetraethylorthosilicate and methyltrimethoxylsilane co-precursor, J. Non. Cryst. Solids. 481 (2018) 622–626. https://doi.org/https://doi.org/10.1016/j.jnoncrysol.2017.12.015

[83] M. Karzar Jeddi, O. Laitinen, H. Liimatainen, Magnetic superabsorbents based on nanocellulose aerobeads for selective removal of oils and organic solvents, Mater. Des. 183 (2019) 108115. https://doi.org/https://doi.org/10.1016/j.matdes.2019.108115

[84] D.A. Osorio, B.E.J. Lee, J.M. Kwiecien, X. Wang, I. Shahid, A.L. Hurley, E.D. Cranston, K. Grandfield, Cross-linked cellulose nanocrystal aerogels as viable bone tissue scaffolds, Acta Biomater. 87 (2019) 152–165. https://doi.org/https://doi.org/10.1016/j.actbio.2019.01.049

[85] K. Rahbar Shamskar, H. Heidari, A. Rashidi, Preparation and evaluation of nanocrystalline cellulose aerogels from raw cotton and cotton stalk, Ind. Crops Prod. 93 (2016) 203–211. https://doi.org/https://doi.org/10.1016/j.indcrop.2016.01.044

[86] N.T. Cervin, C. Aulin, P.T. Larsson, L. Wågberg, Ultra porous nanocellulose aerogels as separation medium for mixtures of oil/water liquids, Cellulose. 19 (2012) 401–410. https://doi.org/10.1007/s10570-011-9629-5

[87] H. Jin, M. Kettunen, A. Laiho, H. Pynnönen, J. Paltakari, A. Marmur, O. Ikkala, R.H.A. Ras, Superhydrophobic and superoleophobic nanocellulose aerogel membranes as bioinspired cargo carriers on water and oil, Langmuir. 27 (2011) 1930–1934. https://doi.org/10.1021/la103877r

[88] F. Jiang, Y.L. Hsieh, Amphiphilic superabsorbent cellulose nanofibril aerogels, J. Mater. Chem. A. 2 (2014) 6337–6342. https://doi.org/10.1039/C4TA00743C

[89] C. Buesch, S.W. Smith, P. Eschbach, J.F. Conley, J. Simonsen, The microstructure of cellulose nanocrystal aerogels as revealed by transmission electron microscope tomography, Biomacromolecules. 17 (2016) 2956–2962. https://doi.org/10.1021/acs.biomac.6b00764

[90] J. Ha, J. Kim, Y. Jung, G. Yun, D.-N. Kim, H.-Y. Kim, Poro-elasto-capillary wicking of cellulose sponges, Sci. Adv. 4 (2018) eaao7051. https://doi.org/10.1126/sciadv.aao7051

[91] Z. Zhang, G. Sèbe, D. Rentsch, T. Zimmermann, P. Tingaut, Ultralightweight and flexible silylated nanocellulose sponges for the selective removal of oil from water, Chem. Mater. 26 (2014) 2659–2668. https://doi.org/10.1021/cm5004164

[92] S. Elazzouzi-Hafraoui, Y. Nishiyama, J.L. Putaux, L. Heux, F. Dubreuil, C. Rochas, The shape and size distribution of crystalline nanoparticles prepared by acid hydrolysis of native cellulose, Biomacromolecules. 9 (2008) 57–65. https://doi.org/10.1021/bm700769p

[93] Y. Zhao, C. Hu, Y. Hu, H. Cheng, G. Shi, L. Qu, A Versatile, Ultralight, nitrogen-doped graphene framework, Angew. Chemie Int. Ed. 51 (2012) 11371–11375. https://doi.org/10.1002/anie.201206554

[94] J. Zou, J. Liu, A.S. Karakoti, A. Kumar, D. Joung, Q. Li, S.I. Khondaker, S. Seal, L. Zhai, Ultralight multiwalled carbon nanotube aerogel, ACS Nano. 4 (2010) 7293–7302. https://doi.org/10.1021/nn102246a

[95] E. Abraham, D.E. Weber, S. Sharon, S. Lapidot, O. Shoseyov, Multifunctional cellulosic scaffolds from modified cellulose nanocrystals, ACS Appl. Mater. Interfaces. 9 (2017) 2010–2015. https://doi.org/10.1021/acsami.6b13528

[96] S.T. Nguyen, J. Feng, S.K. Ng, J.P.W. Wong, V.B.C. Tan, H.M. Duong, Advanced thermal insulation and absorption properties of recycled cellulose aerogels, Colloids Surfaces A Physicochem. Eng. Asp. 445 (2014) 128–134. https://doi.org/https://doi.org/10.1016/j.colsurfa.2014.01.015

[97] H. Huang, P. Chen, X. Zhang, Y. Lu, W. Zhan, Edge-to-edge assembled graphene oxide aerogels with outstanding mechanical performance and superhigh chemical activity, Small. 9 (2013) 1397–1404. https://doi.org/10.1002/smll.201202965

[98] Y. Kharbanda, M. Urbańczyk, O. Laitinen, K. Kling, S. Pallaspuro, S. Komulçainen, H. Liimatainen, V.V. Telkki, Comprehensive NMR analysis of pore structures in superabsorbing cellulose nanofiber aerogels, J. Phisical Chem. C. 123 (2019) 30986–30995. https://doi.org/10.1021/acs.jpcc.9b08339

[99] X. Yang, E.D. Cranston, Chemically cross-linked cellulose nanocrystal aerogels with shape recovery and superabsorbent properties, Chem. Mater. 26 (2014) 6016–6025. https://doi.org/10.1021/cm502873c

[100] W.J. Yang, A.C.Y. Yuen, A. Li, B. Lin, T.B.Y. Chen, W. Yang, H.D. Lu, G.H. Yeoh, Recent progress in bio-based aerogel absorbents for oil/water separation, Cellulose. 26 (2019) 6449–6476. https://doi.org/10.1007/s10570-019-02559-x

[101] J. Huang, X. Wang, Q. Jin, Y. Liu, Y. Wang, Removal of phenol from aqueous solution by adsorption onto OTMAC-modified attapulgite, J. Environ. Manage. 84 (2007) 229–236. https://doi.org/10.1016/j.jenvman.2006.05.007

[102] R.J. Moon, A. Martini, J. Nairn, J. Simonsen, J. Youngblood, Cellulose nanomaterials review: Structure, properties and nanocomposites, Chem. Soc. Rev. 40 (2011) 3941–3994. https://doi.org/10.1039/c0cs00108b

[103] S.J. Eichhorn, Cellulose nanowhiskers: Promising materials for advanced applications, Soft Matter. 7 (2011) 303–315. https://doi.org/10.1039/c0sm00142b

[104] Y. Kharbanda, M. Urbańczyk, O. Laitinen, K. Kling, S. Pallaspuro, S. Komulainen, H. Liimatainen, V.V. Telkki, Comprehensive NMR analysis of pore

structures in superabsorbing cellulose nanofiber aerogels, J. Phys. Chem. C. 123 (2019) 30986–30995. https://doi.org/10.1021/acs.jpcc.9b08339

[105] A. Fakhru'l-Razi, A. Pendashteh, L.C. Abdullah, D.R.A. Biak, S.S. Madaeni, Z.Z. Abidin, Review of technologies for oil and gas produced water treatment, J. Hazard. Mater. 170 (2009) 530–551. https://doi.org/10.1016/j.jhazmat.2009.05.044

[106] B. Liu, L. Zhang, H. Wang, Z. Bian, Preparation of MCC/MC silica sponge and its oil/water separation apparatus application, Ind. Eng. Chem. Res. 56 (2017) 5795–5801. https://doi.org/10.1021/acs.iecr.6b04854

[107] J. Saleem, M. Adil Riaz, M. Gordon, Oil sorbents from plastic wastes and polymers: A review, J. Hazard. Mater. 341 (2018) 424–437. https://doi.org/10.1016/j.jhazmat.2017.07.072

[108] W. Wan, Y. Lin, A. Prakash, Y. Zhou, Three-dimensional carbon-based architectures for oil remediation: from synthesis and modification to functionalization, J. Mater. Chem. A. 4 (2016) 18687–18705. https://doi.org/10.1039/C6TA07211A

[109] J.T. Korhonen, M. Kettunen, R.H.A. Ras, O. Ikkala, Hydrophobic nanocellulose aerogels as floating, sustainable, reusable, and recyclable oil absorbents, ACS Appl. Mater. Interfaces. 3 (2011) 1813–1816. https://doi.org/10.1021/am200475b

[110] T. Ohno, S. Tashiro, Y. Amano, R. Yoshida, H. Abe, Rapid clogging of high-efficiency particulate air filters during in-cell solvent fires at reprocessing facilities, Nucl. Technol. 206 (2020) 40–47. https://doi.org/10.1080/00295450.2019.1620057

[111] I.M. Hutten, Handbook of nonwoven filter media, Elsevier Inc., 2015. https://doi.org/10.1016/C2011-0-05753-8

[112] J. Nemoto, T. Saito, A. Isogai, Simple freeze-drying procedure for producing nanocellulose aerogel-containing, high-performance air filters, ACS Appl. Mater. Interfaces. 7 (2015) 19809–19815. https://doi.org/10.1021/acsami.5b05841

[113] B. Bereiter, S. Eggleston, J. Schmitt, C. Nehrbass-Ahles, T.F. Stocker, H. Fischer, S. Kipfstuhl, J. Chappellaz, Revision of the EPICA Dome C CO_2 record from 800 to 600-kyr before present, Geophys. Res. Lett. 42 (2015) 542–549. https://doi.org/10.1002/2014GL061957

[114] N. Mahfoudhi, S. Boufi, Nanocellulose as a novel nanostructured adsorbent for environmental remediation: a review, Cellulose. 24 (2017) 1171–1197. https://doi.org/10.1007/s10570-017-1194-0

[115] C. Gebald, J.A. Wurzbacher, P. Tingaut, T. Zimmermann, A. Steinfeld, Amine-based nanofibrillated cellulose as adsorbent for CO_2 capture from air, Environ. Sci. Technol. 45 (2011) 9101–9108. https://doi.org/10.1021/es202223p

[116] H. Sehaqui, M.E. Gálvez, V. Becatinni, Y. cheng Ng, A. Steinfeld, T. Zimmermann, P. Tingaut, Fast and reversible direct CO_2 capture from air onto all-polymer nanofibrillated cellulose—polyethylenimine foams, Environ. Sci. Technol. 49 (2015) 3167–3174. https://doi.org/10.1021/es504396v

[117] W.D. Callister Júnior, D.G. Retcwisch, Materials science and engineering: An introduction, 10th ed., WileyPlus, Hoboken, 2018

[118] R.C. Hibbeler, Mechanics of Materials, 10th ed., Pearson, London, 2016

[119] S.K. Goyal, The big book of packaging : science, art & technology, Sanex Packaging Connections Pvt. Ltd., Gurugram, 2016

[120] J. Wang, Z. Yu, P. Li, D. Ding, X. Zheng, C. Hu, T. Hu, X. Gong, Y. Chang, C. Wu, Poly(styrene-: Ran -cinnamic acid) (SCA), an approach to modified polystyrene with enhanced impact toughness, heat resistance and melt strength, RSC Adv. 9 (2019) 39631–39639. https://doi.org/10.1039/c9ra08635h

[121] J. Wiener, F. Arbeiter, A. Tiwari, O. Kolednik, G. Pinter, Bioinspired toughness improvement through soft interlayers in mineral reinforced polypropylene, Mech. Mater. 140 (2020). https://doi.org/10.1016/j.mechmat.2019.103243

[122] M. Mihalic, L. Sobczak, C. Pretschuh, C. Unterweger, Increasing the impact toughness of cellulose fiber reinforced polypropylene composites—influence of different impact modifiers and production scales, J. Compos. Sci. 3 (2019) 82. https://doi.org/10.3390/jcs3030082

[123] I. Nennewitz, W. Nutsch, P. Peschel, S. Schulzig, G. Seifert, T. Strechel, Holztechnik Tabellenbuch, 11th ed., Europa Lehrmittel Verlag, Haan, 2019

[124] A.E. Donius, A. Liu, L.A. Berglund, U.G.K. Wegst, Superior mechanical performance of highly porous, anisotropic nanocellulose-montmorillonite aerogels prepared by freeze casting, J. Mech. Behav. Biomed. Mater. 37 (2014) 88–99. https://doi.org/10.1016/j.jmbbm.2014.05.012

[125] A.B. Perumal, P.S. Sellamuthu, R.B. Nambiar, E.R. Sadiku, O.A. Adeyeye, Biocomposite reinforced with nanocellulose for packaging applications, in: D. Gnanasekaran (Ed.), Green biopolym. their nanocomposites, Springer, Singapore, 2019: pp. 83–123. https://doi.org/10.1007/978-981-13-8063-1_4

[126] P. Komarnicki, P. Lombardi, Z. Styczynski, Electric energy storage systems, Springer, Berlin, 2017. https://doi.org/10.1007/978-3-662-53275-1

[127] X. Yang, K. Shi, I. Zhitomirsky, E.D. Cranston, Cellulose Nanocrystal aerogels as universal 3D lightweight substrates for supercapacitor materials, Adv. Mater. 27 (2015) 6104–6109. https://doi.org/10.1002/adma.201502284

[128] M. Hamedi, E. Karabulut, A. Marais, A. Herland, G. Nyström, L. Wågberg, Nanocellulose Aerogels functionalized by rapid layer-by-layer assembly for high charge storage and beyond, Angew. Chemie Int. Ed. 52 (2013) 12038–12042. https://doi.org/10.1002/anie.201305137

[129] M. Pääkkö, J. Vapaavuori, R. Silvennoinen, H. Kosonen, M. Ankerfors, T. Lindström, L.A. Berglund, O. Ikkala, Long and entangled native cellulose i nanofibers allow flexible aerogels and hierarchically porous templates for functionalities, Soft Matter. 4 (2008) 2492–2499. https://doi.org/10.1039/b810371b

[130] S. Zhou, M. Wang, X. Chen, F. Xu, Facile template synthesis of microfibrillated cellulose/polypyrrole/silver nanoparticles hybrid aerogels with electrical conductive and pressure responsive properties, ACS Sustain. Chem. Eng. 3 (2015) 3346–3354. https://doi.org/10.1021/acssuschemeng.5b01020

[131] J. Fricke, Thermal transport in porous superinsulations, in: J. Fricke (Ed.), Aerogels, Springer, Chan, 1985: pp. 94–103. https://doi.org/10.1007/978-3-642-93313-4_11

[132] J. Zhou, Y. Lo Hsieh, Nanocellulose aerogel coaxial fibers for thermal insulation, Nano Energy. (2019) 104305. https://doi.org/https://doi.org/10.1016/j.nanoen.2019.104305

[133] B. Seantier, D. Bendahou, A. Bendahou, Y. Grohens, H. Kaddami, Multi-scale cellulose based new bio-aerogel composites with thermal super-insulating and tunable mechanical properties, Carbohydr. Polym. 138 (2016) 335–348. https://doi.org/10.1016/j.carbpol.2015.11.032

[134] B. Wicklein, A. Kocjan, G. Salazar-Alvarez, F. Carosio, G. Camino, M. Antonietti, L. Bergström, Thermally insulating and fire-retardant lightweight anisotropic foams based on nanocellulose and graphene oxide, Nat. Nanotechnol. 10 (2015) 277–283. https://doi.org/10.1038/nnano.2014.248

[135] C. He, J. Huang, S. Li, K. Meng, L. Zhang, Z. Chen, Y. Lai, Mechanically resistant and sustainable cellulose-based composite aerogels with excellent flame retardant, sound-absorption, and superantiwetting ability for advanced engineering

materials, ACS Sustain. Chem. Eng. 6 (2018) 927–936.
https://doi.org/10.1021/acssuschemeng.7b03281

[136] B. Yuan, J. Zhang, Q. Mi, J. Yu, R. Song, J. Zhang, Transparent cellulose–silica composite aerogels with excellent flame retardancy via an in situ sol–gel process, ACS Sustain. Chem. Eng. 5 (2017) 11117–11123. https://doi.org/10.1021/acssuschemeng.7b03211

[137] Y. Han, X. Zhang, X. Wu, C. Lu, Flame retardant, heat insulating cellulose aerogels from waste cotton fabrics by in situ formation of magnesium hydroxide nanoparticles in cellulose gel nanostructures, ACS Sustain. Chem. Eng. 3 (2015) 1853–1859. https://doi.org/10.1021/acssuschemeng.5b00438

[138] D.A. Gopakumar, S. Thomas, O. F.A.T, S. Thomas, A. Nzihou, S. Rizal, H.P.S. Abdul Khalil, Nanocellulose based aerogels for varying engineering applications, in: S. Hashmi (Eds.) Reference Module in Materials Science and Materials Engineering, Elsevier, 2019. https://doi.org/10.1016/b978-0-12-803581-8.10549-1

[139] M.L. Workman, L.A. LaCharity, S.L. Kruchko, Understanding pharmacology: Essentials for medication safety., Elsevier, Amsterdan, 2013

[140] U. Pal, S.K. Pramanik, Advances in the application of nanomaterials and nanosacled materials in physiology or medicine: Now and the future, in: B.N. Ganguly (Ed.), Nanomaterials in bio-medical applications; A novel approach, Materials Research Forum, Millersville, 2018: pp. 147–178

[141] V. Gopinath, S. Saravanan, A.R. Al-maleki, M. Ramesh, J. Vadivelu, Biomedicine & Pharmacotherapy A review of natural polysaccharides for drug delivery applications : Special focus on cellulose , starch and glycogen, Biomed. Pharmacother. 107 (2018) 96–108

[142] K. Löbmann, A.J. Svagan, Cellulose nano fi bers as excipient for the delivery of poorly soluble drugs, Int. J. Pharm. 533 (2017) 285–297

[143] H. Valo, S. Arola, P. Laaksonen, M. Torkkeli, L. Peltonen, M.B. Linder, R. Serimaa, S. Kuga, J. Hirvonen, T. Laaksonen, Drug release from nanoparticles embedded in four different nanofibrillar cellulose aerogels, Eur. J. Pharm. Sci. 50 (2013) 69–77

[144] A.J. Svagan, J.W. Benjamins, Z. Al-Ansari, D.B. Shalom, A. Müllertz, L. Wågberg, K. Löbmann, Solid cellulose nanofiber based foams – Towards facile design of sustained drug delivery systems, J. Control. Release. 244 (2016) 74–82. https://doi.org/10.1016/j.jconrel.2016.11.009

[145] J. Zhao, C. Lu, X. He, X. Zhang, W. Zhang, X. Zhang, Polyethylenimine-grafted cellulose nanofibril aerogels as versatile vehicles for drug delivery, ACS Appl. Mater. Interfaces. 7 (2015) 2607–2615. https://doi.org/10.1021/am507601m

[146] M. Bhattacharya, M.M. Malinen, P. Lauren, Y.R. Lou, S.W. Kuisma, L. Kanninen, M. Lille, A. Corlu, C. Guguen-Guillouzo, O. Ikkala, A. Laukkanen, A. Urtti, M. Yliperttula, Nanofibrillar cellulose hydrogel promotes three-dimensional liver cell culture, in: J. Control. Release, 2012: pp. 291–298. https://doi.org/10.1016/j.jconrel.2012.06.039

[147] H. Cai, S. Sharma, W. Liu, W. Mu, W. Liu, X. Zhang, Y. Deng, Aerogel microspheres from natural cellulose nanofibrils and their application as cell culture scaffold, Biomacromolecules. 15 (2014) 2540–2547. https://doi.org/10.1021/bm5003976

[148] E. Carletti, A. Motta, C. Migliaresi, Scaffolds for tissue engineering and 3D cell culture., Methods Mol. Biol. 695 (2011) 17–39. https://doi.org/10.1007/978-1-60761-984-0_2

[149] M. Jorfi, E.J. Foster, Recent advances in nanocellulose for biomedical applications, J. Appl. Polym. Sci. 132 (2015) n/a-n/a. https://doi.org/10.1002/app.41719

[150] A.B. Seabra, J.S. Bernardes, W.J. Fávaro, A.J. Paula, Cellulose nanocrystals as carriers in medicine and their toxicities : A review, Carbohydr. Polym. 181 (2018) 514–527

Aerogels I: Preparation, Properties and Applications
Materials Research Foundations **84** (2020) 34-82

Materials Research Forum LLC
https://doi.org/10.21741/9781644900994-2

Chapter 2

Porous Aerogels

Samad Yaseen[1], Ata-ur-Rehman[1], Ghulam Ali[2], Ghulam Shabbir[3], Syed Mustansar Abbas[4]*

[1]Department of Chemistry, Quaid-e-Azam University, Islamabad, Pakistan

[2]U.S.-Pakistan Center for Advanced Studies in Energy (USPCAS-E), National University of Science and Technology (NUST), Sector H-12, Islamabad, 44000, Pakistan

[3]Department of Chemistry, University of Chakwal, Chakwal Campus, Punjab, Pakistan

[4]Nanoscience and Technology Department, National Centre for Physics, Islamabad, Pakistan

*qau_abbas@yahoo.com (Syed Mustansar Abbas)

Abstract

The current chapter reviews porous aerogels that have been more utilized including silicate and non-silicate (ZrO_2 and TiO_2) aerogels. In addition, the recently focused aerogels like composite/hybrid aerogels, polymeric, carbon-based (mostly CNTs and graphene-based) and biogels have been summarized. The metal chalcogenide aerogels display unique properties, unlike the conventional oxide-based aerogels. The diverse utility of these porous aerogels encompasses environmental, biomedical, catalytic and advanced applications especially graphene and CNTs based aerogels have applications in modern devices primarily due to surpassed surface area, porosity and structural adaptability.

Keywords

Xerogel, Hybrid, Chalcogenide, Graphene, Biogels, Tissue Engineering, Surface Area

Contents

Materials Research Forum LLC
https://doi.org/10.21741/9781644900994-2

1.   Porous aerogel history

Kistler et al. [1] for the first time coined the term of aerogel in 1932, where the gaseous phase takes the place of liquid in the wet gel without compromising on the 3D solid porous network of gel. Kistler utilized the supercritical fluid drying technique (SCF) instead of evaporation to remove the trapped liquid out of gel. In this technique, the liquid found in the pores of gel transforms into the supercritical fluid without disturbing the porous structure. Because of the evacuation of supercritical fluid as a gas, a dry solid network of a wet gel is obtained, keeping the open porous texture intact. Therefore, the aerogels are the materials where the air has replaced the liquid in the wet gel without considerable shrinkage in the solid porous network.

Aerogels have the characteristics of the high specific pore volume. Mostly, the silica-based aerogel has the relative pore volume of around 90% [2]. Different series of aerogel synthesized by Kistler show that the results are less promising than the silica aerogel [1]. Apart from silica-based aerogels, which are easy to handle due to their robust nature, he also synthesized alumina-based aerogels but these proved to be mechanically very fragile.

Initially, the precursors utilized are the metallic salt containing the cations to synthesize the oxide gel. Kistler utilized the sodium metasilicate (Na_2SiO_3), as it was cost-effective [1]. Hence, BASF developed an industrially feasible process benefitting from this compound [3]. Metallic salts also remained important when other precursors are not available.

2. Aerogel pore classification

The aerogel can be categorized based upon their microstructure as given below; Microporous aerogel (less than 2 nm), Mesoporous aerogel (2-50 nm) and mixed porous aerogel with larger and varyingly pore size.

3. Inorganic-silica based aerogels

Silica-based aerogels are usually obtained by the sol-gel method where a porous network of gel is developed due to the polymerization mechanism, where the (-Si–O–Si-) linkage is created between initial precursors. In the first stage, colloidal silica particles are arranged to form linear oligomers with nanosized particles. Then afterwards, these elementary units are linked together to form a 3D network of a gel in the solvent [4,5]. Kistler synthesized the first silica aerogel from sodium metasilicate (termed as waterglass) reaction with HCl as shown in Eq (1). Product of the reaction is salt, so it can easily be removed by acidic ion-exchanger or by dialysis [6].

$$Na_2SiO_3 + 2HCl + (x\text{-}1)H_2O \implies SiO_2.xH_2O + 2NaCl \qquad (1)$$

However, nowadays precursors of Si used are alkoxides of silicon such as $Si(OR)_4$, in which -OR are usually methoxy or ethoxy moities [7].

Other varieties of alkoxide derived precursors include perfluoroalkylsilane (PFAS), methyltrimethoxysilane (MTMS), 3-(2-aminoethylamino) propyltrimethoxysilane (EDAS) [8]. All of these mentioned examples have the polar Si-O bond which permits the Si-O-Si wide-angle distribution values to assist the three-dimensional network like that found in glass [9].

3.1 Properties of silica-based aerogel

3.1.1 Texture

Silica-based aerogel has amorphous nature. They usually have a skeletal density of 2 gcm^{-3} which is near to non-crystalline silica (2.2 g cm^{-3}) having above 90% pore volume [10]. Few ultra-light and -porous silica aerogel are reported with extraordinary low density (0.003 g cm^{-3}) [11].

The mesoporous structure is usually found in silica aerogels, with the pore diameter ranging from 20-40 nm. Microporous aerogel of less than 2 nm can be synthesized by particular conditions like acid catalysed treatment [12]. Silica aerogel has a surface area ranging from 250-800 m^2g^{-1} but could be exceeding 1000 m^2g^{-1} in controlled condition [13].

3.1.2 Thermal properties

Silica aerogel has a very low thermal conductivity in the order of 0.015W m^{-1} K^{-1}. These conductivities are less than that of air (0.025W m^{-1} K^{-1}) at ambient pressure, temperature and relative humidity [14]. Considering these properties, the silica aerogel are the promising candidate for thermal insulation.

3.1.3 Optical properties

The transparent form of silica aerogel can be achieved although their optical quality is reduced due to the scattering of transmitted light to some extent. The heterogenous solid network nature of gel in the nanometer range causes Rayleigh scattering which is responsible for a bluish colouration in reflection and yellowish colouration in the transmission mode [15]. While the micrometre range heterogeneities in the solid network are the reason behind the blurred deformation of optical images [16].

3.1.4 Entrapment, release, sorption, and storage properties

High specific pore volume and somewhat resistant porous SiO_2 network of aerogel can be utilized to trap the nanoparticles and a variety of other molecules [17-20]. Similarly, silica aerogel could be impregnated with nematic crystalline phase. As a result of impregnation, the long-range order of crystal can be destroyed by the randomness of a gel network and liquid crystal changes into the glassy state [21].

Bacteria and viruses can be easily trapped inside aerogels and still remain alive [22]. High specific pore volume and controllable pore size of silica-based aerogel could be utilized to release drugs and different chemicals in a controlled manner [23]. Likewise, aerogel can also adsorb harmful toxic chemicals from wastewater. Silica-based aerogels as suggested by Woignier et al. [24] can be employed for storage of actinides, as they are chemically stable in water. Silica-based aerogel impregnated with different salts like calcium chloride, lithium bromide is checked as low-temperature water sorbents for heat storage. They have the energy storage capacity as measured by differential scanning calorimetry (DSC) up to 4 kJg^{-1}, that is far greater than unimpregnated silica gel and zeolites [25].

4. Inorganic-nonsilicate aerogels

4.1 ZrO_2 aerogels

For the synthesis of zirconia material, a number of experiments are performed to evaluate the textural features like pore size, porosity, surface area, etc. Different methodologies such as electrolysis, sol-gel, etc. are often used. It is an established fact that the supercritical drying (SCD) of zirconia-based gel is the best method to get enhanced textural and structural properties [26]. Recently it has been reported that the zirconia aerogel synthesized by SCD have an increased surface area with large pore volume [27]. Moreover, by the SCD procedure ZrO_2 nanoparticles with better morphological features and high surface reactivity are obtained [28].

4.1.1 ZrO_2 aerogels in catalysis

ZrO_2 aerogels have the numerous properties suitable for catalytic support such as surface having both acidic and basic for redox properties. When textural and structural feature combines with previously mentioned properties, they make a catalytically active phase with increased thermal stability with better resistance to catalytic poisoning [29].

4.1.2 ZrO$_2$ aerogels in ceramics

With exceptional thermal, mechanical and electrical properties, ZrO$_2$ aerogel shows a competing aspect in the ceramic industry for the usage in the fuel cell, thermal barrier coatings [30]. But mostly it is doped with other metal oxides to have the low-temperature stabilization of tetragonal and cubic phase. Low dopant concentration increased the mechanical toughness due to stabilizing tetragonal phase while the high dopant concentration favours the cubic phase with high ionic conductivity [31].

Similarly, if the ZrO$_2$ is doped with Y^{3+}, Ce^{3+}, stabilization of nanoscale tetragonal particles enhanced their fracture toughness. It is seen that the tetragonal phase stabilization is being controlled by two important factors including preparation methodology and particle size [32-35].

4.1.3 ZrO$_2$ aerogels in solid oxide fuel cells

Solid oxide fuel cells (SOFCs) utilizes the yttria-stabilized zirconia (YSZ) due to high-temperature oxide conductor. When 8-10% of yttrium oxide is incorporated in the ZrO$_2$ aerogel, it causes low-temperature stabilization of zirconia in the cubic crystalline phase. In addition to structure and composition, particle size and homogeneity of precursors are the factors to consider while preparing the YSZ. It is noted that SCD is a useful process that will produce the ZrO$_2$ aerogel with enhanced surface area which is desirable for greater conductivity in YSZ [36].

4.2 TiO$_2$ aerogels

Due to the surface properties, titania has been utilized in numerous applications like photocatalysts [37,38], gas sensors [39], solar cell electrodes [40], etc. Structure and controlled porosity of TiO$_2$ materials is important for these applications [41,42].

TiO$_2$ aerogel could be either pure titania or mixed oxides. Mixed oxide aerogel is known to show better photocatalytic activity [43-45]. Porous titania could be made by various processes. Sol-gel [45,46] method is usually employed despite a few drawbacks such as costly raw materials, long process and large shrinkage during the process. But it is still an attractive method because the "tailored materials" can be formed with high surface area and controlled pore structure.

5.   Organic-natural/biogels

5.1   Polysaccharides aerogels

Polysaccharides are the potential candidates for the production of biogels. A vast variety of polysaccharides can be used for aerogel production. Their use in a number of applications might be due to its abundance, biocompatibility, biodegradability and cost-effectiveness [47]. Cellulose, chitin, chitosan, alginate, starch, carrageenan, curdlan are usually considered abundant polysaccharides for the preparation of cost-effective aerogels. Polysaccharides based aerogel can replace the silica-based aerogel. It would be beneficial in pharmaceutical development [48-50]. Few polysaccharides based aerogels are discussed below.

5.2   Chitosan aerogel

Chitosan hydrogel was synthesized from different methods including ionically cross-linked hydrogel, entangled gel, polyelectrolyte complexes, electro-rheological fluids, etc. Usually, an aqueous solution of chitosan was prepared using sodium hydroxide solution for gelation [51]. The chitosan hydrogel formed is then followed by supercritical fluid drying by using CO_2 [52]. The morphological structure of the formed chitosan aerogel was studied and it has been noted that the chitosan hydrogel drying in air tend to form film-like chitosan structure, without significant pores, thereby giving a low surface area. The reason for the less porous film structure is that the drying by air tends to collapse the chitosan network that leads to the enormous contraction of gel [53]. While, the supercritical drying using CO_2, allows the chitosan aerogel with 3D porous networking structure with high porosity and surface area. Therefore, the supercritical drying using CO_2 is the best drying method for chitosan aerogel synthesis.

5.3   Pectin aerogel

Pectin is the polysaccharide obtained directly from the cell wall of the plant. Firstly, the intra-hydrogen bonds are created by acidic or thermal means to form the wet gel structure [54]. The carboxylic groups present on the molecules of pectin are involved in the formation of these hydrogen bonds and the neighbouring molecules also form the hydrogen bonds between their hydroxyl moieties [55]. Then, the solvent from the pectin hydrogel is removed by supercritical drying using CO_2 to form pectin aerogel with high surface area and high porosity [56].

5.4   Alginate aerogel

The alginate aerogel can be synthesized by various routes; (1) the wet gel of alginate is produced by adding the dissolved sodium alginate solution dropwise to the calcium chloride solution. Then the synthesized hydrogel is dried by freeze-drying technique. (2) Similarly, the hydrogel of alginate can be transformed into alcogel by substituting the water present in the pores with alcohol. Then, the alcogel is dried by supercritical drying (SCF) using CO_2 [57].

Recently, the alginate/clay aerogel was synthesized by taking the equal mass of clay and alginate in water. The resultant solution is completely mixed. The pH of the resulting mixture is maintained at 6 or 8 by adding the toluene sulfonic acid monohydrate. The resulting mixture is frozen for 6 hr in a freezer at -80°C and then dried at low pressure using the freeze-drying technique. It was noted that the alginate colour changes to dark brown by adding the clay in alginate/clay aerogel.

It is also noted that the specific density of alginate/clay-based aerogel is around 400 cm^3 g^{-1} at a pH of ~6, while the specific density of the alginate/clay aerogel is around 232 $\pm$ 20 (cm^3g^{-1}) at the pH of about ~ 8 [58].

5.5   κ –Carrageenan aerogel

The κ-carrageenan based aerogel was synthesized by the following steps. Briefly, the κ-carrageenan was dissolved in water at 80°C by magnetic stirring for 12 hours. Then the κ-carrageenan based hydrogel was prepared by the emulsion method using the potassium carbonate (K_2CO_3) [59]. Finally, the κ-carrageenan based aerogel was prepared by freeze-drying the κ-carrageenan hydrogel [60]. The resultant porous κ-carrageenan aerogel has a high surface area in the order of 167 m^2g^{-1} and the pore volume of around 0.54 cm^3g^{-1}. The pore diameter of the aerogel is around 13 nm [59].

5.6   Starch aerogel

Amylose and amylopectin are the two components which comprise starch. The composition of starch varies in the relative amount of amylose and amylopectin depending upon the source material. The ratio of amylose to amylopectin affect the crystallinity and molecular order of starch [61,62]. Gelation process of starch is a thermally assisted three-step mechanism where the plasticization and hydration of the network take place at the same time. Firstly, the starch granules get swollen by adsorption of water due to hydrophilicity [63,64]. Secondly, the amylose molecules are discharged by elevating the temperature of the starch solution. The removal of amylose molecules destroys the granular structure of starch and this irreversible change leads to the gelatinization of starch. Finally, the ageing process is carried out with cooling to room temperature that leads to the gelatinized starch. This retrogradation step leads to the formation of the starch hydrogel. Here, the re-crystallization and reorganization of starch structure can take place. It can be inferred that the gelation process in starch is mainly affected by gelation temperature and amylose composition.

In the gelation process of starch, the retrogradation step involves the reorganization of the amylose molecules which are dissociated from starch granules. These reorganized amylose fractions then get deposited on the amylopectin portion. The starch gel is somewhat amorphous due to the linear amylose portion incorporation. This will create the mesoporosity in the starch gel structure. During the retrogradation step, the low cooling temperatures are maintained to get a high surface area aerogel. The hydrogel of starch is transformed to aerogel by drying through the water-ethanol solvent exchange, to avoid the shrinkage of 3D porous aerogel structure [65, 66]. This water-ethanol based system is also critical to prevent the particle coalescence during the gelation.

5.7   Curdlan aerogel

D-glucose monomers are linked in $(1{\rightarrow}3)$ manner to create the glycosidic linkage, which leads to a polymer called curdlan (CURD) as shown in Fig.1 [67]. Curdlan is employed for various applications such as medicinal purposes or as food because it is biodegradable, biocompatible and safe. In the medicine field, it is especially utilized as a drug loading, for tissue repairing, anti-tumour, etc. but it is insoluble in water which restricts its medical domain. The insolubility of CURD lies inherently in the structure where triple-helix is formed and a large number of hydrogen bonds are associated with it. The solubility of CURD can be improved by sodium hydroxide solution which helps in the straightening of coiled structure [68-74]. Therefore, the solubility of CURD in cold water in enhanced and the curdlan can be utilized to form the aerogel [75-77]. Unfortunately, the curdlan based cryogenic fragile and mechanically weak. Mostly, the polysaccharides possess a huge number of pores but with the fragile characteristics [73]. To improve the mechanical strength, polyetheneoxide (PEO) is usually added to the curdlan in a different ratio to support the aerogel formation [73, 67]. Similarly, the mechanical features of aerogel are improved by reinforcing with cellulose nanofibrils (CNF).

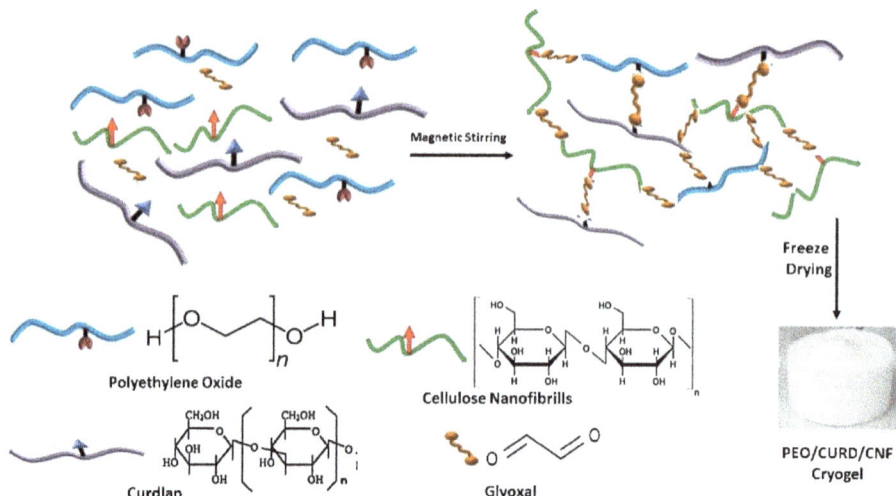

Fig. 1 *Graphical representation for the synthesis of CURD aerogel. Reproduced with permission from [67] Copyright (2017) Elsevier*

The morphological features of CURD aerogel owe the porous aerogel network in the structure which contains a huge number of pores on the basis of CURD to PEO composition. Similarly, it was noticed the curd cryogel swelling phenomenon when the different concentration of PEO is mixed.

5.8 Cellulose aerogels

Cellulose is a naturally found macromolecule synthesized by the plants. It is a polymer of β-D-glucopyranose having the 1-4 linkage and can exist in four polymorphic forms in a crystalline state [46]. Plant cells synthesize cellulose from glucose monomers which are prepared by plants in the photosynthesis. Plants cell wall carries the hemicellulose and lignin along with cellulose, forming three-phase composite material for mechanical strength to plants especially the big trees.

To make the cellulose fibre, hemicelluloses and lignin should be separated from the cellulose matrix. This process of separation is done chemically to make the viscous liquid that goes through the spinning process. Then the regeneration of pure cellulose fibrils and fringes is completed [78,79]. Cellulose usually has organized fibrils and fringes in a way that there is an area of crystalline ordered phase being mixed with amorphous disordered phase [46].

Cellulose is hydrophilic in nature and swells considerably in an aqueous medium but it does not dissolve in water along with other organic solvents. Because cellulose is a very stable material, the suitable process should be adopted for the breakdown of cellulose into its fibrils and then reconstruct these elementary units into a porous gel with less density that could be supercritically dried to get the three-dimensional structure of aerogel.

5.8.1 Cellulose aerogel monoliths

Weatherwax et al. [80] for the first time successfully preserved the bulged structure of cellulose. They utilized pulp from different sources such as cotton, rayon, etc. These pulpy substances were swollen in water, aq. NaOH and ethylenediamine [78,80] dried through an exchange of solvent and left for slow evaporation. They were able to make porous material but considerable shrinkage was also observed. Weatherwax and Caulfield [80] later used supercritical drying along with solvent exchange and were able to make cellulose aerogel having a surface area of around 200 m^2g^{-1} [81].

First cellulose aerogel that gains the attention and becomes well known were synthesized by Tan et al. [82]. They de-esterified the cellulose acetate and crosslinked with toluene-2,4-diisocyanate. They observed that cellulose gel could be formed with the cellulose concentration between 5-30% by weight. Their supercritically dried aerogel has a surface area of around 400 m^2g^{-1}.

Jin et al. [83] developed a method to make a good quality cellulose aerogel in which they avoided the toxic isocyanates with a lesser amount of cellulose. They utilized the semi-crystalline cellulose in which highly ordered crystalline areas are interconnected with unordered ones.

Ishida and co-workers [84] prepared carbon aerogel from cellulose precursors by dissolving it in H_2SO_4 or NaOH aqueous solution. The resultant suspension of cellulose was sprayed onto a Cu plate. After the supercritical or freeze-drying process, the powder form is achieved having the surface area in order of 32 to 178 m^2g^{-1}. Pyrolysis can suitably enhance the surface area up to 500 m^2g^{-1} consisting of carbon ribbons.

5.8.2 Nanostructured cellulose filaments in textile

Among the earliest material known to make the fibre for the fabrics includes the cellulose. Nowadays, fibre, yarn is produced both naturally and synthetically [85]. Fibre and filamenttoday have a large aspect ratio with compact microstructure. Schmenk et al. [86] and Hacker et al. [87] for the first time produced the cellulose aerogel by sol-gel method and spinning techniques, having the open porous structure with a nanostructured filament of cellulose. The synthesized fibres are usually characterized based on their pore size distribution, specific surface area, microstructure, tensile strength, density, etc.

6. Resorcinol–formaldehyde aerogels

These aerogels are classified as open-celled foam having low density due to the large internal void space up to 98% by volume that's why these materials have low thermal properties, enhanced acoustic impedance along with high surface area [88,89]. These low-density open-celled foams are produced from the wet gel by supercritical fluid drying (SCF) such as CO_2 [2,90]. These aerogels can be utilized in thermal and acoustic insulation, gas storage material and gas filter, conducting and dielectric materials [91].

Pekala and co-workers [92] for the first time synthesized purely organic aerogel comprises of resorcinol–formaldehyde (RF) resin. These aerogels were made by resorcinol and formaldehyde using the condensation mechanism which utilized the Na_2CO_3 as a catalyst in an aqueous medium. The wet gel was supercritically dried using CO_2 [90]. In recent literature, RF synthesis in both acidic and alkaline catalytic conditions have been explored [92,93]. Studies reflect that the variation in the process conditions such as the resorcinol to catalyst ratios (R/C ratio) in RF aerogel make the variety of nanostructures. Aerogel proceeds with higher R/C ratio obtained in a mesoporous skeletal network, having the secondary particle in the 40-70nm range which in turn formed by primary particles of 10-12nm [94-98].

7. Composite aerogels

7.1 Polymer-crosslinked aerogels

It was noticed that the silica-based aerogel has some issues which hinder in their widespread commercial applications like fragility, hydrophilicity, drying by SCF route (leads to high expense and hazard issue as SCF drying is a high-pressure method). Hydrophilicity and drying issues are correlated and addressed by the surface tailoring of silica with alkyl moiety to make it water repellent in nature. This modification makes the surface of wet gel, void of -OH groups present on it.

In the process of drying, siloxane linkages are developed by the reaction of –OH functional groups when they come close to one another and cause irreversible shrinkage. While, the wet gel with alkyl modified surface also shrink but after the solvent evaporation in the pores, alkyl moieties repel each other and consequently aerogel regain the original pore size of the wet gel. This type of spring-back effect was first introduced

Materials Research Forum LLC
https://doi.org/10.21741/9781644900994-2

by Prakash et al. [99]. Similarly, Rao et al. [100,101] has noted that silica-based gel modified with hexamethyldisilazane cause only 2% volume shrinkage.

Despite the fact that the issue of drying and hydrophilicity (stability) was addressed, the commercialization of aerogel is slow and the issue of fragility need to be further investigated.

7.2   Effect of polymer addition on aerogel fragility

As silica-based aerogel has pearl-necklace like structure in which the interparticle necks are vulnerable to stress, reflecting their fragility [102-105]. It was observed that the dissolution and again solidification of silica can enhance the mechanical features of silica aerogel due to the appearance of negative curvature at the surface of the interparticle neck but the strength is achieved at the cost of skeletal nanoparticles [106, 107]. Another approach to strengthening the silica is by treating it with tetraethylorthosilicate (TEOS) after gelation [108].

Aerogel can be compounded with polymers to enhance their tensile strength. The polymers are being incorporated into silica-based aerogel to solve the problem of cracking and shrinkage during drying. These incorporations of polymer also create transparency in the aerogels which make them a suitable candidate for utilization in optics.

8.   Exotic aerogels

8.1   Chalcogenide aerogels

Recent effort to make the aerogel on the chalcogenide framework has been tried out. Chalcogenide (i.e., S_2, Se_2, or Te_2), have the physicochemical properties distinct from traditional carbon/organic framework or oxides aerogel. These include bandgap semiconductivity, redox stability that controls catalytic and acid/base behaviour. Three different approaches are adopted to form these aerogels. (1) thiolytic cleavage of metal precursor (2) condensing the anionic chalcogenide cluster with cation (3) condensing the metal chalcogenide nano-assemblies [109-111].

8.1.1 Chalcogenide aerogels formation by thiolysis: GeS_2

Brock and co-workers [112] for the first time attempted to form chalcogenide aerogel using GeS_x compositions. They have chosen the system that is already known for making xerogel with high surface area (500 m^2g^{-1}), showing high porosity. When H_2S reacts with $Ge(OEt)_4$ in an inert atmosphere, followed by supercritical drying using CO_2 creates a porous, white-coloured chalcogenide aerogel having the composition $GeS_{2.4}$ like GeS_2 glass.

XRD of these chalcogenide aerogels confirms the amorphous nature and surface area analysis along with transmission electron microscopy (TEM) show porous aerogel network like silica aerogel. Brunauer–Emmett–Teller (BET) analysis gives the surface area of around 755 m^2g^{-1} which is greater than GeS_x xerogels previously formed. This method has some drawbacks in synthetic handling like the wet GeS_x on exposure to air during gelation will form crystalline GeO_2 rendering poor sulphur aerogels.

8.1.2 Chalcogenide aerogels formation by cluster-linking

Bag et al. [113] adopted a different methodology for chalcogen formation in which the anionic clusters like $[Ge_4S_{10}]^{4-}$ reacted with Pt^{2+} in a metathesis reaction. Previous attempts have formed mesostructured chalcogenide aerogels along with surfactant as template [114-116]. But the effort for the removal of surfactant causes the destruction of the internal porosity. New sol-gel methods create the aerogel with intact pore volume but disordered structures. This approach has quite a flexibility to the nature of building blocks. Different geometries of chalcogenide cluster along with varied coordination number of transition metal effects the properties of chalcogenide aerogels like specific surface area, pore size, optical, catalytic, etc.

8.1.3 Chalcogenide aerogels formation by nanoparticle assembly

In this method, chalcogenide aerogel formation is achieved by starting with nanoassembly of metal chalcogenide followed by condensing them to create a gel. Gacoin and co-workers for the [117-120] first time used the nanoparticles of CdS tailored with 4-fluorophenylthiolate to create a gel network on prolong standing. They noticed that surface thiolate groups are air oxidized forming disulphide linkages which create aggregate leading to porous aerogel as shown in Fig. 2 [110]. The gel has optical properties that rely on the size of initial nanoparticles and similar band gaps to founding building blocks.

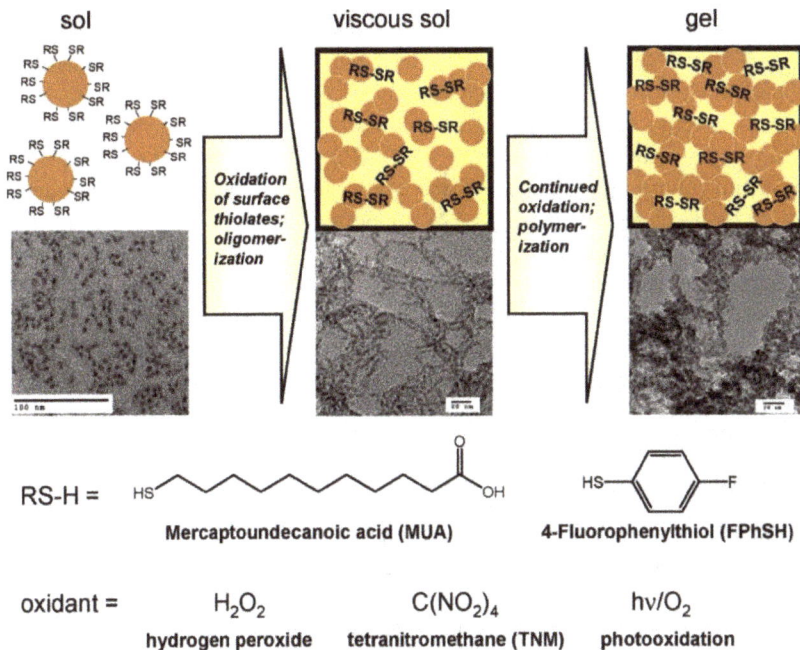

Fig. 2 *Schematic representation for the preparation of Chalcogenide based aerogel using nanoparticle assembly via Sol-gel synthetic route. (Reproduced with permission from [110] Copyright (2007) The American Chemical Society*

9. Conducting polymer aerogel

Conducting polymers are the polymer with highly conjugated chains. Because of their useful redox and electronic properties, they have been studied extensively. Commonly employed conducting polymers include polyaniline (PANi), polypyrrole (PPy), polyacetylene, polyfuran (PF), etc. There can be two types of aerogel based on their electric features; non-conducting and conducting aerogel. Conducting aerogel also exhibit good magnetic performance along with electric conductivity. Therefore, these aerogels have the huge potential to be utilized in catalysis, energy storage, gas sensing, etc.

Conducting aerogel obtained from conducting polymers is named as conducting polymer aerogel (CPAs). Rigidity associated with most of the conducting polymer helps in maintaining the microporosity (<2 nm). Conducting polymer aerosols are formed by the

Materials Research Forum LLC
https://doi.org/10.21741/9781644900994-2

coupling of superior properties of conducting polymers with aerogel, providing high specific surface area, more target sites for faster functionalization for a host-guest relationship, enhanced energy storage capacity, etc. But few challenges are there to be addressed. Firstly, the inherent poor solubility of conducting polymer poses great difficulty in large scale manufacturing of CPAs. Secondly, the inherent rigidity beneficial in maintaining porosity, on the other hand cause elasticity problems and new strategies needs to be developed for making CPAs strong and elastic. Recently, Zhang and co-workers [121] have worked out the improvement in CPAs regarding their rigidity and brittle character. Their success could lead to the development of flexible electronics. It is important to discuss three CPAs including polypyrrole (PPy), polyaniline (PANi), poly(3,4-ethylenedioxythiophene) (PEDOT) related to their porosity and other physicochemical properties.

9.1   Conducting polymer aerogels- A property prospective

9.1.1 PEDOT aerogels

The PEDOT along with poly(styrenesulfonate) ((abbreviated as PSS) form a complex PEDOT-PSS aerogel dried by SCD using CO_2 show enhanced surface area (170–370 m^2 g^{-1}) accompanying the density in the range of (0.138–0.232 g cm^{-3}) keeping the porous structure intact [121]. PEDOT-PSS aerogels have the electrical conductivity of around 10^{-1} S cm^{-1}. PEDOT aerogels prepared by emulsion polymerization and dried by SCF using CO_2 has the conductivity level as high as up to 101 S m^{-1} [122]. Compared to the conventional aerogel, PEDOT aerogels demonstrate excellent adsorption capability to different guest substances like heavy metal, dyes, etc. Embedding of PEDOT-PSS aerogel matrix with multiwalled carbon nanotubes [123] also improves different properties including surface area, thermal stability, conductivity, etc. The resulting composite aerogel have shown increased surface area (280–400 m^2 g^{-1}) and conductivity (1.2–6.9 × 10^{-2} S cm^{-1}) which are quite lightweight (0.044–0.062 g cm^{-3}) keeping the porous structure intact.

9.1.2 Polypyrrole (Ppy) aerogels

Lu's group [124] for the first time synthesize PPy aerogel through SCF method, having electrical conductivity (0.5 S m$^{-1)}$, low density (0.07 g cm^{-3}) still keeping the elasticity intact in aerogel initially present in hydrogel i.e. $\geq$70% compressibility regains its original shape in 30 as seconds shown in Fig. 3 [124]. Similarly, PPy-Ag NW aerogels exhibit excellent compressive elasticity which can bear up to 90% stress deformation and return to the original shape in few seconds.

Materials Research Forum LLC
https://doi.org/10.21741/9781644900994-2

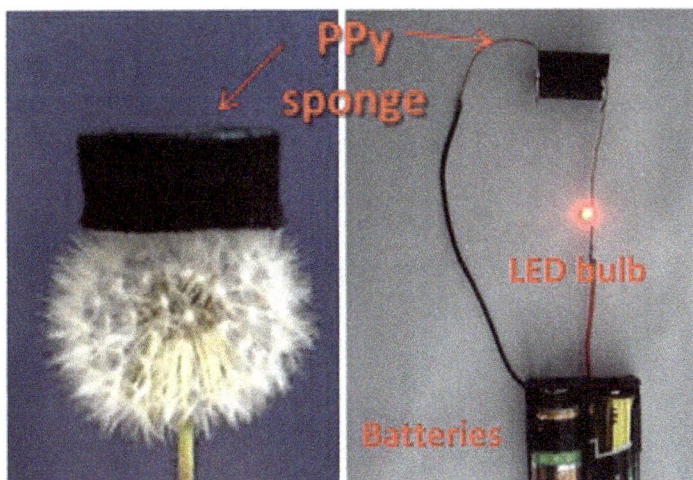

Fig. 3 *Digital Photographs of lightweight and conductive PPy based aerogel connected to a LED bulb (Reproduced with permission from [124] Copyright (2014) Springer Nature*

9.1.3 Polyaniline (PANi) aerogels

PANi aerogels can be prepared from their particles hydrogel by SCF drying method [125] which exhibit densities in range 0.03–0.15 g cm^{-3}, micromorphology just similar to corals, large BET surface area (39.54 m^2 g^{-1}) along with high mechanical endurance. Interesting coral-like morphology of PANi aerogels can be achieved by nanofiber having the diameter around 50-150 nm and length in few microns showing in Fig. 4 [125].

Due to this unique morphology, electrochemical performance is enhanced as the coral arrangement having the three-dimensional porous network help in an effective transport of charges. SEM analysis shows high porosity in PANi aerogels. Brunauer-Emmett-Teller (BET) analysis shows the high mesoporosity (2–50 nm). BET also indicates that the PANi aerogel has the pore size of 28 nm and an enhanced surface area of 39.54 m^2 g^{-1}. The mesoporous structure along with large specific surface area enhances the electrolyte-electrode contact and consequently increases the electrolyte transfer. PANi aerogels have the integrated hydrogel network shown in their density which in turn enhanced their mechanical strength and help in charge transfer in electrochemical tests.

Fig. 4 *SEM image of PANI hydrogels by showing different porosity (Reproduced with permission from [125] Copyright (2015) Elsevier*

10. Sonogels

Tarasevich et al. [126] for the first time reported the use of ultrasound to the sol-gel method. In the acid catalysed process, inside the cavitation bubbles, the alkoxides undergo hydrolysis to produce alcohol as a by-product of reaction which acts as a solvent. So no use of an additional solvent is required. Afterwards, the condensation reaction takes place to form the metal oxide. Therefore, we get the product just like the

Materials Research Forum LLC
https://doi.org/10.21741/9781644900994-2

conventional method (in the absence of ultrasound). But the difference lies in the structural and textural properties of aerogel produced by sonification method. Zarzycki and Blanco et al. [127, 128] have also focused on the textural and kinetic properties of sonogel. They have investigated several systems including SiO_2–Al_2O_3–MgO, SiO_2–TiO_2, ZrO_2, etc.

11. Graphene aerogel

Aerogel comprising of reduced graphene oxide (rGO) are currently much explored owing to the low density, high surface area, high porosity, good conductivity of rGO. These aerogel materials are a good prospect to be utilized as a gas sensor, batteries, supercapacitors, etc. Different efforts are being made to prepare aerogel based on reduced graphene oxide to enhance their surface area, conductivities, elasticity, and strength. So, they can be utilized especially in portable electronic devices [129].

11.1 Preparation of reduced graphene oxide aerogels

Reduced graphene oxide aerogel has a unique structure and morphological properties which makes it a good candidate to be synthesized by different approaches in controlled structure. Main methods employed for the fabrication of such structures include the self-linkage of graphene oxide (GO) systems, cross-linking of such structures, hydrothermal assisted aerogel. All the mentioned methods require the prevention of stacking of GO/rGO layers during the drying process which can be achieved using the freeze-drying or supercritical drying technique. Fixation of the dispersion structure and less mobile layers of GO/rGO makes it feasible to get material with enhanced surface area and porosity. This generates the exfoliation of graphene-like layers in the finished product.

If the freeze-drying method is opted it will lead to a large number of macropores in the structure because the ice crystals are formed during the freezing, which will dislocate some material part from the whole volume. It will eventually lead to structural flaws and the creation of macropores in the drying process as the speed of freezing controls the size of an ice crystal. While the supercritical drying leads towards the microporous structure. Both drying approaches form the ultralight and porous structure with different advantages due to their highly developed morphology.

For example, the large number of molecules and ions can diffuse in various pathways if macropores are present in the structure. This is beneficial in numerous applications. Reducing species of different nature are employed. Zhang and co-workers [129] prepared the robust and electrically conductive aerogel using L-ascorbic acid. They showed that the different by-products which are formed with other reducing agents like $NaBH_4$,

LiAlH₄, hydrazine are absent in case of L-ascorbic acid [130,131]. So, the uniformity of structure in the aerogel is not disturbed. Aerogel based on reduced graphene oxide is also formed with vitamin C, NaHSO₃, Na₂S, hydroquinone, sodium ascorbate, etc. Recently, aerogel based on rGO is also synthesized with an environment-friendly and cheap reducing medium like the combination of sodium iodide with oxalic acid [132]. These materials have shown high porosity, lightweight and enhanced electrical conductivities. Yang et al. [133] have prepared the rGO based aerogel by incorporating the weak alkali (NaHCO₃) into graphene oxide suspension using the thermal treatment. This method employed the in-situ reduction to prepare the rGO aerogel. Likewise, hypophosphorous acid and I_2 reduction process also synthesized the rGO based aerogel. These aerogels have shown the enhanced surface area of around 830 m^2g^{-1} [134]. Mostly, the reduction process takes place in the liquid medium. Interestingly, the hydrazine vapours are utilized for the reduction of rGO based aerogel at room temperature conditions [135]. The author reported the highly vigorous nature of the reduction process. The obtained material is quite similar to the material produced by the thermal process.

12.   Carbon nanotubes (CNTs) aerogel

Carbon nanotubes (CNTs) possess the extraordinary electrical and mechanical properties which impart promise in their usage in actuators, energy storage devices, interfaces, artificial membranes and composite materials [136,137]. However, the synthesis of CNTs aerogel with the inherent properties of CNTs is a challenging task. Mostly, the synthesis process of CNTs aerogel involves the dispersion of CNTs but the growth of CNTs within the aerogel matrix is also reported [138-141]. Dispersion of CNTs in aqueous medium require surfactant due to its hydrophobic nature. Then, gelation followed by drying for post-treatment. Bryning and co-workers [137] for the first time synthesized the CNTs aerogel using a surfactant to disperse the single-walled CNTs (SWCNTs) in aqueous medium. The resultant suspension would be changed to wet gel and then washed by polyvinylalcohol (PVA) solution to clear out the remaining surfactant. Both SCF drying using CO_2 and freeze-drying can be utilized to get the final aerogel.

CNTs aerogel properties are varied with the concentration of CNTs, the concentration of PVA and drying method employed. The increased concentration of CNTs gave the aerogel with enhanced electrical conductivity up to1S cm^{-1}as compared to traditional carbon aerogels (CA) of the same density exhibit much lower conductivity. Few limitations need to be addressed including the fragility of CNTs aerogel. Mechanical features of CNTs aerogel can be improved by reinforcing with PVA. So, these reinforced aerogel has shown excellent mechanical properties as they can bear the weight 8000 times to their own weight [142-144]. But these reinforced aerogel has lost the

Materials Research Forum LLC
https://doi.org/10.21741/9781644900994-2

conductivity due to the interference by the reinforced binder in the conduction of CNTs. This has led the researchers to reinforce the CNTs aerogel with a conductive binder like resorcinol-formaldehyde (RF) instead of traditional binders to get better mechanical properties along with improved conductivity.

Worsley et al. [145] for the first time added the evenly distributed CNTs in surfactant to the RF solution to get the CNTs-carbon aerogel (CNTs-CA) composite. SEM analysis of CNTs-CA composite was shown to have the CNTs bundles being dispersed in the CA matrix. They have better electrical properties along with mechanical ones but still less improved because the content of carbon derived from RF is still high as compared to CNTs. So, there is a requirement to reduce the RF-derived carbon percentage in the CNTs aerogel to just coat and crosslink the CNTs network. Worsley et al. [146] decreased the RF amount by weight from 12 to 4% while keeping the number of CNTs the same as before. It has been observed that the reduced amount of RF sol-gel initiates polymerization process at the sidewalls of CNT and at the junction of CNTs, bundles to act as a binder. Scanning electron microscopy (SEM) analysis confirms the CNTs based foams are formed having the filament-like struts randomly connected to each other of diameter between 5-40 nm and length between 500-1000 nm. While the TEM reveals the layer of carbon on the CNTs, confirming the RF-derived carbon growth and nucleation on the surface of CNTs. The use of carbonaceous binder prepares the CNTs aerogel exhibiting the excellent electrical and mechanical properties. In Fig. 5 [146], the CNTs aerogel are showing comparable properties to conventional CA and other porous carbon material like the modulus of elasticity. Similarly, the density of aerogel showing that with over 16% by weight coating of CNTs make them twelve times stiffer than SiO_2 and three times more stiffer than CA aerogel [147-149]. Likewise, CNTs aerogel is three times stronger than alumina nanofoams in which morphology of struts are like curled nano-leaflets [150]. In Fig.5 [146] also shows that aerogel with 55% by weight CNTs loading has a of density 30 mg cm^{-3} and exhibit strain of 76%. This confirms that aerogel is super elastic due to fully strain recovery that is being noted for nanotube-based aerogel with a density < 50 mg cm^{-3}. This low-density CNTs aerogel has also shown 5 times higher electrical conductivity (>1 S cm^{-1}) as compared to conventional CAs.

Gutiérrez et al. [151] synthesized high-performance CNTs aerogel by utilizing the deep eutectic solvents (DESs) to disperse the CNTs. DES catalyses the polycondensation of furfuryl alcohol to make the porous network of colloidal particles. On the addition of multiwalled carbon nanotubes (MWCNTs), furfuryl alcohol act as conductive glue between the (MWCNTs) and coat them. These CNTs aerogel exhibit the high electrical conductivities in order of 4.8 S cm^{-1} and elastic modulus of up to 24 MPa [152].

Materials Research Forum LLC
https://doi.org/10.21741/9781644900994-2

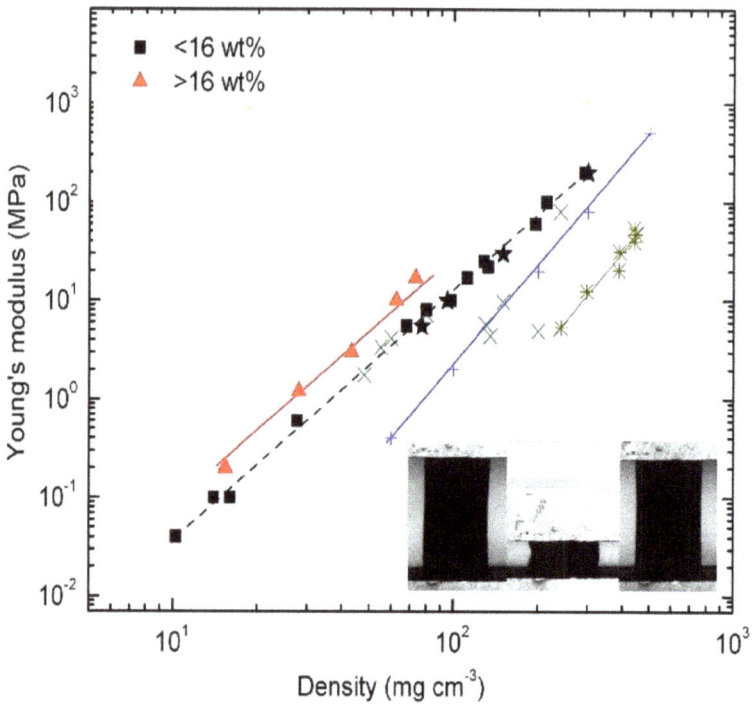

Fig. 5 *Comparison of Young's modulus of monolithic CNTs foam, carbon, silica and alumina aerogels. (Reproduced with permission from [146]Copyright (2009) AIP Publishing*

13. Hybrid aerogel

It has been observed that the reinforcing the silica backbone with the polymer can be useful to increase the mechanical strength [153]. Different types of bonding of silica particles with the organic functionalities of the polymer are observed. Depending upon the chemical interactions, hybrid aerogel can be divided into the following classes; Class-I hybrid composite aerogels and class-II hybrid composite aerogels [154-156].

13.1 Class-I hybrid composites

Class-I hybrid aerogel includes composite materials where the hydrogen bonding and van der Waals type interactions are observed in the organic and inorganic phases [157, 158]. In this type, the organic polymer is used to reinforce the inorganic matrix and both are independent of each other. Various routes are opted to prepare these materials. The most important examples of class-I are the insertion of the organic dyes into the sol-gel matrix [159,160]. The organic dyes are physically dissolved with the inorganic precursors (e.g., TEOS or TMOS) or introduced at the sol stage to get trapped in the gel during the gelation and drying of the mixture. Composite materials having the organic group with the polar end, usually form the hydrogen bonds between different components. The repeating unit of the polyethene glycol [161] or PVA [162, 163] carries the hydroxyl group. Silanol groups of the silica matrix create a strong bonding with them. These hybrid materials are usually employed to enhance the mechanical characteristics of an organic polymer like inorganic filler SiO_2 incorporated into the polysiloxanes polymer (PDMS) [164,165]. Class-I hybrid aerogels are seldom studied because of the post-gelation washing of these aerogel cause percolation of the polymer out of the pores, as the weak bonding at the interphase between the two phases unable to hold them together. However, class-I hybrid composite aerogel can be straightforward to make xerogel composite by the monomer induction in the sol or by monomer polymerization in the post-gelation step. The hybrid composite produced by this method shows the less elastic modulus along with 3 times increase in the compressive strength [159, 160]. The ambient dried class-I hybrid composite aerogel has cost-effectivity and easy to produce. So, the removal of post-gelation step is important for the large scale manufacturing of these materials [166].

13.2 Class-II hybrid composite

Class-II hybrid composites aerogel comprises of strong covalent bonds network between the interfaces of the organic phase and silica matrix. Hu et al. [167] studied the various types of covalent bonding among the phases. This methodology requires the reactants that contain the stable chemical bond between the inorganic network and an organic phase. Class-II hybrid composite aerogel possesses unique properties compared to initial precursors. In this regard, the polymer cross-linkage, composition of aerogel, and desired properties with simple and scalable processing are the points of attention for many researchers [153, 168, 169].

14. Application of porous aerogel

14.1 Thermal insulation

Aerogel consists of a trapped gas phase in the solid phase so the heat transfer in aerogel could be of two types, heat conduction in solid and gas phase. Heat conduction in gas phase involves various mechanism like a collision between gas molecules [170], by radiative conduction of gas contained in pores to a solid surface, by convection from the pores [171]. The average pore diameter in an aerogel is 20-40 nm and the mean free path of gas under STP is typically 70 nm. Microporous and mesoporous structure in aerogel due to small diameter hamper the free movement of air, it would lead to a reduction in heat conduction through gaseous medium [172]. A different mechanism of heat transfer in aerogel has a coupling effect with each other [173]. The solid heat conduction usually proceeds by the reticular vibration of molecular solid. Solid heat conduction has been significantly reduced in such aerogels because of the small particle size of the porous network [174,175].

In general, the aerogel based insulation product could be divided into the three following types [171]:

i.   Monolithic aerogels: Homogeneous blocks of aerogel having the dimensions in few cms.

ii.   Granules or powders: fine fragment of aerogel of diameter less than 1 cm and 1 mm in case of powders.

iii.   Composites: Homo or heterogeneous aerogel in which one of the adding substance is inserted in a gel matrix or by surface modification inserted as the second phase.

14.2 Removal of pollutants

The separation process in the gas phase can be done in three different ways; particle separation by filtration, gas sorption and destroying the compound by chemical reaction. Silica-based aerogels are commonly employed when the pollutant particles in the gas are to be removed. These pollutant particles are captured in the small size pores of aerogel while the gas passes through them. The process of sorption is commonly employed where molecular separation in the gas phase is required and surface modification by the relevant functional group is a necessary condition for the sorption process. The separating molecules are trapped in pores and react with the appropriate functional group on the surface of the pores.

14.3 Elimination of solid particle from gases

Silica-based aerogels are used in the elimination of solids from gases due to its porous structure. The silica aerogel shows the excellent separation efficiency for solid particles having the size <70 nm when employed in the form of powder or granules having the micro or mesoporous structure. Oil particles could be separated with greater efficiency of around 2.3 cm s^{-1} [176]. More selective filtration is attainable by controlling the pore size during synthesis. Hybrid aerogels can be employed where the flow rate of gases are high because they possess the reinforced structure with mechanical strength to bear the high pressure of gas [177]. The hybrid composition also gives the advantage of both mesoporous and microporous structures [178].

14.4 CO_2 capture

To purify and separate the CO_2, aerogels are used with few amino-functionalities on the solid surface [179]. To introduce the amino functionality on the surface of aerogel different strategies are used such as co-condensation in gel preparation and post-synthesis functionalization. Another method to introduce amino group is the usage of macromolecules with a high content of the amino group. Similarly, amino-functionalized silane can be used to introduce chemically linked amino group in the porous network during the sol-gel process. The CO_2 is captured due to the affinity of CO_2 towards the amino group in a humid atmosphere at 100°C as given by the reaction below [180]:

$$CO_2 + RNH_2 + RNH_2 \rightleftharpoons RNHCOO^- + RNH_3^+ \tag{2}$$

The physically adsorbed amines to aerogel surface are advantageous than chemically linked amines to aerogel surface because of the greater CO_2 absorption capacity. But they have little cyclic stability in the thermal regeneration process. On the other hand, the chemically linked amines (e.g., mono-, di-, and trialkylaminotrimethoxysilane) have excellent cyclic stability. Among the chemically linked amines, the di-, and tri-amino groups provide the surpassed performance because ample space is available for the gaseous exchange.

To capture the CO_2 in the industrial equipment at a large scale required to have the low water sorption capability along with good thermal regeneration capacity. Lin et al. [181] has developed the CO_2 capture system based on the fluidized-bed bench reactors in which silica aerogel is incorporated with amino functionality with a hydrophobic feature.

14.5 Volatile organic compounds/catalysis

Silica aerogel has (-OH) functionalities on the surface where the organic compound can be easily adsorbed, leads to the degradation of hydrocarbon due to surface chemistry. Titina doped silica aerogel is used for adsorption of chlorinated hydrocarbon. Shengli et al. [182] has employed the TiO_2-SiO_2 aerogel having the high TiO_2 content for adsorption and photodegradation of trichloroethylene.

14.6 Water treatment

Oil and intoxicating organic contaminants in the water is an important factor regarding water pollution [183, 184]. Because the uncontrolled drainage of industrial and municipal waste in water bodies without treatment is the cause of environmental concern and hazard for water flora and fauna [185].

14.6.1 Oils in water

Oils could be classified in different types such as light and heavy hydrocarbon, emulsified and non-emulsified oil, plant and animal fat, lubricant and cutting fluid, etc. [185]. Many techniques are used to separate these oil fractions from water. The main mechanisms are filtration and adsorption, which are used for capturing and then biological or chemical degradation. Industrial sorbents employed for the adsorption have the issues of low selectivity, slow kinetics and low elimination rate. Adsorption of a pollutant can be improved by using the hydrophobic aerogel with the oleophilic surface [186-189]. Hydrophobic silica aerogel is used for efficient elimination of non-water-soluble organic compounds and the hydrophilic silica aerogel is used for efficient elimination of water-soluble organic compounds [190]. The Cabot corporation has prepared the hydrophobic silica aerogel called nanogel. They have modified the surface of aerogel with trimethylsilyl (TMS) moiety to impart the hydrophobic characteristics. These nanogels are used for the removal of emulsions by reverse fluidization-bed configuration. They have noted the different factors affecting the absorption capacity of aerogel in reverse fluidization-bed systems; including bed height, size of the granule, water flow rate, oil/water ratio in the emulsion. This nanogel has particles with an absorption capability of 2.8 times of their mass [190].

14.6.2 Wastewater and brackish water treatment

Treatment of brackish water near coastline and cleaning of wastewater is the requirement for the progressive usage of natural resources for drinking purpose. So, the capacitive deionization (CDI) systems are used where the carbon aerogel as an electrode is utilized due to high capacitance and high conductivity. CDI based desalination is the electricity-based technique [191]. This technique has the capability to remove the divalent ions up to 85% [192]. The CDI system has the cyclic process of the loading, purification involving the ion removal and regeneration (Fig. 6) [193] involving ion discharge to generate the two streams: one is desalinated water and another one is brine.

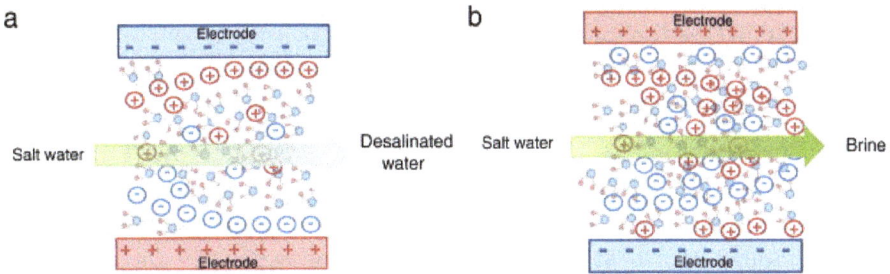

Fig. 6 a) Deionization and b) Regeneration step in CDI system (Reproduced with permission from [193] Copyright (2014) Elsevier

Many articles have been reported regarding the synthesis of high-performance electrode material of CDI; including mostly carbon-based materials like Al_2O_3 and SiO_2 nanocomposite [194], activated carbons [195], carbide-derived carbon (CDCs) [196], carbon nanotubes (CNTs) [197], carbon aerogel [198], mesoporous carbon [199], nano carbon fiber [200]. Farmer et al. [201], during 1990 made advancement in the CDI system by using carbon aerogel based electrode [201]. He has published three patents on CDI systems [202]. The carbon aerogel based electrode exhibit the high electrical conductivity, low resistivity of around 40 mΩ cm^{-1}, enhanced surface area between 400–1100 m^2 g^{-1} and pore size of 50 nm [203].

14.7   Biomedical applications

14.7.1 Aerogels for the administration of medicines

Aerogels are also capable to be utilized to release the drug in a controlled manner in a drug delivery system (DDS). The aerogel composition should be such that it could carry the drug without major changes during storage and then offer a safer release. For cytotoxic drugs, it is necessary to release the drug at the target point without changes to have a reduced dosage and fewer side effects. These drugs can be employed by aerogel in a better and controlled manner because of higher surface area, controlled pore size and the more surface functionalities in aerogel enhanced their utilization [203, 204]. The incorporation of drugs in aerogel is carried out in the sol-gel synthesis by the solvent exchange during gelation or supercritical drying using CO_2. The extent of loading depends upon the adsorption, drug stability in aqueous medium and other organic media [63, 66, 205, 206].

14.7.2 Tissue engineering

The tissue engineering is focused on implantable porous material development for bone tissue repairing regarding aerogel [207,208]. The porous aerogel structure resembles with natural porous structure found in the calcareous bone tissue. Aerogels can enhance cell-binding due to porosity in the structure in a way that the interconnected pores in aerogels provide the basis for essential nutrient and oxygen supply to the cell and the excretion of harmful wastes of cells. Mostly, aerogel lack macroporosity, which play a vital role in adhesion and proliferation of bone cells along with vascularization for tissue growth. The mechanical strength of aerogel is not enough to sustain the stress required for large bone structure implantation. So, many efforts are done to increase the aerogel strength by making their composite with an appropriate reinforcing agent. The reinforced aerogel is usually strengthened with inorganic fillers or biopolymers, which increase their density along with enhanced rigidity and ductility [209]. The reinforcement of aerogel is usually carried out employing the gel in the solution of a reinforcing material, which deposited on the gel and then dried [210].

14.7.3 Biosensing

The interconnected porous structure of aerogels can host various compounds and act as a material for biological sensing [211, 212]. Power et al. [21] reported that the silica-based aerogel can be used for colonizing and growth of bacteria. The pores of silica-based aerogel are increased between 10–100 μm on the treatment with a water-soluble polymer which results in successful cell growth and colonization. When bacteria are infected by bacteriophage virus they emit the fluorescent light. This fluorescent light is detected without disturbing the other cell of the colony. The emitted light is due to the expression of green fluorescence protein (GFP) on the bacteriophage infection.

References

[1]   S.S. Kistler, Coherent expanded-aerogels, J. Phys. Chem. 36 (2002) 52-64. https://doi.org/10.1021/j150331a003

[2]   C.J. Brinker, G.W. Scherer, Sol-gel science: the physics and chemistry of sol-gel processing, Academic press, California, 2013

[3]   F. Broecker, W. Heckmann, F. Fischer, M. Mielke, J. Schroeder, A. Stange, Structural analysis of granular silica aerogels, in: J. Fricke (Ed.) Aerogels, Springer, Berlin, Heidelberg, 1986, pp. 160-166. https://doi.org/10.1007/978-3-642-93313-4_21

[4]   S.W. Hwang, H.H. Jung, S.H. Hyun, Y.S. Ahn, Effective preparation of crack-free silica aerogels via ambient drying, J. Sol-Gel Sci.Technol. 41 (2007) 139-146. https://doi.org/10.1007/s10971-006-0513-y

[5]   C. Lee, G. Kim, S. Hyun, Synthesis of silica aerogels from waterglass via new modified ambient drying, J.Mater.Sci. 37 (2002) 2237-2241. https://doi.org/10.1023/A:1015309014546

[6]   K. Nakanishi, H. Minakuchi, N. Soga, N. Tanaka, Structure design of double-pore silica and its application to HPLC, J. Sol-Gel Sci. Technol. 13 (1998) 163-169. https://doi.org/10.1023/A:1008644514849

[7]   L. Pauling, The Nature of the Chemical Bond, Cornell university press Ithaca, Newyork, 1960

[8]   R. Mozzi, B. Warren, The structure of vitreous silica, J. Appl. Crystallogr. 2 (1969) 164-172. https://doi.org/10.1107/S0021889869006868

[9]   A. Ayral, J. Phalippou, T. Woignier, Skeletal density of silica aerogels determined by helium pycnometry, J. Mater. Sci. 27 (1992) 1166-1170. https://doi.org/10.1007/BF01142014

[10]   L. Kocon, F. Despetis, J. Phalippou, Ultralow density silica aerogels by alcohol supercritical drying, J. Non-Cryst. Solids. 225 (1998) 96-100. https://doi.org/10.1016/S0022-3093(98)00322-6

[11]   G. Reichenauer, Thermal aging of silica gels in water, J. Non-Cryst. Solids. 350 (2004) 189-195. https://doi.org/10.1016/j.jnoncrysol.2004.07.073

[12]   B. Zhou, J. Shen, Y. Wu, G. Wu, X. Ni, Hydrophobic silica aerogels derived from polyethoxydisiloxane and perfluoroalkylsilane, Mater. Sci. Eng., C. 27 (2007) 1291-1294. https://doi.org/10.1016/j.msec.2006.06.032

[13]   B. Yoldas, M. Annen, J. Bostaph, Chemical engineering of aerogel morphology formed under nonsupercritical conditions for thermal insulation, Chem. Mater. 12 (2000) 2475-2484. https://doi.org/10.1021/cm9903428

[14]   K. Duer, S. Svendsen, Monolithic silica aerogel in superinsulating glazings, Sol. Energy. 63 (1998) 259-267. https://doi.org/10.1016/S0038-092X(98)00063-2

[15]   N. Hüsing, U. Schubert, Aerogels-airy materials: chemistry, structure, and properties, Ang. Chem. Int. Ed. 37 (1998) 22-45. https://doi.org//10.1002/(SICI)1521-3773(19980202)37:1/2<22::AID-ANIE22>3.0.CO;2-I

[16]   A. Charlton, I. McKinnie, M. Meneses-Nava, T. King, A tunable visible solid state laser, J. Mod. Opt. 39 (1992) 1517-1523. https://doi.org/10.1080/09500349214551531

[17]   S. Jun, Z. Lei, W. Jue, W. Xiang, L. Yufen, Preparation of fullerence doped silica aerogels and the study of photoluminescence-properties, J. Inorg. Mater.11 (1997) 371-374

[18]   J. Shen, J. Wang, B. Zhou, Z. Deng, Z. Weng, L. Zhu, L. Zhao, Y. Li, Photoluminescence of fullerenes doped in silica aerogels, J. Non-Cryst. Solids. 225 (1998) 315-318. https://doi.org/10.1016/S0022-3093(98)00050-7

[19]   N. Leventis, I.A. Elder, D.R. Rolison, M.L. Anderson, C.I. Merzbacher, Durable modification of silica aerogel monoliths with fluorescent 2, 7-diazapyrenium moieties. Sensing oxygen near the speed of open-air diffusion, Chem. Mater. 11 (1999) 2837-2845. https://doi.org/10.1021/cm9901966

[20]   D. Feldman, Quasi-long-range order in nematics confined in random porous media, Phys.Rev.Lett. 84 (2000) 4886. https://doi.org/10.1103/PhysRevLett.84.4886

[21]   M. Power, B. Hosticka, E. Black, C. Daitch, P. Norris, Aerogels as biosensors: viral particle detection by bacteria immobilized on large pore aerogel, J. Non-Cryst. Solids. 285 (2001) 303-308. https://doi.org/10.1016/S0022-3093(01)00471-9

[22]   D. L. Bernik, Silicon based materials for drug delivery devices and implants, Recent patents on nanotechnology, 1 (2007) 186-192. https://doi.org/10.2174/187221007782360402

Materials Research Forum LLC
https://doi.org/10.21741/9781644900994-2

[23]   A. Buzykaev, A. Danilyuk, S. Ganzhur, E. Kravchenko, A. Onuchin, Measurement of optical parameters of aerogel, nuclear instruments and methods in physics research section A: Accelerators, Spectrometers, Res., Sect. A. 433 (1999) 396-400. https://doi.org/10.1016/S0168-9002(99)00325-3

[24]   T. Woignier, J. Reynes, J. Phalippou, Sintering of silica aerogels for glass synthesis: application to nuclear waste containment. In Aerogels Handbook, Springer, New York, 2011, pp. 665-680. https://doi.org/10.1007/978-1-4419-7589-8_29

[25]   I. Mejri, M. Younes, A. Ghorbel, P. Eloy, E. M. Gaigneaux, Comparative study of the sulfur loss in the xerogel and aerogel sulfated zirconia calcined at different temperatures: effect on n-hexane isomerization, in Studies in surface science and catalysis, Elsevier. 2006, pp. 953-960. https://doi.org/10.1016/S0167-2991(06)81002-5

[26]   D.A. Ward, E.I. Ko, One-step synthesis and characterization of zirconia-sulfate aerogels as solid superacids, J. Catal. 150 (1994) 18-33. https://doi.org/10.1006/jcat.1994.1319

[27]   J. V. Stark, D.G. Park, I. Lagadic, K.J. Klabunde, Nanoscale metal oxide particles/clusters as chemical reagents. Unique surface chemistry on magnesium oxide as shown by enhanced adsorption of acid gases (sulfur dioxide and carbon dioxide) and pressure dependence, Chem. Mater. 8 (1996) 1904-1912. https://doi.org/10.1021/cm950583p

[28]   I. Ferino, M.F. Casula, A. Corrias, M.G. Cutrufello, R. Monaci, G. Paschina, 4-Methylpentan-2-ol dehydration over zirconia catalysts prepared by sol-gel, Phys. Chem. Chem. Phys. 2 (2000) 1847-1854. https://doi.org/10.1039/A908992F

[29]   D.W. Richerson, Modern ceramic engineering: properties, processing, and use in design, CRC press, 2005

[30]   J. F. Shackelford, R.H. Doremus, Ceramic and glass materials, Structure, properties and processing, Springer, Boston MA, 2008. https://doi.org/10.1007/978-0-387-73362-3

[31]   D.J. Suh, T.J. Park, Sol-gel strategies for pore size control of high-surface-area transition-metal oxide aerogels, Chem. Mater. 8 (1996) 509-513. https://doi.org/10.1021/cm950407g

[32]   A. Bedilo, K. Klabunde, Synthesis of high surface area zirconia aerogels using high temperature supercritical drying, Nanostruct.Mater. 8 (1997) 119-135. https://doi.org/10.1016/S0965-9773(97)00011-1

[33]   G. Štefanić, S. Musić, Factors influencing the stability of low temperature tetragonal ZrO_2, Croa.Chem.Acta, 75 (2002) 727-767

[34]   C.N. Chervin, B.J. Clapsaddle, H.W. Chiu, A.E. Gash, J.H. Satcher, S.M. Kauzlarich, Role of cyclic ether and solvent in a non-alkoxide sol-gel synthesis of yttria-stabilized zirconia nanoparticles, Chem. Mater. 18 (2006) 4865-4874. https://doi.org/10.1021/cm061258c

[35]   R.W. Matthews, Photooxidation of organic impurities in water using thin films of titanium dioxide, J. Phys. Chem. 91 (1987) 3328-3333. https://doi.org/10.1021/j100296a044

[36]   L.R. Matthews, D. Avnir, A.D. Modestov, S. Sampath, O. Lev, The incorporation of titania into modified silicates for solar photodegradation of aqueous species, J. Sol-Gel Sci. Technol. 8 (1997) 619-623. https://doi.org/10.1007/BF02436911

[37]   E. Traversa, M. L. Di Vona, S. Licoccia, M. Sacerdoti, M. C. Carotta, M. Gallana, G. Martinelli, Sol-gel nanosized semiconducting titania-based powders for thick-film gas sensors, J. Sol-Gel Sci. Technol. 19 (2000) 193-196. https://doi.org/10.1023/A:1008723902604

[38]   P. Wang, S.M. Zakeeruddin, J.E. Moser, M.K. Nazeeruddin, T. Sekiguchi, M. Grätzel, A stable quasi-solid-state dye-sensitized solar cell with an amphiphilic ruthenium sensitizer and polymer gel electrolyte, Nature Mat. 2 (2003) 402. https://doi.org/10.1038/nmat904

[39]   H. Hirashima, C. Kojima, H. Imai, Application of alumina aerogels as catalysts, J. Sol-Gel Sci. Technol. 8 (1997) 843-846. https://doi.org/10.1023/A:1018310829773

[40]   H. Hirashima, H. Imai, V. Balek, Characterization of alumina gel catalysts by emanation thermal analysis (ETA), J. Sol-Gel Sci. Technol. 19 (2000) 399-402. https://doi.org/10.1023/A:1008731026228

[41]   R.J. Willey, C.T. Wang, J.B. Peri, Vanadium-titanium oxide aerogel catalysts, J. Non-Cryst. Solids. 186 (1995) 408-414. https://doi.org/10.1016/0022-3093(95)00063-1

[42]   C. Hoang-Van, O. Zegaoui, P. Pichat, Vanadia–titania aerogel deNOx catalysts, J. Non-Cryst. Solids. 225 (1998) 157-162. https://doi.org/10.1016/S0022-3093(98)00036-2

[43]   K. Shimizu, H. Imai, H. Hirashima, K. Tsukuma, Low-temperature synthesis of anatase thin films on glass and organic substrates by direct deposition from aqueous solutions, Thin Solid Films, 351 (1999) 220-224. https://doi.org/10.1016/S0040-6090(99)00084-X

[44] H. Imai, Y. Takei, K. Shimizu, M. Matsuda, H. Hirashima, Direct preparation of anatase TiO_2 nanotubes in porous alumina membranes, J. Mater. Chem. 9 (1999) 2971-2972. https://doi.org/10.1039/A906005G

[45] P. Zugenmaier, Crystalline cellulose and derivatives: characterization and structures, Springer, 2008

[46] J. T. Marsh, F. C. Wood, Introduction to the Chemistry of Cellulose, AGRIS, Chapman & Hall Ltd, 1942

[47] H. Maleki, L. Durães, C. A. García-González, P. del Gaudio, A. Portugal, M. Mahmoudi, Synthesis and biomedical applications of aerogels: Possibilities and challenges, Adv. ColloidInterface Sci. 236 (2016) 1-27. https://doi.org/10.1016/j.cis.2016.05.011

[48] K.S. Mikkonen, K. Parikka, A. Ghafar, M. Tenkanen. Prospects of polysaccharide aerogels as modern advanced food materials, Trends Food Sci.Technol. 34 (2013) 124-136. https://doi.org/10.1016/j.tifs.2013.10.003

[49] H. Derakhshankhah, M.J. Hajipour, E. Barzegari, A. Lotfabadi, M. Ferdousi, A.A. Saboury, E. P. Ng, M. Raoufi, H. Awala, S. Mintova, Zeolite nanoparticles inhibit Aβ–fibrinogen interaction and formation of a consequent abnormal structural clot, ACS Appl.Mater.Interfaces. 8 (2016) 30768-30779. https://doi.org/10.1021/acsami.6b10941

[50] E. Poorakbar, A. Shafiee, A.A. Saboury, B.L. Rad, K. Khoshnevisan, L. Ma'mani, H. Derakhshankhah, M.R. Ganjali, M. Hosseini, Synthesis of magnetic gold mesoporous silica nanoparticles core shell for cellulase enzyme immobilization: Improvement of enzymatic activity and thermal stability, Process Biochem. 71 (2018) 92-100. https://doi.org/10.1016/j.procbio.2018.05.012

[51] M. Robitzer, F. Di Renzo, F. Quignard, Natural materials with high surface area. Physisorption methods for the characterization of the texture and surface of polysaccharide aerogels, Microporous Mesoporous Mater. 140 (2011) 9-16. https://doi.org/10.1016/j.micromeso.2010.10.006

[52] J. Singh, P. Dutta, J. Dutta, A. Hunt, D. Macquarrie, J. Clark, Preparation and properties of highly soluble chitosan–l-glutamic acid aerogel derivative, Carbohydr. Polym. 76 (2009) 188-195. https://doi.org/10.1016/j.carbpol.2008.10.011

[53] J. Radwan-Pragłowska, M. Piątkowski, Ł. Janus, D. Bogdał, D. Matysek, V. Cablik, Microwave-assisted synthesis and characterization of antioxidant chitosan-based aerogels for biomedical applications, Int. J. Polym. Anal. Charact. 23 (2018) 721-729. https://doi.org/10.1080/1023666X.2018.1504471

[54]   A. Veronovski, G. Tkalec, Ž. Knez, Z. Novak, Characterisation of biodegradable pectin aerogels and their potential use as drug carriers, Carbohydr. Polym. 113 (2014) 272-278. https://doi.org/10.1016/j.carbpol.2014.06.054

[55]   C. A. García-González, E. Carenza, M. Zeng, I. Smirnova, A. Roig, Design of biocompatible magnetic pectin aerogel monoliths and microspheres, RSC Adv. 2 (2012) 9816-9823. https://doi.org/10.1039/C2RA21500D

[56]   G. Horvat, T. Fajfar, A.P. Uzunalić, Ž. Knez, Z. Novak, Thermal properties of polysaccharide aerogels, J.Therm. Anal. Calorim. 127 (2017) 363-370. https://doi.org/10.1007/s10973-016-5814-y

[57]   M. Pantić, Ž. Knez, Z. Novak, Supercritical impregnation as a feasible technique for entrapment of fat-soluble vitamins into alginate aerogels, J. Non-Cryst. Solids. 432 (2016) 519-526. https://doi.org/10.1016/j.jnoncrysol.2015.11.011

[58]   X.L. Li, M.J. Chen, H.B. Chen, Facile fabrication of mechanically-strong and flame retardant alginate/clay aerogels, Compos., B. 164 (2019) 18-25. https://doi.org/10.1016/j.compositesb.2018.11.055

[59]   M. Alnaief, R. Obaidat, H. Mashaqbeh, Effect of processing parameters on preparation of carrageenan aerogel microparticles, Carbohydr. Polym., 180 (2018) 264-275. https://doi.org/10.1016/j.carbpol.2017.10.038

[60]   K. Ganesan, L. Ratke, Facile preparation of monolithic κ-carrageenan aerogels, Soft Matter. 10 (2014) 3218-3224. https://doi.org/10.1039/C3SM52862F

[61]   J. Blazek, H. Salman, A.L. Rubio, E. Gilbert, T. Hanley, L. Copeland, Structural characterization of wheat starch granules differing in amylose content and functional characteristics, Carbohydr. Polym. 75 (2009) 705-711. https://doi.org/10.1016/j.carbpol.2008.09.017

[62]   P. Chen, L. Yu, G.P. Simon, X. Liu, K. Dean, L. Chen, Internal structures and phase-transitions of starch granules during gelatinization, Carbohydr. Polym. 83 (2011) 1975-1983. https://doi.org/10.1016/j.carbpol.2010.11.001

[63]   T. Mehling, I. Smirnova, U. Guenther, R. Neubert, Polysaccharide-based aerogels as drug carriers, J. Non-Cryst. Solids. 355 (2009) 2472-2479. https://doi.org/10.1016/j.jnoncrysol.2009.08.038

[64]   E.D. Barker, Starch-based hydrogel for biomedical applications, U.S. Patent Application No. 12/459,123, 2010

[65]   G.M. Glenn, A.P. Klamczynski, D.F. Woods, B. Chiou, W.J. Orts, S.H. Imam, Encapsulation of plant oils in porous starch microspheres, J. Agric. Food. Chem. 58 (2010) 4180-4184. https://doi.org/10.1021/jf9037826

[66] C. García-González, M. Alnaief, I. Smirnova, Polysaccharide-based aerogels-Promising biodegradable carriers for drug delivery systems, Carbohydr. Polym. 86 (2011) 1425-1438. https://doi.org/10.1016/j.carbpol.2011.06.066

[67] M.E. El-Naggar, A.M. Abdelgawad, A. Tripathi, O. J. Rojas, Curdlan cryogels reinforced with cellulose nanofibrils for controlled release, J. Environ. Chem. Eng. 5 (2017) 5754-5761. https://doi.org/10.1016/j.jece.2017.10.056

[68] G. El-Feky, M. El-Rafie, M. El-Sheikh, M. E. El-Naggar, A.V. Hebeish, Utilization of crosslinked starch nanoparticles as a carrier for indomethacin and acyclovir drugs, J. Nanomed. Nanotechnol. 6 (2015) 1-8. https://doi.org/10.4172/2157-7439.1000254

[69] M.E. El-Naggar, T.I. Shaheen, M.M. Fouda, A.A. Hebeish, Eco-friendly microwave-assisted green and rapid synthesis of well-stabilized gold and core–shell silver–gold nanoparticles, Carbohydr. Polym. 136 (2016) 1128-1136. https://doi.org/10.1016/j.carbpol.2015.10.003

[70] A. Hebeish, M. El-Rafie, M. El-Sheikh, M.E. El-Naggar, Ultra-fine characteristics of starch nanoparticles prepared using native starch with and without surfactant, J. Inorg. Organomet. Polym. Mater. 24 (2014) 515-524. https://doi.org/10.1007/s10904-013-0004-x

[71] T.I. Shaheen, M.E. El-Naggar, J.S. Hussein, M. El-Bana, E. Emara, Z. El-Khayat, M.M. Fouda, H. Ebaid, A. Hebeish, Antidiabetic assessment; in vivo study of gold and core-shell silver-gold nanoparticles on streptozotocin-induced diabetic rats, Biomed. Pharmacother. 83 (2016) 865-875. https://doi.org/10.1016/j.biopha.2016.07.052

[72] A.M. Abdelgawad, M.E. El-Naggar, W.H. Eisa, O.J. Rojas, Clean and high-throughput production of silver nanoparticles mediated by soy protein via solid state synthesis, J.Cleaner Prod. 144 (2017) 501-510. https://doi.org/10.1016/j.jclepro.2016.12.122

[73] M.E. El-Naggar, A.M. Abdelgawad, C. Salas, O.J. Rojas, Curdlan in fibers as carriers of tetracycline hydrochloride: Controlled release and antibacterial activity, Carbohydr. Polym. 154 (2016) 194-203. https://doi.org/10.1016/j.carbpol.2016.08.042

[74] A. Hebeish, T. I. Shaheen, M. E. El-Naggar, Solid state synthesis of starch-capped silver nanoparticles, Int. J. Biol.Macromol. 87 (2016) 70-76. https://doi.org/10.1016/j.ijbiomac.2016.02.046

[75] J. Hussein, M.E. El Naggar, Y. A. Latif, D. Medhat, M. El Bana, E. Refaat, S. Morsy, Solvent-free and one pot synthesis of silver and zinc nanoparticles: activity

toward cell membrane component and insulin signaling pathway in experimental diabetes, Colloids Surf. B. 170 (2018) 76-84. https://doi.org/10.1016/j.colsurfb.2018.05.058

[76] D. Medhat, J. Hussein, M.E. El-Naggar, M.F. Attia, M. Anwar, Y.A. Latif, H.F. Booles, S. Morsy, A.R. Farrag, W.K. Khalil, Effect of Au-dextran NPs as anti-tumor agent against EAC and solid tumor in mice by biochemical evaluations and histopathological investigations, Biomed. Pharmacother. 91 (2017) 1006-1016. https://doi.org/10.1016/j.biopha.2017.05.043

[77] M.E. El-Naggar, M. El-Rafie, M. El-Sheikh, G.S. El-Feky, A. Hebeish, Synthesis, characterization, release kinetics and toxicity profile of drug-loaded starch nanoparticles, Int. J. Biol. Macromol. 81 (2015) 718-729. https://doi.org/10.1016/j.ijbiomac.2015.09.005

[78] D. Klemm, B. Philpp, T. Heinze, U. Heinze, W. Wagenknecht, Comprehensive cellulose chemistry. Volume 1: Fundamentals and analytical methods, Wiley-VCH Verlag GmbH, Weinheim, 1998

[79] J. Hearle, A fringed fibril theory of structure in crystalline polymers, J. Polym. Sci. 28 (1958) 432-435. https://doi.org/10.1002/pol.1958.1202811722

[80] R. Weatherwax, D. Caulfield, Cellulose aerogels: An improved method for preparing a highly expanded form of dry cellulose, Tappi, 54 (1971) 985-986

[81] B. Alinče, Porosity of swollen solvent-exchanged cellulose and its collapse during final liquid removal, Colloid Polym. Sci. 253 (1975) 720-729. https://doi.org/10.1007/BF02464455

[82] C. Tan, B.M. Fung, J.K. Newman, C. Vu, Organic aerogels with very high impact strength, Adv.Mater. 13 (2001) 644-646. https://doi.org/10.1002/1521-4095(200105)13:9<644::AID-ADMA644>3.0.CO;2-%23

[83] H. Jin, Y. Nishiyama, M. Wada, S. Kuga, Nanofibrillar cellulose aerogels, Colloids Surf. A. 240 (2004) 63-67. https://doi.org/10.1016/j.colsurfa.2004.03.007

[84] O. Ishida, D.Y. Kim, S. Kuga, Y. Nishiyama, R.M. Brown, Microfibrillar carbon from native cellulose, Cellulose, 11 (2004) 475-480. https://doi.org/10.1023/B:CELL.0000046410.31007.0b

[85] J.C. Intyre, Synthetic Fibres: Nylon, Polyester, Acrylic, Polyolefin,. Woodhead Publishing Limited, Cambridge. England. 2004

[86] B. Schmenk, L. Ratke, T. Gries. Solution spinning process for porous cellulose aerogel filaments. in Proceedings of the 2nd Aachen-Dresden International Textile Conference, Dresden. 2008

[87]   C. Hacker, T. Gries, C. Popescu, L. Ratke, Solution spinning process for highly porous, nanostructured cellulose fibers, Chem. Fibers Int. 59 (2009) 85-87

[88]   J. Fricke, Aerogels-highly tenuous solids with fascinating properties, J. Non-Cryst. Solids. 100 (1988) 169-173. https://doi.org/10.1016/0022-3093(88)90014-2

[89]   S. Mulik, C. Sotiriou-Leventis, Resorcinol–formaldehyde aerogels, in Aerogels handbook, Springer, New York, 2011, pp. 215-234. https://doi.org/10.1007/978-1-4419-7589-8_11

[90]   L.L. Hench, J.K. West, The sol-gel process, Chem.Rev. 90 (1990) 33-72

[91]   G. Carlson, D. Lewis, K. McKinley, J. Richardson, T. Tillotson, Aerogel commercialization: technology, markets and costs, J. Non-Cryst. Solids., 186 (1995) 372-379. https://doi.org/10.1016/0022-3093(95)00069-0

[92]   R. Pekala, Organic aerogels from the polycondensation of resorcinol with formaldehyde, J. Mater. Sci. 24 (1989) 3221-3227. https://doi.org/10.1007/BF01139044

[93]   S.A. Al-Muhtaseb, J.A. Ritter, Preparation and properties of resorcinol–formaldehyde organic and carbon gels, Adv. Mater. 15 (2003) 101-114. https://doi.org/10.1002/adma.200390020

[94]   S. Mulik, C. Sotiriou-Leventis, N. Leventis, Time-efficient acid-catalyzed synthesis of resorcinol–formaldehyde aerogels, Chem. Mater. 19 (2007) 6138-6144. https://doi.org/10.1021/cm071572m

[95]   O. Barbieri, F. Ehrburger-Dolle, T.P. Rieker, G.M. Pajonk, N. Pinto, A.V. Rao, Small-angle X-ray scattering of a new series of organic aerogels, J. Non-Cryst. Solids. 285 (2001) 109-115.https://doi.org/10.1016/S0022-3093(01)00440-9

[96]   R. Brandt, J. Fricke, Acetic-acid-catalyzed and subcritically dried carbon aerogels with a nanometer-sized structure and a wide density range, J. Non-Cryst. Solids. 350 (2004) 131-135. https://doi.org/10.1016/j.jnoncrysol.2004.06.039

[97]   S. Mulik, L. Sotiriou-Leventis, N. Leventis, Acid-catalyzed time-efficient synthesis of resorcinol-formaldehyde aerogels and crosslinking with isocyanates, Polym. Preprints.47 (2006) 364-365

[98]   F. Conceição, P. Carrott, M.R. Carrott, New carbon materials with high porosity in the 1–7 nm range obtained by chemical activation with phosphoric acid of resorcinol–formaldehyde aerogels, Carbon. 47 (2009) 1874-1877. https://doi.org/10.1016/j.carbon.2009.03.026

[99]   S.S. Prakash, C.J. Brinker, A.J. Hurd, S.M. Rao, Silica aerogel films prepared at ambient pressure by using surface derivatization to induce reversible drying shrinkage, Nature, 374 (1995) 439-443

[100] A.P. Rao, A.V. Rao, G. Pajonk, Hydrophobic and physical properties of the ambient pressure dried silica aerogels with sodium silicate precursor using various surface modification agents, Appl.Surf.Sci. 253 (2007) 6032-6040. https://doi.org/10.1016/j.apsusc.2006.12.117

[101] A.P. Rao, A.V. Rao, Microstructural and physical properties of the ambient pressure dried hydrophobic silica aerogels with various solvent mixtures, J. Non-Cryst. Solids. 354 (2008) 10-18. https://doi.org/10.1016/j.jnoncrysol.2007.07.021

[102] J. She, T. Ohji, S. Kanzaki, Oxidation bonding of porous silicon carbide ceramics with synergistic performance, J. Eur. Ceram. Soc. 24 (2004) 331-334. https://doi.org/10.1016/S0955-2219(03)00225-5

[103] S.T. Oh, K.I. Tajima, M. Ando, T. Ohji, Strengthening of porous alumina by pulse electric current sintering and nanocomposite processing, J. Am. Ceram. Soc. 83 (2000) 1314-1316. https://doi.org/10.1111/j.1151-2916.2000.tb01380.x

[104] H.S. Ma, A.P. Roberts, J.H. Prévost, R. Jullien, G.W. Scherer, Mechanical structure–property relationship of aerogels, J. Non-Cryst. Solids. 277 (2000) 127-141. https://doi.org/10.1016/S0022-3093(00)00288-X

[105] T. Woignier, J. Phalippou, Mechanical strength of silica aerogels, J. Non-Cryst. Solids. 100 (1988) 404-408. https://doi.org/10.1016/0022-3093(88)90054-3

[106] S. Hæreid, J. Anderson, M. Einarsrud, D. Hua, D. Smith, Thermal and temporal aging of TMOS-based aerogel precursors in water, J. Non-Cryst. Solids. 185 (1995) 221-226. https://doi.org/10.1016/0022-3093(95)00016-X

[107] E.M. Lucas, M.S. Doescher, D.M. Ebenstein, K.J. Wahl, D.R. Rolison, Silica aerogels with enhanced durability, 30-nm mean pore-size, and improved immersibility in liquids, J. Non-Cryst. Solids. 350 (2004) 244-252. https://doi.org/10.1016/j.jnoncrysol.2004.07.074

[108] M.A. Einarsrud, M.B. Kirkedelen, E. Nilsen, K. Mortensen, J. Samseth, Structural development of silica gels aged in TEOS, J. Non-Cryst. Solids. 231 (1998) 10-16. https://doi.org/10.1016/S0022-3093(98)00405-0

[109] S. Bag, I. U. Arachchige, M.G. Kanatzidis, Aerogels from metal chalcogenides and their emerging unique properties, J. Mater. Chem. 18 (2008) 3628-3632. https://doi.org/10.1039/B804011G

[110] I.U. Arachchige, S.L. Brock, Sol–gel methods for the assembly of metal chalcogenide quantum dots, Acc.Chem.Res. 40 (2007) 801-809. https://doi.org/10.1021/ar600028s

[111] S. L. Brock, I.U. Arachchige, K.K. Kalebaila, Metal chalcogenide gels, xerogels and aerogels, Comments Inorg. Chem. 27 (2006) 103-126. https://doi.org/10.1080/02603590601084434

[112] K.K. Kalebaila, D.G. Georgiev, S.L. Brock, Synthesis and characterization of germanium sulfide aerogels, J. Non-Cryst. Solids. 352 (2006) 232-240. https://doi.org/10.1016/j.jnoncrysol.2005.11.035

[113] S. Bag, P.N. Trikalitis, P.J. Chupas, G.S. Armatas, M.G. Kanatzidis, Porous semiconducting gels and aerogels from chalcogenide clusters, Science. 317 (2007) 490-493. https://doi.org/10.1126/science.1142535

[114] P.N. Trikalitis, K.K. Rangan, T. Bakas, M.G. Kanatzidis, Varied pore organization in mesostructured semiconductors based on the $[SnSe_4]^{4-}$ anion, Nature. 410 (2001) 671-675. https://doi.org/10.1038/35070533

[115] M.J. MacLachlan, N. Coombs, G.A. Ozin, Non-aqueous supramolecular assembly of mesostructured metal germanium sulphides from $(Ge_4S_{10})^{4-}$ clusters, Nature. 397 (1999) 681-684. https://doi.org/10.1038/17776

[116] S.D. Korlann, A.E. Riley, B.L. Kirsch, B.S. Mun, S.H. Tolbert, Chemical tuning of the electronic properties in a periodic surfactant-templated nanostructured semiconductor, J. Am. Chem. Soc. 127 (2005) 12516-12527. https://doi.org/10.1021/ja045446k

[117] T. Gacoin, L. Malier, J.P. Boilot, New transparent chalcogenide materials using a Sol-Gel process, Chem. Mater., 9 (1997) 1502-1504. https://doi.org/10.1021/cm970103p

[118] T. Gacoin, L. Malier, J.P. Boilot, Sol–gel transition in CdS colloids, J. Mater. Chem.7 (1997) 859-860. https://doi.org/10.1039/A701035D

[119] T. Gacoin, K. Lahlil, P. Larregaray, J. Boilot, Transformation of CdS colloids: sols, gels, and precipitates, J. Phys. Chem., B. 105 (2001) 10228-10235. https://doi.org/10.1021/jp011738l

[120] L. Malier, J. Boilot, T. Gacoin, Sulfide gels and films: Products of non-oxide gelation, J. Sol-Gel Sci. Technol. 13 (1998) 61-64. https://doi.org/10.1023/A:1008695003946

[121] X. Zhang, D. Chang, J. Liu, Y. Luo, Conducting polymer aerogels from supercritical CO_2 drying PEDOT-PSS hydrogels, J. Mater. Chem. 20 (2010) 5080-5085. https://doi.org/10.1039/C0JM00050G

[122] Y. Xu, Z. Sui, B. Xu, H. Duan, X. Zhang, Emulsion template synthesis of all conducting polymer aerogels with superb adsorption capacity and enhanced

electrochemical capacitance, J. Mater. Chem. 22 (2012) 8579-8584.
https://doi.org/10.1039/C2JM30565H

[123] X. Liang, M. Zeng, C. Qi, One-step synthesis of carbon functionalized with sulfonic acid groups using hydrothermal carbonization, Carbon. 48 (2010) 1844-1848. https://doi.org/10.1016/j.carbon.2010.01.030

[124] Y. Lu, W. He, T. Cao, H. Guo, Y. Zhang, Q. Li, Z. Shao, Y. Cui, X. Zhang, Elastic, conductive, polymeric hydrogels and sponges, Sci.Rep. 4 (2014) 5792. https://doi.org/10.1038/srep05792

[125] W. He, G. Li, S. Zhang, Y. Wei, J. Wang, Q. Li, X. Zhang, Polypyrrole/silver coaxial nanowire aero-sponges for temperature-independent stress sensing and stress-triggered joule heating, ACS Nano. 9 (2015) 4244-4251. https://doi.org/10.1021/acsnano.5b00626

[126] M. Tarasevich, Ultrasonic hydrolysis of a metal alkoxide without alcohol solvents, Am. Cer. Bull. 63 (1984) 500

[127] J. Zarzycki, Sonogels, Hetero. Chem. Rev. 1 (1994) 243-253

[128] E. Blanco, L. Esquivias, R. Litrán, M. Piñero, M. R.D. Solar, N.D.L. Rosa-Fox, Sonogels and derived materials, Appl.Organomet.Chem. 13 (1999) 399-418. https://doi.org/10.1002/(SICI)1099-0739(199905)13:5<399::AID-AOC825>3.0.CO;2-A

[129] X. Zhang, Z. Sui, B. Xu, S. Yue, Y. Luo, W. Zhan, B. Liu, Mechanically strong and highly conductive graphene aerogel and its use as electrodes for electrochemical power sources, J. Mater. Chem. 21 (2011) 6494-6497. https://doi.org/10.1039/C1JM10239G

[130] W. Chen, L. Yan, In situ self-assembly of mild chemical reduction graphene for three-dimensional architectures, Nanoscale. 3 (2011) 3132-3137. https://doi.org/10.1039/C1NR10355E

[131] Z. S. Wu, A. Winter, L. Chen, Y. Sun, A. Turchanin, X. Feng, K. Müllen, Three-dimensional nitrogen and boron Co-doped graphene for high-performance all-solid-state supercapacitors, Adv. Mater. 24 (2012) 5130-5135. https://doi.org/10.1002/adma.201201948

[132] L. Zhang, G. Chen, M. N. Hedhili, H. Zhang, P. Wang, Three-dimensional assemblies of graphene prepared by a novel chemical reduction-induced self-assembly method, Nanoscale. 4 (2012) 7038-7045. https://doi.org/10.1039/C2NR32157B

[133] S. Yang, L. Zhang, Q. Yang, Z. Zhang, B. Chen, P. Lv, W. Zhu, G. Wang, Graphene aerogel prepared by thermal evaporation of graphene oxide suspension

containing sodium bicarbonate, J. Mater. Chem., A. 3 (2015) 7950-7958.
https://doi.org/10.1039/C5TA01222H

[134] W. Si, X. Wu, J. Zhou, F. Guo, S. Zhuo, H. Cui, W. Xing, Reduced graphene
       oxide aerogel with high-rate supercapacitive performance in aqueous electrolytes,
       Nanoscale Res.Lett. 8 (2013) 247. https://doi.org/10.1186/1556-276X-8-247

[135] M. Gudkov, A. Y. Gorenberg, A. Shchegolikhin, D. Shashkin, V. Mel'nikov.
       Explosive reduction of graphite oxide by hydrazine vapor at room temperature,
       Doklady Phys. Chem.478 (2018) 11-14.
       https://doi.org/10.1134/S0012501618010037

[136] S. Nardecchia, D. Carriazo, M. L. Ferrer, M. C. Gutiérrez, F. del Monte, Three
       dimensional macroporous architectures and aerogels built of carbon nanotubes
       and/or graphene: synthesis and applications, Chem. Soc. Rev. 42 (2013) 794-830.
       https://doi.org/10.1039/C2CS35353A

[137] M. Bryning, D. MilNie, M. Islam, J. KiNNawa, Yodh, Carbon nanotube aerogels.
       Adv. Mater. 19 (2007) 661-664. https://doi.org/10.1002/adma.200601748

[138] X. Gui, J. Wei, K. Wang, A. Cao, H. Zhu, Y. Jia, Q. Shu, D. Wu, Carbon nanotube
       sponges, Adv. Mater. 22 (2010) 617-621. https://doi.org/10.1002/adma.200902986

[139] M. A. Worsley, M. Stadermann, Y. M. Wang, J. H. Satcher Jr, T. F. Baumann,
       High surface area carbon aerogels as porous substrates for direct growth of carbon
       nanotubes, Chem. Commun. 46 (2010) 9253-9255.
       https://doi.org/10.1039/C0CC03457F

[140] B. Lee, S. Lee, M. Lee, D. H. Jeong, Y. Baek, J. Yoon, Y. H. Kim, Carbon
       nanotube-bonded graphene hybrid aerogels and their application to water
       purification, Nanoscale. 7 (2015) 6782-6789.
       https://doi.org/10.1039/C5NR01018G

[141] C. Hoecker, F. Smail, M. Pick, A. Boies, The influence of carbon source and
       catalyst nanoparticles on CVD synthesis of CNT aerogel, Chem. Eng. J. 314
       (2017) 388-395. https://doi.org/10.1016/j.cej.2016.11.157

[142] S. M. Jung, H. Y. Jung, M. S. Dresselhaus, Y. J. Jung, J. Kong, A facile route for
       3D aerogels from nanostructured 1D and 2D materials, Sci.Rep. 2 (2012) 849.
       https://doi.org/10.1038/srep00849

[143] K. H. Kim, Y. Oh, M. Islam, Graphene coating makes carbon nanotube aerogels
       superelastic and resistant to fatigue, Nature Nanotechnol. 7 (2012) 562.
       https://doi.org/10.1038/nnano.2012.118

[144] Z. Lin, X. Gui, Q. Gan, W. Chen, X. Cheng, M. Liu, Y. Zhu, Y. Yang, A. Cao, Z.
       Tang, In-situ welding carbon nanotubes into a porous solid with super-high

Materials Research Forum LLC
https://doi.org/10.21741/9781644900994-2

compressive strength and fatigue resistance, Sci. Rep. 5 (2015) 11336.
https://doi.org/10.1038/srep11336

[145] M.A. Worsley, J.H. Satcher Jr, T.F. Baumann, Synthesis and characterization of monolithic carbon aerogel nanocomposites containing double-walled carbon nanotubes, Langmuir, 24 (2008) 9763-9766. https://doi.org/10.1021/la8011684

[146] M.A. Worsley, S.O. Kucheyev, J.H. Satcher Jr, A.V. Hamza, T.F. Baumann, Mechanically robust and electrically conductive carbon nanotube foams, Appl. Phys. Lett. 94 (2009) 073115. https://doi.org/10.1063/1.3086293.

[147] R. Pekala, C. Alviso, J. LeMay, Organic aerogels: microstructural dependence of mechanical properties in compression, J. Non-Cryst. Solids. 125 (1990) 67-75. https://doi.org/10.1016/0022-3093(90)90324-F

[148] T. Woignier, J. Reynes, A.H. Alaoui, I. Beurroies, J. Phalippou, Different kinds of structure in aerogels: relationships with the mechanical properties, J. Non-Cryst. Solids. 241 (1998) 45-52. https://doi.org/10.1016/S0022-3093(98)00747-9

[149] N. Leventis, C. Sotiriou-Leventis, G. Zhang, A.M. M. Rawashdeh, Nanoengineering strong silica aerogels, Nano Lett. 2 (2002) 957-960. https://doi.org/ 10.1021/nl025690e

[150] S. Kucheyev, T. Baumann, C. Cox, Y. Wang, J. Satcher Jr, A. Hamza, J. Bradby, Nanoengineering mechanically robust aerogels via control of foam morphology, Appl. Phys. Lett. 89 (2006) 041911. https://doi.org/10.1063/1.2236222

[151] M. C. Gutiérrez, D. Carriazo, A. Tamayo, R. Jiménez, F. Picó, J. M. Rojo, M.L. Ferrer, F. del Monte, Deep-eutectic-solvent-assisted synthesis of hierarchical carbon electrodes exhibiting capacitance retention at high current densities, Chem. Eur. J. 17 (2011) 10533-10537. https://doi.org/10.1002/chem.201101679

[152] M.A. Worsley, P.J. Pauzauskie, S.O. Kucheyev, J.M. Zaug, A.V. Hamza, J.H. Satcher Jr, T. F. Baumann, Properties of single-walled carbon nanotube-based aerogels as a function of nanotube loading, Acta Mater. 57 (2009) 5131-5136. https://doi.org/10.1016/j.actamat.2009.07.012

[153] M.M. Koebel, L. Huber, S. Zhao, W.J. Malfait, Breakthroughs in cost-effective, scalable production of superinsulating, ambient-dried silica aerogel and silica-biopolymer hybrid aerogels: from laboratory to pilot scale, J. Sol-Gel Sci. Technol. 79 (2016) 308-318. https://doi.org/10.1007/s10971-016-4012-5

[154] C. Sanchez, B. Julián, P. Belleville, M. Popall, Applications of hybrid organic–inorganic nanocomposites, J. Mater. Chem. 15 (2005) 3559-3592. https://doi.org/10.1039/B509097K

[155] C. Sanchez, F. Ribot, B. Lebeau, Molecular design of hybrid organic-inorganic nanocomposites synthesized via sol-gel chemistry, J. Mater. Chem. 9 (1999) 35-44. https://doi.org/10.1039/A805538F

[156] C. Sanchez, P. Belleville, M. Popall, L. Nicole, Applications of advanced hybrid organic–inorganic nanomaterials: from laboratory to market, Chem. Soc. Rev. 40 (2011) 696-753. https://doi.org/10.1039/C0CS00136H

[157] B.M. Novak, D. Auerbach, C. Verrier, Low-density, mutually interpenetrating organic-inorganic composite materials via supercritical drying techniques, Chem. Mater. 6 (1994) 282-286. https://doi.org/10.1021/cm00039a006

[158] G. Gould, D. Ou, R. Begag, W. Rhine, Highly-transparent polymer modified silica aerogels, Polym. Prepr, 49 (2008) 534-535.

[159] R. Reisfeld, Spectroscopy and applications of molecules in glasses, J. Non-Cryst. Solids. 121 (1990) 254-266. https://doi.org/10.1016/0022-3093(90)90141-8

[160] J. McKiernan, E. Simoni, B. Dunn, J.I. Zink, Proton diffusion in the pores of silicate sol-gel glasses, J. Phys. Chem. 98 (1994) 1006-1009. https://doi.org/10.1021/j100054a043

[161] R. Takahashi, S. Sato, T. Sodesawa, M. Suzuki, K. Ogura, Preparation of microporous silica gel by sol-gel process in the presence of ethylene glycol oligomers, Bull. Chem. Soc. Jpn. 73 (2000) 765-774. https://doi.org/10.1246/bcsj.73.765

[162] K. Nakane, T. Yamashita, K. Iwakura, F. Suzuki, Properties and structure of poly (vinyl alcohol)/silica composites, J. Appl. Polym. Sci. 74 (1999) 133-138. https://doi.org/10.1002/(SICI)1097-4628(19991003)74:1<133::AID-APP16>3.0.CO;2-N

[163] A. Bandyopadhyay, M. De Sarkar, A. Bhowmick, Poly (vinyl alcohol)/silica hybrid nanocomposites by sol-gel technique: Synthesis and properties, J. Mater. Sci. 40 (2005) 5233-5241. https://doi.org/10.1007/s10853-005-4417-y

[164] D. Fragiadakis, P. Pissis, L. Bokobza, Modified chain dynamics in poly (dimethylsiloxane)/silica nanocomposites, J. Non-Cryst. Solids. 352 (2006) 4969-4972. https://doi.org/10.1016/j.jnoncrysol.2006.02.159

[165] D. Fragiadakis, P. Pissis, Glass transition and segmental dynamics in poly (dimethylsiloxane)/silica nanocomposites studied by various techniques, J. Non-Cryst. Solids. 353 (2007) 4344-4352. https://doi.org/10.1016/j.jnoncrysol.2007.05.183

[166] N. Leventis, A. Palczer, L. McCorkle, G. Zhang, C. Sotiriou-Leventis, Nanoengineered silica-polymer composite aerogels with no need for supercritical

Materials Research Forum LLC
https://doi.org/10.21741/9781644900994-2

fluid drying, J. Sol-Gel Sci. Technol. 35 (2005) 99-105.
https://doi.org/10.1007/s10971-005-1372-7

[167] Y. Hu, J. Mackenzie, Rubber-like elasticity of organically modified silicates, J. Mater. Sci. 27 (1992) 4415-4420. https://doi.org/10.1007/BF00541574

[168] N. Leventis, Three-dimensional core-shell superstructures: mechanically strong aerogels, Acc. Chem. Res. 40 (2007) 874-884. https://doi.org/10.1021/ar600033s

[169] J. P. Randall, M.A. B. Meador, S.C. Jana, Tailoring mechanical properties of aerogels for aerospace applications, ACS Appl. Mater. Interfaces. 3 (2011) 613-626. https://doi.org/10.1021/am200007n.

[170] Y.L. He, T. Xie, Advances of thermal conductivity models of nanoscale silica aerogel insulation material, Appl. Therm. Eng. 81 (2015) 28-50. https://doi.org/10.1016/j.applthermaleng.2015.02.013.

[171] M. Koebel, A. Rigacci, P. Achard, Aerogel-based thermal superinsulation: an overview, J. Sol-Gel Sci. Technol. 63 (2012) 315-339. https://doi.org/10.1007/s10971-012-2792-9

[172] H. Yang, H. Zhao, Z. Li, K. Zhang, X. Liu, C. Tang, Microstructure evolution process of porous silicon carbide ceramics prepared through coat-mix method, Ceramics Int. 38 (2012) 2213-2218. https://doi.org/10.1016/j.ceramint.2011.10.069

[173] C. Bi, G. Tang, Z. Hu, H. Yang, J. Li, Coupling model for heat transfer between solid and gas phases in aerogel and experimental investigation, Int. J. Heat Mass Transfer. 79 (2014) 126-136. https://doi.org/10.1016/j.ijheatmasstransfer.2014.07.098

[174] Y. Liu, Heat transfer mechanism and thermal design of nanoporous insulating materials, School of Mechanical Engineering, PhD Thesis, University of Science and Technology Beijing, Beijing, 2007

[175] C. Bi, G. Tang, Effective thermal conductivity of the solid backbone of aerogel, Int. J. Heat Mass Transfer. 64 (2013) 452-456. https://doi.org/10.1016/j.ijheatmasstransfer.2013.04.053

[176] J. Quevedo, G. Patel, R. Pfeffer, R. Dave, Agglomerates and granules of nanoparticles as filter media for submicron particles, Powder Technol. 183 (2008) 480-500. https://doi.org/10.1016/j.powtec.2008.01.020

[177] M. Guise, B. Hosticka, B. Earp, P. Norris, An experimental investigation of aerosol collection utilizing packed beds of silica aerogel microspheres, J. Non-Cryst. Solids. 285 (2001) 317-322. https://doi.org/10.1016/S0022-3093(01)00473-2

[178] S. Deville, Freeze-casting of porous ceramics: a review of current achievements and issues, Adv. Eng. Mater 10 (2008) 155-169. https://doi.org/10.1002/adem.200700270

[179] J.E. Amonette, J. Matyáš, Functionalized silica aerogels for gas-phase purification, sensing, and catalysis: A review, Microporous Mesoporous Mater. 250 (2017) 100-119. https://doi.org/10.1016/j.micromeso.2017.04.055

[180] C. Gebald, J. A. Wurzbacher, P. Tingaut, T. Zimmermann, A. Steinfeld, Amine-based nanofibrillated cellulose as adsorbent for CO_2 capture from air, Environ.Sci.Technol. 45 (2011) 9101-9108. https://doi.org/10.1021/es202223p

[181] Y.F. Lin, C.C. Ko, C.H. Chen, K.L. Tung, K.S. Chang, Reusable methyltrimethoxysilane-based mesoporous water-repellent silica aerogel membranes for CO_2 capture, RSC Adv. 4 (2014) 1456-1459. https://doi.org/10.1039/C3RA45371E

[182] S. Cao, N. Yao, K.L. Yeung, Synthesis of freestanding silica and titania-silica aerogels with ordered and disordered mesopores, J. Sol-Gel Sci. Technol. 46 (2008) 323-333. https://doi.org/10.1007/s10971-008-1701-8

[183] P. Hu, B. Tan, M. Long, Advanced nanoarchitectures of carbon aerogels for multifunctional environmental applications, Nanotech. Rev. 5 (2016) 23-39. https://doi.org/ 10.1515/ntrev-2015-0050

[184] E. Unur, Functional nanoporous carbons from hydrothermally treated biomass for environmental purification, Microporous Mesoporous Mater. 168 (2013) 92-101. https://doi.org/10.1016/j.micromeso.2012.09.027

[185] D. Wang, T. Silbaugh, R. Pfeffer, Y. Lin, Removal of emulsified oil from water by inverse fluidization of hydrophobic aerogels, Powder Technol. 203 (2010) 298-309. https://doi.org/10.1016/j.powtec.2010.05.021

[186] M.O. Adebajo, R.L. Frost, J.T. Kloprogge, O. Carmody, S. Kokot, Porous materials for oil spill cleanup: a review of synthesis and absorbing properties, J. Porous Mater. 10 (2003) 159-170. https://doi.org/10.1023/A:1027484117065

[187] N. Chen, Q. Pan, Versatile fabrication of ultralight magnetic foams and application for oil–water separation, ACS Nano. 7 (2013) 6875-6883. https://doi.org/10.1021/nn4020533

[188] A. Pasila, A biological oil adsorption filter, Mar.Pollut.Bull. 49 (2004) 1006-1012. https://doi.org/10.1016/j.marpolbul.2004.07.004

[189] M. Hartmann, S. Kullmann, H. Keller, Wastewater treatment with heterogeneous Fenton-type catalysts based on porous materials, J. Mater. Chem. 20 (2010) 9002-9017. https://doi.org/10.1039/C0JM00577K

[190] H. Liu, W. Sha, A.T. Cooper, M. Fan, Preparation and characterization of a novel silica aerogel as adsorbent for toxic organic compounds, Colloids Surf. A. 347 (2009) 38-44. https://doi.org/10.1016/j.colsurfa.2008.11.033

[191] M.A. Anderson, A.L. Cudero, J. Palma, Capacitive deionization as an electrochemical means of saving energy and delivering clean water. Comparison to present desalination practices: Will it compete?, Electrochim. Acta. 55 (2010) 3845-3856. https://doi.org/10.1016/j.electacta.2010.02.012

[192] S.J. Seo, H. Jeon, J.K. Lee, G.Y. Kim, D. Park, H. Nojima, J. Lee, S.H. Moon, Investigation on removal of hardness ions by capacitive deionization (CDI) for water softening applications, Water Res. 44 (2010) 2267-2275. https://doi.org/10.1016/j.watres.2009.10.020

[193] F. A. AlMarzooqi, A.A. Al Ghaferi, I. Saadat, N. Hilal, Application of capacitive deionisation in water desalination: a review, Desalination, 342 (2014) 3-15. https://doi.org/10.1016/j.desal.2014.02.031

[194] L. Han, K. Karthikeyan, M. Anderson, J. Wouters, K.B. Gregory, Mechanistic insights into the use of oxide nanoparticles coated asymmetric electrodes for capacitive deionization, Electrochim. Acta. 90 (2013) 573-581. https://doi.org/10.1016/j.electacta.2012.11.069

[195] J.Y. Lee, S.J. Seo, S.H. Yun, S.H. Moon, Preparation of ion exchanger layered electrodes for advanced membrane capacitive deionization (MCDI), Water Res. 45 (2011) 5375-5380. https://doi.org/10.1016/j.watres.2011.06.028

[196] S. Porada, L. Weinstein, R. Dash, A. Van Der Wal, M. Bryjak, Y. Gogotsi, P. Biesheuvel, Water desalination using capacitive deionization with microporous carbon electrodes, ACS Appl. Mater. Interfaces. 4 (2012) 1194-1199. https://doi.org/10.1021/am201683j

[197] H. Li, L. Pan, T. Lu, Y. Zhan, C. Nie, Z. Sun, A comparative study on electrosorptive behavior of carbon nanotubes and graphene for capacitive deionization, J. Electroanal. Chem. 653 (2011) 40-44. https://doi.org/10.1016/j.jelechem.2011.01.012

[198] I. Villar, D. J. Suarez-De la Calle, Z. González, M. Granda, C. Blanco, R. Menéndez, R. Santamaría, Carbon materials as electrodes for electrosorption of NaCl in aqueous solutions, Adsorption. 17 (2011) 467-471. https://doi.org/10.1007/s10450-010-9296-0

[199] L. Li, L. Zou, H. Song, G. Morris, Ordered mesoporous carbons synthesized by a modified sol–gel process for electrosorptive removal of sodium chloride, Carbon. 47 (2009) 775-781. https://doi.org/10.1016/j.carbon.2008.11.012

[200] G. Wang, Q. Dong, Z. Ling, C. Pan, C. Yu, J. Qiu, Hierarchical activated carbon nanofiber webs with tuned structure fabricated by electrospinning for capacitive deionization, J. Mater. Chem. 22 (2012) 21819-21823. https://doi.org/10.1039/C2JM34890J

[201] J.C. Farmer, J.H. Richardson, D.V. Fix, S.L. Thomson, S.C. May, Desalination with carbon aerogel electrodes, Lawrence Livermore National Laboratory Report No. UCRL-ID-125298, 1996

[202] M. E. Suss, T.F. Baumann, W.L. Bourcier, C.M. Spadaccini, K.A. Rose, J.G. Santiago, M. Stadermann, Capacitive desalination with flow-through electrodes, Energy Environ. Sci. 5 (2012) 9511-9519. https://doi.org/10.1039/C2EE21498A

[203] C. Hou, C. Huang, C. Hu, Application of capacitive deionization technology to the removal of sodium chloride from aqueous solutions, Int. J. Environ. Sci. Technol. 10 (2013) 753-760. https://doi.org/10.1007/s13762-013-0232-1

[204] C. García-González, I. Smirnova, Use of supercritical fluid technology for the production of tailor-made aerogel particles for delivery systems, J. Supercrit. Fluids. 79 (2013) 152-158. https://doi.org/10.1016/j.supflu.2013.03.001

[205] M. Betz, C. García-González, R. Subrahmanyam, I. Smirnova, U. Kulozik, Preparation of novel whey protein-based aerogels as drug carriers for life science applications, J. Supercrit. Fluids. 72 (2012) 111-119. https://doi.org/10.1016/j.supflu.2012.08.019

[206] A. Veronovski, Ž. Knez, Z. Novak, Comparison of ionic and non-ionic drug release from multi-membrane spherical aerogels, Int.J.Pharma. 454 (2013) 58-66. https://doi.org/10.1016/j.ijpharm.2013.06.074

[207] C.A. Garcia-Gonzalez, A. Concheiro, C. Alvarez-Lorenzo, Processing of materials for regenerative medicine using supercritical fluid technology, Bioconjugate Chem. 26 (2015) 1159-1171. https://doi.org/10.1021/bc5005922

[208] L. Servat-Medina, A. Gonzalez-Gomez, F. Reyes-Ortega, I.M.O. Sousa, N.D.C.A. Queiroz, P.M.W. Zago, M.P. Jorge, K.M. Monteiro, J.E. de Carvalho, J. San Román, Chitosan–tripolyphosphate nanoparticles as Arrabidaea chica standardized extract carrier: synthesis, characterization, biocompatibility, and antiulcerogenic activity, Int.J.Nanomed. 10 (2015) 3897. https://doi.org/10.2147/IJN.S83705

[209] E. Reverchon, P. Pisanti, S. Cardea, Nanostructured PLLA-hydroxyapatite scaffolds produced by a supercritical assisted technique, Ind. Eng. Chem.Res. 48 (2009) 5310-5316. https://doi.org/10.1021/ie8018752

[210] N. Pircher, S. Veigel, N. Aigner, J.M. Nedelec, T. Rosenau, F. Liebner, Reinforcement of bacterial cellulose aerogels with biocompatible polymers,

Carbohydr. Polym. 111 (2014) 505-513.
https://doi.org/10.1016/j.carbpol.2014.04.029

[211] X. Chen, G.S. Wilson, Electrochemical and spectroscopic characterization of surface sol-gel processes, Langmuir. 20 (2004) 8762-8767.
https://doi.org/10.1021/la034940j

[212] J.M. Wallace, J.K. Rice, J.J. Pietron, R.M. Stroud, J.W. Long, D.R. Rolison, Silica nanoarchitectures incorporating self-organized protein superstructures with gas-phase bioactivity, Nano Lett. 3 (2003) 1463-1467.
https://doi.org/10.1021/nl034646b

Aerogels I: Preparation, Properties and Applications   Materials Research Forum LLC
Materials Research Foundations **84** (2020) 83-108   https://doi.org/10.21741/9781644900994-3

Chapter 3

Hybrid Silica Aerogel

Matheus Costa Cichero[1], João Henrique Zimnoch Dos Santos*[1]

[1]Instituto de Química - Universidade Federal do Rio Grande do Sul, Av. Bento Gonçalves 9500, CEP 91501-970, Porto Alegre, RS, Brasil

*jhzds@iq.ufrgs.br

Abstract

Silica aerogel is one of most promising material in many application segments, such as insulator, adsorbents, catalyst, batteries, sensing and drug delivery devices and many more. All due its impressive properties as low thermal conductivity, low density and high surface area. In spite of the potential, silica aerogel is associated with some drawbacks as cost, time-consuming process and especially the inherent brittleness that limits its full applicability. To overpass these disadvantages, one of the most common procedures is to incorporate an organic compound to the silica backbone – enhancing its mechanical strength and potentially decreasing processing time – and thus creating a hybrid silica aerogel. This chapter presents an overview of the recent strategies adopted in the literature that utilizes polymers, biomolecules and graphene when composing hybrid silica aerogel.

Keywords

Hybrid Silica Aerogel, Silica, Polymer, Biomolecule, Graphene

Contents

Materials Research Forum LLC
https://doi.org/10.21741/9781644900994-3

1. Introduction

As defined by the International Union for Pure and Applied Chemistry (IUPAC), aerogel is a gel that comprises a microporous solid in which the dispersed phase is a gas [1]. Aerogel was first introduced in the study done by Kistler in 1931. He reported the possibility to replace the liquid contained in jelly by a gas without the collapse or shrinkage of the initial structure [2]. He successfully preserved the network of the initial wet jelly exchanging the pore filling solvent (water) to an adequate solvent (ethanol) that possessed a lower critical temperature and submitted to its critical point, creating a supercritical fluid and venting off the gas. This process is now referred as supercritical fluid (SCF) drying. Kistler did not only report the jelly study, but other successfully attempts with nickel, alumina, some oxides as stannic and tungstic, and the most recognized one: silica aerogel [2,3].

The strategy to take the liquid solvent to the critical point aims at minimizing the capillary forces that occurs when gels are drying at ambient pressure. The capillary force associated with high tensile strength present in liquids at ambient pressure drying act upon the silica framework. These forces crush the structure causing the shrinkage and increase density of the final material. These materials are referred to as xerogels [2,4]. Taking the solvent to its critical point, no liquid-solid interface is ever formed, and no tensile forces are exerted upon silica structure, thus preserving it in shape and size of the initial wet gel with minimal or no shrinkage. Since Kistler's report in the 1930's, silica aerogel properties have been thoroughly investigated. Many reviews about history, synthesis, properties and applications of silica aerogel are available [5–11], probably the most recognized characteristic of silica aerogel is its low thermal conductivity. Silica aerogel possesses lower thermal conductivity than air at the same conditions of temperature, pressure and humidity which make it an excellent insulating material being used in aerospace applications [12]. Furthermore, other known properties of silica aerogel comprise acoustic insulation, optically transparency, non-flammability, amorphous structure and high porosity (usually around 90%). These properties make silica aerogel a promising material in different applications, such as catalysis, adsorbents, thermal insulator, for instance [12,13].

On the other hand, silica aerogels exhibit low mechanical properties which make handling them difficult, as they are very brittle. It is worth mentioning that although SCF drying is a well-known procedure, it has some drawbacks such as size restriction and time consumption, which in turn have undermined industrial scalability and practical application of silica aerogel. Others drying procedures have been investigated in attempt to overcome SCF drying issues, and namely there are three drying procedures options: (i) SCF drying, (ii) freeze-drying and (iii) subcritical drying (also known as ambient pressure drying) [12–14].

There are presently two types of SCF drying procedures: The hot SCF drying procedure consists in raising the solvent in an autoclave to its critical temperature and pressure point, followed by gas evacuation; and the cold SCF drying procedure involves solvent exchange to liquid CO_2 as the solvent and followed to submission to its critical point temperature and pressure. The advantage of the cold procedure, comparing to the hot one, is that CO_2 critical conditions are milder since critical temperature is around 31°C. Furthermore, liquid CO_2 is non-flammable, non-toxic and can be recovered to be re-used. On the other hand, the cold process is time consuming, especially for large monoliths production, as the solvent exchange is dictated by diffusion and multiple solvent exchanges (and washes) may be necessary. Another imposed difficult is that size restriction is still present since the drying station determines the gel size [12].

In the freeze-drying procedure, it is mandatory to freeze the gel before the drying step, which, depending on solvent, can be easily done: Commercial refrigerator can be employed if the solvent is water, for example, or liquid nitrogen is necessary for solvents with lower freezing point (such as alcohols and others organic solvents). Evaporation of the frozen solvent is done in low pressure and although this process reduces the capillary forces, it usually results in cracked monoliths, besides being timing consuming depending on the size of the gel [12].

Finally, subcritical drying procedure presents as an alternative to the before mentioned ones. At ambient pressure (AP), drying is necessary to turn a conventional hydrophilic silica aerogel in a hydrophobic one. In the hydrophobic silica aerogel, the usual hydroxyl (-OH) groups are substituted by alkyl groups. In the case of hydrophilic silica gel, during AP drying process, the surface silanols are close enough to react in a condensation reaction, forming siloxane (Si-O-Si) bridges, which are responsible to the irreversible shrinkage at ambient pressure drying. Although shrinkage may be still present when hydrophobic silica wet gel is dried at AP conditions, as the solvent is complete evaporated from the monolith, the surface alkyl groups repel one to each other and the monolith recovers the shape and size closer to that present on the former wet gel [15]. This concept was introduced by Brinker in 1995 by preparing silica films (of tetraethoxysilane, TEOS) that were organosilyl derivatized with trimethyl chlorosilane

(TMCS) [16]. This was the first report on the silica functionalization (or hybridization) that allows obtaining silica aerogel under subcritical procedure.

Stergar et al. [17] classify aerogel in terms of appearance, microstructure and composition. From the appearance point of view, aerogels may be presented as monolith, powders or films, bearing microstructure that can be microporous, mesoporous or mixed porous. From the compositional point of view, it can be organic, inorganic or mixed (hybrid) in nature. A hybrid material, as defined by IUPAC, is a material that is composed of an intimate mixture of inorganic components, organic components or both types of components [1]. In sol-gel composites, it is considered that a hybrid system is a mixture of organic and inorganic components, which may be classified in two classes, according with the nature of the interface between the components: Class I and II [18].

In Class I systems there are no bonds (covalent nor iono-covalent) involved connecting the organic and inorganic components. The interaction occurs via weak forces, such as electrostatic forces, van der Waals forces and hydrogen bonds. Class II systems present a strong bond between the components (covalent or iono-covalent) - Class II having more sub classifications, in models 1,2 and 3 as described by Mackenzie [19] corresponding the way the particles arrange themselves: if the particle of the same nature (organic or inorganic) are connected, separate by the other matrix or form clusters.

From the practical point of view, materials classified as Class I are usually associated with xerogels characteristics (although it is not a rule), as the organic counterpart is either incorporated during the sol stage or added after gelation being completed. As for aerogels, as previously described, the procedure for obtaining silica aerogel is usually associated with multiple washes steps and solvent exchanges, which makes leaching of the organic counterpart concern for Class I materials. This issue can be avoided upon fixation of the two components by a covalent bond, which is capable to endure the aerogel multiple washes procedures. Naturally, this configuration is no longer categorized as a Class I, but a Class II material. It is worth mentioning that, although Class II are more common, hybrid aerogel in both classification do exist [15].

For the sake of clarity, we classify hybrid silica aerogel according to their composition (using Stergar's classification) [17]. In addition, we are considering hybrid material to be a combination of an inorganic component, silica in this case, with an organic one. This chapter is organized according to the nature of the hybrid components, the strategies adopted for the cited aerogels development, as well as their properties or application are here described and discussed.

2. Hybrid silica aerogel

As stated before, silica aerogels possess unique properties, in particular low-density (approximated 0.003-0.5 g. cm^{-3}), high degree of porosity (between 80 and 99.8%), high surface area (varies from 500 to 1000 m^2. g^{-1}), low index of refraction (n = 1.05), thermal conductivity (0.005-0.1 W/mK) and dielectric constant (k = 1.0 − 2.0) [20,21]; that qualifies aerogel applications in numerous areas such as an insulator component for windows [22], acrylic-based paints [23], catalyst support [24], absorbent materials [25], photocatalyst degradation support [26], gas-phase purifications and humidity sensing devices [27], enzyme and protein entrapment matrices [8,28], batteries [7], even on space missions [29,30] and many more [20,31–33]. In spite all the potential, silica aerogel has two main drawbacks that limit its practical application: Hydrophilicity and poor mechanical properties.

Silica aerogel is obtained from sol-gel process, where a solution of silica precursor undergoes a polymerization reaction and creates a solid network. At this point, the solution is no longer a liquid but a gel, which gelation or sol-gel transition occurred. After a period of aging the gel is dried to obtain silica aerogel [35]. The first silica precursor, used by Kistler, was sodium metasilicate (Na$_2$SiO$_3$) [3]; but there are many type of silica precursors such as salt, oxides, complexes, alkoxides. Alkoxides, like TEOS and tetramethyl orthosilicate (TMOS) are the most popular silica precursor, as they are relatively low cost and easily available [20]. Sol-gel chemistry is a well-known process and has been thoroughly described by Brinker et al. [35,36]. Without delving into the subject, it is possible to summarize the process in two reactions: Hydrolysis and condensation reactions.

Using an alkoxide as example − Si(OR)$_4$, where R = alky group, the reaction between water and alkoxide (hydrolysis reaction) forms silicic acid (Si(OH)$_4$); concomitantly, condensation reaction occurs, which generates the after mentioned siloxane groups (Si-O-Si). As time passes and the two reactions continue to occur, siloxane bridges (or connections) numbers grows, and a 3D polymeric network is formed. This polymeric structure is responsible for the sol-gel transition and also is the aerogel backbone after solvent removal [12,20,34–36]. In this usual sol-gel procedure, not only siloxane is generated but also silanol groups (Si-OH) that are responsible for the hydrophilicity in silica aerogel. Silica polymeric network is a particulate backbone, which it consists of particles interconnected; although each particle are strong the interconnection between each of these particles (also referred as "neck") are weaker, resulting in the fragility of silica-based aerogel [12,20,34].

Materials Research Forum LLC
https://doi.org/10.21741/9781644900994-3

Some techniques have been reported to overcome these limitations, such as functionalization to remove hydroxyl from the silanol groups and turn the aerogel hydrophobic, as contemplated in the previous section when ambient pressure drying was discussed [16,34]. There are a few strategies reported in the literature to enhance aerogel mechanical strength, from simple approaches such as increasing period of aging, in order to maximize the number of siloxane bridges and thus fortifying the interconnections, to incorporating components of different nature (as biomolecules, graphene and polymers) to the silica backbone structure to be able to withstand tensile forces or even add flexibility and elastic recovery to the aerogels [14,15,19,34].

Although the prolonged aging period improves the mechanical strength of the silica aerogel, the increase is not desirable and associated with the fact that the procedure is time consuming and requires greater amounts of solvent. This procedure alone does not solve the problem as the system remains a particulate structure [34]. Thereby, most studies are based on the addition of a second organic component to the system, either incorporating through or between crosslinks in the silica structure (Class I and Class II, respectively). Table 1 [6,37–46] depicts some recent examples (2019-2020) of these strategies to produce hybrid silica aerogel using several types of organic components.

As shown in table 1 [6, 37-46], there are different types of organic molecules that can be used in the production of hybrid silica aerogel, either as integral part of the aerogel structure (Class II) or as reinforced molecule that act as filler (Class I). The most used silica precursors are alkoxides, most have a modification with a react group or are used in association of organically-modified silica (or ORMOSIL) precursor. Table 1 also demonstrated that hybrid silica aerogel can be obtained by any of the drying procedures but SCF and ambient pressure are the most used ones, whether because it is a well-known procedure like SCF drying or it is a desirable protocol the reduces aerogel overall costs (AP drying). We can create a classification based on the type of precursors adopted: Polymer-silica aerogels, biomolecules-silica aerogel and graphene-silica aerogels. There are several studies reported and dedicated to some of the components mentioned, such as polymers [10,15,47–52], a diverse of biomolecules alternatives [9, 53–57] and many more [58–61], but there are few studies focused on the strategies to create different hybrid silica aerogels, from the processing of the materials and its applications.

Table 1: Recent publications onto the hybrid aerogel, research on Scopus (www.scopus.com) with "Hybrid Silica Aerogel" keywords. Accessed on March 2020.

Silica precursors	Organic Precursors	Dying Technique	Applications	Hybrid Classification	References
Sodium Silicate/APTES	Dextran and Dextran aldehyde	Spray drying	Drug Delivery	Class II - Biomolecule- Silica	[47]
GPTMS	Polyether-based precursor	SCF Drying	Insulator	Class II - Polymer- Silica	[38]
TEOS	Cellulose diacetate (CDA)	Freeze drying	Flame retardant and oil separation	Class I – Biopolymer- Silica	[6]
TMOS	Gelatine	SCF Drying	Selective adsorption	Class II – Biomolecule - Silica	[44]
TEOS and APS	Chitosan	AP Drying	Selective adsorption	Class I- Biomolecule- Silica	[43]
TMOS	Polysaccharides	SCF Drying	Insulator	Class I- Biomolecule- Silica	[42]
Triethoxy(1-phenylethenyl) silane	Polyethylene-based precursor	SCF Drying	Insulator	Class II – Polymer- Silica	[40]
TEOS, DMDES and HDMZ	ORMOSIL	AP Drying	Hydrophobic Insulator	Class II	[39]
OTS and MPTMS	Thiol-ene click chemistry	AP Drying	Oil-Water Separation	Class II	[41]
VTMS	Polyvinly-based percursor and graphene nanoplatelets	SCF Drying	Insulator and Oil-Water Separation	Class II – Polymer- Silica (Class I for graphene)	[45]
TEOS	Reduced graphene oxide (rGO)	AP Drying	Oil (Solvents)- Water Separation	Class I and Class II	[46]

*Acronyms: APS - Trimethoxyaminopropylsilane, APTES - aminopropyltriethoxy silane , DMDES - Dimethlydiethoxysilane, GPTMS - (3- Glycidyloxypropyl) trimethoxysilane, HDMZ – hexamethyldisilazene, MPTMS - γ-Mercaptopropyltrimethoxysilane , OTS - (Octa[2-((3-(trimethoxysilyl)propyl)thio)ethyl] silsesquioxane, TEOS - tetraethoxysilane , TMOS - tetramethoxysilane,VTMS – Vinlytrimethoxysilane.

Materials Research Forum LLC
https://doi.org/10.21741/9781644900994-3

The focus of this chapter is to in depth discuss some of the recent examples in the literature, that are contemplated in Table 1 and more, aiming to do an overview of the strategies adopted, from the precursor to synthesis procedure, the proprieties of the developed materials and their applications.

2.1   Polymer-silica aerogel

One of the earliest attempts to increase silica aerogels mechanical properties was to cross-link silica aerogel to a polymeric chain known as cross-aerogel (or crosslinked aerogels). Usually, the monomer reacts on the surface of the silica network or a modify silica network (as amine modified using APTES, for example), creating polymer chain between the particles and reinforcing the structure. Several crosslinkers have been reported: epoxides [62], isocyanates [63], acrylonitrile [47] and more [64]. Although reinforced, this approach maintains the particulate nature structure which does not respond for issues like solvent usage and processing time as it uses common alkoxides and ORMOSIL; this method also has its disadvantages as density and thermal conductive increases considerably [34]. This approach is known as monomeric silica precursor that renders particulate systems and can be further modify by ORMOSIL or polymerization reactions.

A modern method is to reorder the aerogel creation procedure going on the contrary of the monomeric approach: Substituting monomer and ORMOSIL precursor to either high or low molecular weight polymers that possess silica functional groups which in turn can undergo sol-gel process in a second step.

As Rezaei et al. [65] reported obtaining hybrid silica aerogel with a continuous structure in oppose to the naturally particulate structure of silica aerogel. The experimental procedure began with pre-polymerizing step of the vinyl trimethoxy silane (VTMS) precursor by radical polymerization which creates a polyethylene (PE) framework structure that possesses trimethoxy silane side groups for each monomeric unit. The PE backbones were then connected by hydrolytically sol-gel cross-link reactions of the side groups, creating a Class II material – as illustrated in Fig. 1 [65]. As water does not act as a solvent to the overall structure, during the gelation period, a spinodal decomposition process occurs (a mechanism of fast separation of mixture of liquids). The group suggests that this procedure causes effects on the final structure and in the final properties of the aerogel and controlling the phase-separation mechanism (amount of water) is responsible for the continuous backbone structure and the small pore size aerogels. Aging processing is essential to particulate silica network structure as it enhances the connection between the silica particles (in this case, silica is already connected by the polymeric backbone). The spinodal decomposition brings the advantages of having a fast gelation time and not

Materials Research Forum LLC
https://doi.org/10.21741/9781644900994-3

requiring an aging time period. The final aerogel monolith is obtained by CO_2 SCF drying procedure, and the resulting aerogel exhibits low density (0.05 -0.22 g cm^{-3}), transparency, flexibility, high elastic modulus, and ultra-low thermal conductivity (10.34-17.21 mWm^{-1}K^{-1}).

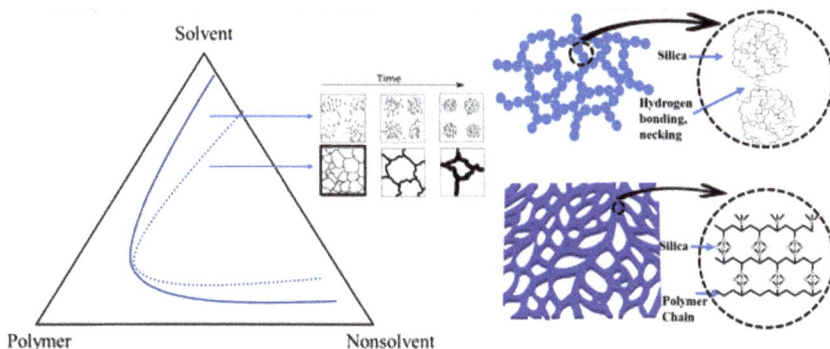

Fig. 1 The effect of the phase decomposition on the final structure aerogel, representation of conventional silica network and the continuous structure backbone obtained by the polymeric precursors and spinodal decomposition. Reprinted from [66]. Copyright (2020), with permission from Elsevier.

In a more recent report done by the same group, the concept was further expanded. Utilizing the pre-polymerization process, followed by hydrolytically cross-linked reaction by the sol-gel process approach, the researchers developed polyether-based silica, instead of that previously describe PE-based silica. They used (3-Glycidyloxypropyl) trimethoxysilane (GPTMS) as precursor that possesses an epoxide ring and reacts via ring opening polymerization. Following the same approach, after polymerization the sol-gel process was carried out employing ammonium hydroxide (NH₄OH) as the catalyst in ethanol. After gelation and aging for 10 hours, the final aerogel is obtained by CO_2 SCF drying method. As mentioned previously, water content during the sol-gel process also induces phase-separation mechanisms which will affect the final properties of the aerogel. The silica modified precursor have an ether function group, which is more flexible when compared to common carbon-carbon bond in alkyl groups – guaranteeing to the aerogel high porosity and good flexibility with an elastic deformation of 15% to its original size [37].

Choi and collaborators [39] reported the first triethoxy(1-phenylethenyl)silane (TPS) crosslinked silica aerogel. TPS was synthetized from the hydrosiliylation of triethoxysilane with phenylacetylene using Karstedt's catalyst (an organoplatinum coordination complex), the resulting TPS molecule possesses two functional groups: phenyl groups that can provide hydrophobic characteristics to the final aerogel; and a vinyl group that can be polymerized and enhance mechanical strength. Following the polymeric approach, the researchers submitted TPS to radical polymerization and after a purification step, PTPES was obtained. The PTPES was mixed with pre-hydrolysed TEOS (TEOS diluted in isopropanol and further catalysed with acid and base subsequently), the condensation of the two species (PTPES and TEOS) occur in the presence of NH_4OH (10 M) creating a Class II network of polymer and silica; the gel was aged in methanol at 50 °C and then dried under SCF procedure. The author's studied different molar ratios of TEOS/PTPES, the optimal hybrid aerogel that presented the most promising properties was at 5% mol of PTPES. The overall samples exhibit properties of regular silica aerogel such as low density (0.03–0.05 g/cm^3), high pore volume (16.5–34.5 cm^3/g) and surface area (520–900 m^2/g). But also, good hydrophobicity degree (120°) and good mechanical strength capable to withstand up to 10 times higher the compressive strength than a regular silica aerogel.

In an opposed direction to the polymer precursors approaches previously described, Jung et al. [52] prepared a Cblass I hybrid polymer-silica aerogel by firstly performing sol-gel process and then proceeding to the polymerization (or the monomer approach). The sol-gel process was carry out by mixing three silica precursors: Tetramethylortosilicate (TMOS), VTMS and Bis[3-(trimethoxysily)-propyl]amine (BTMSPA) were dissolved in methanol under alkaline conditions (using NH_4OH as the catalyst). After an ageing period, the solvent was exchanged to toluene – that was also used as the solvent for the monomers. Styrene and methyl methacrylate (MMA) were used in the study. Three toluene solutions were prepared: Styrene, MMA and styrene-MMA solutions. The synthesized silica mixture and the monomers solutions were combined with a thermal radical initiator, and then submitted to heat treatment. The hybrids systems presented better thermal stabilities than neat polymers counterparts and increasing in the static dielectric constant when compared to bare silica aerogel.

A different approach on the ORMOSIL variable was done by Wang et al. [40]. The authors utilized a polyhedral oligomericsilsesquioxane (POSS) as building block, which has a cage-like structure that can be further functionalized. The resulting POSS was (Octa[2-((3-(trimethoxysilyl)propyl)thio)ethyl] silsesquioxane (OTS), which was synthesized through thiol-ene click reaction, using γ-Mercaptopropyltrimethoxysilane (MPTMS) that possesses a thiol group and Octavinylsilsesquioxane (OVS) that have a vinyl moieties. The resulting OTS have a nano cage-like structure composed of SiO_2 with trimethoxysilyl side groups attaches, which can further react via sol-gel process. The gel was produced via two steps sol-gel procedure - where the process is catalysed first by an acid to promote the hydrolysis of the alkoxide component; after a period of time in a second step a second catalyst (base) is added to favour the condensation reaction of the hydrolysed alkoxide. After the gel formation, the aging period was two days at 40 °C in ethanol. Thereafter, the aged gel was further modified with trimethylchlorosilane (TMCS) ORMOSIL to create hydrophobic characteristics. The modified gel was washed in n-hexane and dried by AP during procedure to obtain the final aerogel. The hybrid aerogels present high surface area (542–834 $m^2 g^{-1}$) and high hydrophobicity degree (with contact angle higher than 150°). The cage-like structure improves the mechanical performance in compression tests, with compression strength of 4.96–6.48 MPa, high compression modulus (18.79–25.84 MPa) and even flexibility.

Polyurethane foams (PUF) are commonly employed as insulating material in edifications, and, despite the thermal insulating capacity, these foams are highly flammable. In order to overcome these limitations, Li and collaborators fabricated PUF-Silica aerogel hybrid system, by immersing the foam in hydrolytic TEOS solution. TEOS solution was submitted to a two-step procedure, where hydrolysis was catalysed by both acidic and alkaline conditions. By dipping the PUF into the solution, TEOS was infused in the PUF and gelation occurs inside the cell walls of the polymeric structure. After aged for 24 hours, the system was soaked in water and submitted to freeze-drying technique: Aerogel produced inside the polymeric cell structure formed a hierarchically porous structure. The final hybrid material probably configures as Class I. Interactions between the matrices seems to occur on the surface of the cell walls. The final composite demonstrated remarkable insulating properties: Reduced thermal conductivity, flammability and smoke release. PUF-Silica aerogel also exhibited an enhanced compressive strength (220%) if compared to neat PUF, and self-extinguishing capacity in vertical burning tests [66].

The examples described in this section, show both "monomeric" and "polymeric" precursors strategies. The monomeric is the most known strategy and has some flaws as mentioned before. The examples here cited of this approach [40,66], overcome this limitations by introducing a building block to the structure of the designed aerogel – through this way the final hybrids maintain the characteristic desirable in a aerogel as low density and thermal conductivity with increased mechanical properties. The polymeric strategy is a recent adaptation for hybrid polymer-silica aerogel and it generates a new opening to hybrid silica aerogel design considering the variability in polymeric structure, although most of the precursors have to be synthesized, they are relatively easy to prepare and more polymeric precursors shall become available as the study goes on. The combination of polymeric precursor with the spinodal decomposition have gained great attention of the community, as there are several advantages such as no aging period requirement and fast gelation time. Although being a well-studied process in the polymer field, there are relatively few studies of spinodal decomposition in the aerogel one, but there is a consensus that aerogels made by this technique offers better mechanical properties without compromise density and thermal conductivity when compared to other methods [34].

2.2   Biomolecules-silica aerogel

Aerogel based on biomolecules, also recognized as "bio-aerogels", are a relatively new branch of materials, especially when compared to silica aerogel. As Kistler's debut silica aerogel in 1931 [2]: The first report of a bio-aerogel by Ookuma [67] dated from 1989 describes cellulose particles with an open porous structure. Recently, the construction of materials based on renewable and bio-resources components as an alternative to petrol-based materials have been a major pursuit in both commercial and research segments. Proteins and polysaccharides are mainly used in bio-aerogel production such as cellulose, alginate, chitosan, lignin, pectin and others. Bio-aerogel shares some similarities to silica aerogel and also has many applications, as thermal and acoustic insulators, for catalyst and food applications where biocompatibility and biodegradability are a major concern [68]. Perhaps, one of the distinct differences between biomolecules and inorganic aerogel is the synthesis process. While silica aerogel (formed by monomeric units that undergo hydrolysis and condensation reactions until gelation), biomolecules in general tend to assemble their constituting molecules by inter- and intramolecular forces, creating a (bio)polymeric network. Thereby, bio-aerogels start with the dissolution of these (bio)polymeric networks in appropriated solvent – allowing to the polymer chains to rearrange, which causes the gelation process in an open network structure. Although it seems simple, gelation process is quite complicated process as it is different for each biomolecule. Starch gelation, for instance, undergoes by thermal mechanism, after being dissolved in hot water and then cooled. The same does not work to alginate or pectin, for example, requiring the presence of ions to induce gel formation [57]. In this section, we highlight a few recent works associated with biomolecules in combination with silica to create bio-aerogel composites with different characteristics. The synthesis and properties or applications are here discussed. For clarity and abbreviation, from this point forward in this section the term "bio-aerogels" will be referred to the hybrid composed by the biocomponent and silica.

Horvat et al. [41] reported an interesting new approach to prepare polysaccharide-silica aerogels. As stated before, one of the difficulties to incorporate a polysaccharide into the inorganic matrix is that the gelation process is different for each biomolecule. To overcome this limitation, the authors proposed that adding alcohol to a polysaccharide solution favour gelation. The elegancy of this approach resides in the mutuality concept in the preparation of the hybrid itself: Tetramethylorthosilicate (TMOS) was used as the silica precursor and when it is hydrolysed, methanol is released, which in turn promotes polysaccharide gelation. Concomitantly, polysaccharides possess carboxylic groups in their structure that acts as acidic catalyst that promotes TMOS hydrolysis. Therefore, there is no need to add alcohol or catalyst to the system. The polysaccharides utilized in the study were: pectin, xanthan, alginate and guar. The final aerogels were obtained by CO_2 SCF drying procedure, being the first report on xanthan-silica and guar-silica hybrid systems. The resulting materials seems to be an interpenetrating network, which is classified as Class I hybrid system. Authors reported that addition of silica enhanced the structure properties, when compared to the blank polysaccharide aerogels, exhibiting surface area of 679 m^2g^{-1} and low thermal conductivity of 19 $mWm^{-1}K^{-1}$.

Hybrid adsorbents, composed of silica aerogel, magnetite (Fe_3O_4) and chitosan (CS), for removal of cadmium II from aqueous solution was designed by Shariatinia et al. [42]. The silica aerogel was prepared by basic catalyst (NH_4OH) route. Trimethoxyaminopropylsilane (APS) was used to functionalize silica surface and enabling ambient pressure drying. The authors elaborated two hybrid systems: The first one is adding Fe_3O_4 nanoparticles (NP) to the sol-gel procedure described above, creating a core-shell with hybrid silica aerogel (HSA) and Fe_3O_4(Fe_3O_4@HSA); and a second one, adding Fe_3O_4@HSA to CS solution, thus creating HSA-Fe_3O_4-CS composite. The four samples, including a bare CS film, were subjected to absorption tests (50 mL of 0.1000 g L^{-1} Cd (II) solution). Tests revealed that HSA-Fe_3O_4-CS system exhibited high sorption capacity (71.9 mg g^{-1}), comparable with other adsorbents in the literature, and the materials maintained up to 95% of their initial sorption capacities after cycles of adsorption and desorption reactions.

Still in the context of sorbent materials, Herman and co-workers [43] created a hybrid bio-aerogels based on gelatine and silica which is selective for Hg (II) ions. The hybrid bio-aerogel was constructed by the cogelation method, in which first each precursor was separately dissolved: Gelatine and ammonium carbonate were dissolved in hot water, whilst TMOS was properly dissolved in methanol. After cooling, both solutions were mixed under stirring, where $(NH_4)_2CO_3$ acts as the basic catalyst for the sol-gel process. After gelation, the gels were aged in methanol for 24 hr, where the solvent was exchange for acetone and aged for 168 hours, in a process where fresh acetone was replaced each 24 hourly. It seems that gelatine is covalently bonded to silica network structure as represented in Fig. 2 [43], since no leaching has been observed after seven days of solvent exchanging process. The authors suggest that the serine and threonine amino acids side chains present in the gelatine are responsible to bonding to SiO_2. After a week, the aged gels were SCF dried with CO_2. The different hybrid aerogels were obtained based on the gelatine content (4–24 wt%). The resulting systems exhibited great selectivity towards aqueous Hg(II) even in the presence of other metal ions (Co(II), Pb(II), Zn(II), Ni(II) and more). The reason of the selectivity is the presence of soft Lewis bases in the side groups of the proteins in the gelatine (-SH, -NH$_2$, -S-S- functions groups).They form stable complex to soft ions such Hg(II) that is only disrupted when strong chelates (EDTA, for example) is added. The best result was presented by 24wt% of gelatine content hybrid bio-aerogel, capable to 91% of Hg(II) removal (adsorption of 200 mg. g^{-1}) with no loss in the adsorption capacity after five cycles of adsorption and desorption.

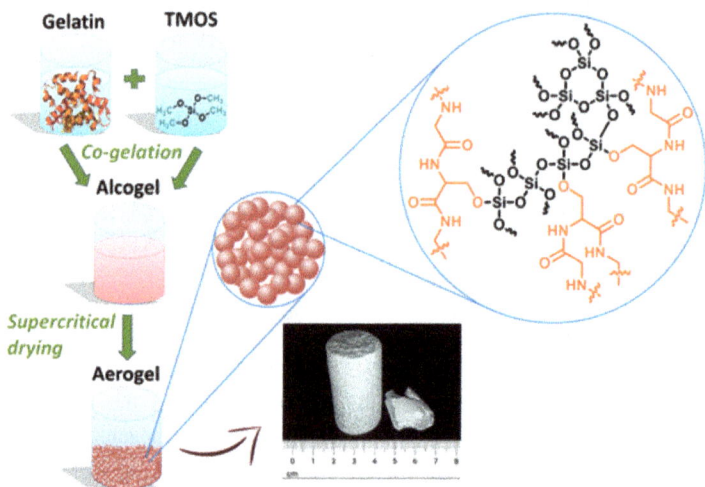

Fig. 2 Schematic of process and sctructure representation of the gelatin-silica hybrid aerogel. Adapted with permission [44]. Copyright (2020) American Chemical Society.

Pirzada et al. [6] used thermal treatment to fix silica and cellulose diacetate (CDA) together, creating nanofibers aerogels that are thermally and mechanically stable. The procedure was carried out by mixing an acid catalysed solution of TEOS in dimethylformamide to a 11 wt% solution of CDA (in a dimethylacetamide/acetone solution). The mixture was poured into a syringe were the needle was attached to a positive electrode. The fibres were spun at a constant flow, current and distance to aluminium foil connected to a grounded electrode. The nanofibers gels were submitted to freeze-drying technique and afterward towards a thermal procedure where the final aerogels were heated at 180 °C for two hours followed by 1 hr at 240 °C under vacuum for the two components glue together. As the thermal treatment induces the crosslinking of the two parts (protein and silica) there are no hydroxyl groups to interact with water molecule. Hence the fibres exhibit a super-hydrophobic behaviour, with water contact angle of 165.7°, and due to the nature of fibres cells structure of the proteins that induce a capillary movement, they also possess oil affinity and work as an excellent oil adsorbent for oil-water separation. Albeit the hybrid fibres also possess excellent mechanical properties, withstanding up to 56.3 kPa in strain tests, recoverability and flexibility up to 75 % of strain stress.

The main method of bio-aerogels synthesis is via sol-gel process where the bio counterpart is normally added during the gelation step, resulting in Class I hybrid materials. The examples here discussed have shown some different approaches by modifying the traditional wet sol-gel process, namely by introducing a self-catalyst among the precursors, the effect of thermal treatment process and complexation. This kind of materials are particularly interesting for the sorbent segment, since such materials exhibit characteristics such as low density, high porosity (as conventional silica aerogel) with improved mechanical properties and adsorption capacities, produced from a variety of different raw materials (agar, alginate, chitosan, chitin, cellulose, gelatine just to mention a few) that are plenty available making it environmentally and economically interesting.

2.3 Graphene-silica aerogel

Graphene is a two-dimension material that possesses extraordinary characteristics such as high strength, low density and high surface area – due its unique honeycomb-like structure monolayer consisting of carbons with sp^2 hybridization. This makes graphene an interesting component for hybrid aerogels [69,70]. Nevertheless, graphene in its pristine form possesses only carbon atoms, which hinders its dissolution in solvents and renders its processability difficult. The majority of the reported literature applying graphene as one of the components is using it as graphene oxide (GO) [61]. The modified Hummer's method is the main procedure to obtaining GO: The original procedure consists in a mixture of potassium permanganate ($KMnO_4$), sodium nitrate ($NaNO_3$) and concentrated sulfuric acid (H_2SO_4) added to a graphite solution [71]. The modified version is normally associated with an exfoliation process [61]. During the oxidation process, epoxy, hydroxyl, carbonyl, and carboxylic function groups are added to graphene sheets. These functionalities not only considerably enhance solubility in aqueous environment, but also contribute to interaction with other materials such as silica network [61,70]. In this section, some of the recent reports on hybrid aerogels based on graphene and silica will be discussed.

Karami kamkar et al. [44] exploited the strategies discussed in the 2.1 polymer-silica section. They studied the effects of adding graphene nanoplatelets (GnPs) as filler to silica based aerogel using the polymeric precursor approach. The polymeric silica precursor used was P-VTMS (VTMS was polymerized via radical polymerization) – a silane functionalized polyethylene (PE).P-VTMS dissolved in ethanol was then submitted to gelation process, where ammonium hydroxide was used as the basic catalyst to the sol-gel process. At this point different amounts of GnPs were added to the mixture and allowed to crosslinking/gelation in vacuum oven at 40 °C for 3 hours. After the gelation, the solvent was exchanged for fresh ethanol before being submitted for supercritical drying, without aging period. As mentioned in section 2.2, one advantage of the polymeric precursors relies on the fact that they render continuous network instead of the conventional silica particulate backbone, i.e., continuous polymeric structure does not possess the fragile neck (Si-O-Si bridges) and therefore no aging is required. The authors observed that GnPs incorporation in the system reduces the time to gel formation in 25%, as they increase the viscosity in the system and limits the effects from shear from break any crosslinked components. The incorporation of graphene does not interfere with the silica structure integrity, but FTIR spectra indicates a decrease in the oxygen related bands (Si-OH and Si-O-Si bands), which suggest an intense interaction between silica and graphene through hydroxyl groups, indicating a Class I type hybrid material. The graphene addition also reduces the shrinkage in the aerogels from 15 % (blank without GnPs) to 5 % with GnPs, while increasing their pore surface area in 21 % (from 570 to 694 m^2 g^{-1}). The combination of both materials also increased the mechanical strength from 7.2 MPa to 68.7 MPa in compressive mechanical tests, maintaining the pore sizes of 32 nm grating excellent thermal isolating properties.

Wang and collaborators [72], created a solid adsorption material assembling 2D graphene and 3D silica in one pot-synthesis. Three surfactants bearing different length alkyl chains were tested in order to increase the pore volumes of the final aerogel: Tetramethylammoniumbromide (C1), triethylhexyl ammonium bromide (C6) and trimethyloctadecyl ammonium bromide (C18). The synthesis started with the oxidation of graphite via a Hummer's method to obtain GO. The produced GO was dispersed in deionized water creating a suspension. Each surfactant (C1, C6 or C18) was dissolved in water containing nitric acid (HNO_3), which acts as the acid catalyst for sol-gel process, and TEOS was dropwise added to acidic solution containing one of the surfactants. After acid solution homogenization occurs, the GO suspension is mixed, and gelation occurs in autoclave at 60°C for two days. The hydrogel formed is frozen in liquid nitrogen and

Materials Research Forum LLC
https://doi.org/10.21741/9781644900994-3

submitted to freeze-drying procedure for another two days. The obtained aerogel has the surfactants removed under furnace tube at 450°C and functionalized with an amine sorbent, tetraethylenepentamine (TEPA). One can assume that the final aerogel does not possess any covalent bonds between silica and graphene components, and the interaction occurs via weak van der Waals force between silanol groups (Si-OH) of silica with any oxygen related groups of GO, hence characterizing a Class I hybrid material. The properties of the final hybrid aerogel composite show an increase in the surface area from 448 to 734 m^2 g^{-1} as the chain length of the surfactants increases (C1 to C18, respectively), maximum pore volume obtained was 0.42 cm^3 g^{-1} (also for C18).Similarly, CO_2 sorption capacity was higher (4.9 mmol g^{-1}) when C18 surfactant was employed in the aerogel synthesis.

Zhao et al. [45] created an interesting hybrid silica-graphene sponge for oil-water separation. The novelty of this material is the integration of graphene sponge with silica. GO was prepared by a modified Hummer's method – the obtained GO was then reduced using an ammonium sulfide solution (($NH_4)_2S$) as reducing agent and transforming GO into graphene hydrogel. Graphene hydrogel was immersed in $NH_3·H_2O$ (14%, v/v) solution and submitted to freeze-drying procedure to finally obtain a compressible graphene sponge. To incorporate de silica counterpart, TEOS was first dissolved in ethanol using HCl as the catalyst. Then ammonia solution was added to elevate pH to 7.0 (two steps catalyst procedure). Before gelation, the silicon solution was mixture drop wise into the graphene sponge. After gelation, the gel was aged in ethanol for six hr to strength the silica network. Solvent was changed to n-hexane were surface modification with TMCS was carried and the final hybrid sponge was dried at AP. The authors suggested that two interactions take place in the hybrid sponge: physical adsorption of silica within the graphene porous sponge (Class I) and, as FTIR Si-O-Si band monitoring suggests, a chemical interaction between the components (Class II). The final structure was able to withstand 146 times their weight without any structural damage. It has a high specific area (803.351 m^2. g^{-1}), and the modification created good hydrophobicity properties(with water contact angle of 129°), which also make an excellent adsorbent capable of fully adsorbed 1.5 g of dyed dichloromethane in water in six seconds, as illustrated in Fig. 3 [45]. The hybrid materials present good adsorption capacities for other five oils (as toluene, ethanol, motor oil and plant oil) up to 10 times its own weight and 10 cycles of adsorption and desorption after heat treatment – making an interesting material for oil spills applications.

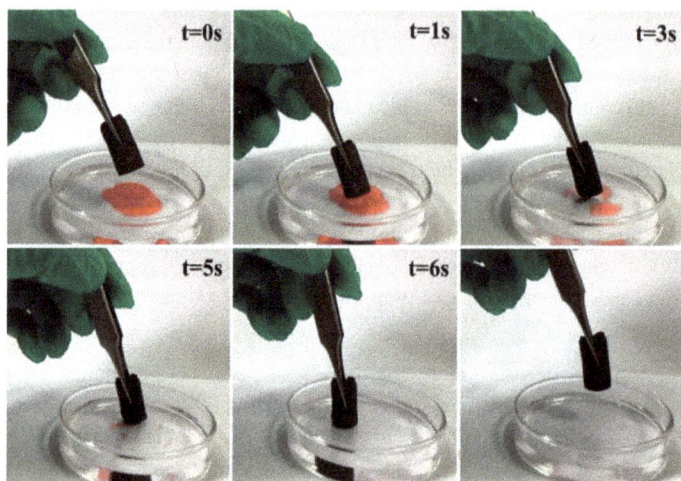

Fig. 3 Fast adsorption of dyed dichloromethane in water by silica-graphene sponge hybrid. Reprinted from [46]. Copyright (2020), with permission from Elsevier.

The strategies described here are the recent approaches on graphene-silica aerogel, but it represents only a fraction of methods and techniques that cannot only be employed to graphene but also to other carbon structures (such carbon nanotubes and nanofibers). As stated in the beginning of this section, the form of graphene mostly employed is GO instead of its pristine condition, due its insolubility in water, as the most common techniques uses aqueous sol-gel process using alkoxides and ORMILS. In the studies here demonstrated, there is a trend to use GO as basis structure for the hybrid, normally as a sponge – where flexibility and adsorptions characteristics are highlighted in the final aerogel.

3. Final remarks

There is a growing interest in obtaining materials that possess characteristic similar or better than those of silica aerogels and that can be mechanical processed to numerous practical applications or even that permits industrial scalability. Hybrid composites that are obtained combining inorganic and organic components, combining advantages from both counterparts is one alternative to overcome the notorious silica aerogels limitations. As show in Fig. 4 there is an increase in the number of publications regarding hybrid silica aerogel subject over the last two decades, especially in the last five years.

In this chapter, three approaches to prepare hybrid silica aerogels by modification of silica aerogel including polymeric structures, biomolecules and graphene have been addressed. The recent strategies for incorporating each component to silica, how they interact, their characteristics and applications have been described. The use of different techniques (as polymeric and monomeric precursors) and components direct affect the final characteristics and properties of the hybrids. Polymeric precursor with addition of spinodal decomposition renders continuous networks instead of fragile particulate that are associated with silica aerogel; the addition of biomolecule greatly improves the strength and flexibility; whilst the addition of graphene greatly increases adsorption capacities of the hybrids. The advances on the hybrid silica are still in its infancy and there is still room for further investigation. The reported studies show that there has been a quest for new approaches that may bring new design, morphological structures and mechanical reinforcement in hybrid silica aerogel. The polymeric precursor with the spinodal decomposition method is a power technique that opens a diversity of potential for new materials. The integration of these procedures with different components (as variety of biomolecules available and carbon-based fillers) represents a promising future for hybrid silica aerogels.

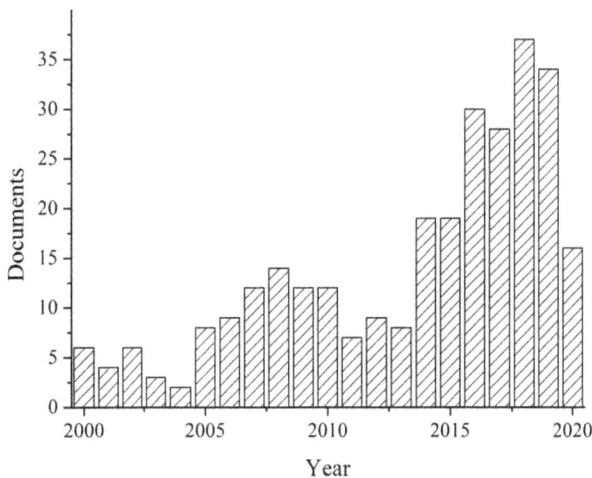

Fig 4 Number of documents published into the hybrid silica aerogel subect along tewnty years. Source: Scopus, search date: 04-17-2020.

Aerogels I: Preparation, Properties and Applications   Materials Research Forum LLC
Materials Research Foundations **84** (2020) 83-108   https://doi.org/10.21741/9781644900994-3

Acknowledgements

Matheus Costa Cichero thanks CAPES for the grants. FAPERGS (16/2551-0000470-6) is also thanked.

References

[1]   J.V. Alemán, A.V. Chadwick, J.He, M.Hess, K.Horie, R.G. Jones, P.Kratochvíl, I. Meisel, I. Mita, G. Moad, S. Penczek and R.F.T.Stepto, Definitions of terms relating to the structure and processing of sols, gels, networks, and inorganic-organic hybrid materials (IUPAC Recommendations 2007), Pure Appl. Chem., 79(2007) 1801-1829. doi:10.1351/pac200779101801.

[2]   S.S. Kistler, Coherent expanded aerogels and jellies, Nature. 127 (1931) 741. doi:10.1038/127741a0.

[3]   S.S. Kistler, Coherent expanded-aerogels, J. Phys. Chem. 36 (1932) 52–64. doi:10.1021/j150331a003.

[4]   U. Schubert, Chemistry and fundamentals of the Sol–Gel process, in: U.Schubert, and N. Hüsing,Synthesis of Inorganic Materials, third ed., VCH Wiley Verlag GmbH, Weinheim, 2012, pp.3–27.

[5]   A. Hulanicki, S. Glab F.Ingman, Chemical sensors: definitions and classification, PureAppl. Chem. 63 (1991) 1247-1250, doi:10.1351/pac199163091247

[6]   T. Pirzada, Z. Ashrafi, W. Xie, S.A. Khan, Cellulose silica hybrid nanofiber aerogels: from sol–gel electrospun nanofibers to multifunctional aerogels. Adv. Funct. Mater. 2020, 30, 1907359.doi:10.1002/adfm.201907359

[7]   X. Zou, K. Liao, D. Wang, Q. Lu, C. Zhou, P. He, R. Ran, W. Zhou, W. Jin, Z. Shao, Water-proof, electrolyte-nonvolatile, and flexible Li-Air batteries via O_2-Permeable silica-aerogel-reinforced polydimethylsiloxane external membranes, Energy Storage Mater. 27 (2020) 297–306. doi:10.1016/j.ensm.2020.02.014.

[8]   A.S. Harper-Leatherman, E.R. Pacer, N.D. Kosciuszek, Encapsulating cytochrome c in silica aerogel nanoarchitectures without metal nanoparticles while retaining gas-phasebioactivity, J.Vis. Exp. (2016) e53802–e53802. doi:10.3791/53802.

[9]   W.-J. Yang, A.C.Y. Yuen, A. Li, B. Lin, T.B.Y. Chen, W. Yang, H.D. Lu, G.H. Yeoh, Recent progress in bio-based aerogel absorbents for oil/water separation, Cellulose. 26 (2019) 6449–6476. doi:10.1007/s10570-019-02559-x.

[10]   S. Salimian, A. Zadhoush, M. Naeimirad, R. Kotek, S. Ramakrishna, A review on aerogel: 3D nanoporous structured fillers in polymer-based nanocomposites,

Materials Research Forum LLC
https://doi.org/10.21741/9781644900994-3

Polym. Compos. 39 (2018) 3383–3408. doi:10.1002/pc.24412.

[11]   A.C. Pierre, History of aerogels, in: M. Aegerter, N. Leventis, M. Koebel (Eds.), Aerogels handbook. Advances in sol-gel derived materials and technologies. Springer, New York, 2011, pp. 3-18.

[12]   A.C. Pierre, A. Rigacci, SiO_2 aerogels, in: M. Aegerter, N. Leventis, M. Koebel (Eds.), Aerogels handbook. Advances in sol-gel derived materials and technologies. Springer, New York, 2011, pp. 21-45.

[13]   K.Kanamori, Hybrid aerogels, in: L. Klein, M. Aparicio, A. Jitianu (Eds.), Handbook of sol-gel science and technology. Springer, Cham, 2016, pp. 3317-3338.

[14]   A.M. Anderson, M.K.Carroll, Hydrophobic silica aerogels: review of synthesis, properties and applications, in: M. Aegerter, N. Leventis, M. Koebel (Eds.), Aerogels handbook. Advances in sol-gel derived materials and technologies. Springer, New York,2011, pp. 47-77.

[15]   N. Leventis, H. Lu, Polymer-crosslinked aerogels, in: M. Aegerter, N. Leventis, M. Koebel (Eds.), Aerogels handbook. Advances in sol-gel derived materials and technologies. Springer, New York, 2011, pp. 251-285.

[16]   S.S. Prakash, C.J. Brinker, A.J. Hurd, S.M. Rao, Silica aerogel films prepared at ambient pressure by using surface derivatization to induce reversible drying shrinkage, Nature 374 (1995) 439–443. doi:10.1038/374439a0.

[17]   J. Stergar, U. Maver, Review of aerogel-based materials in biomedical applications, J. Sol-Gel Sci. Technol. 77 (2016) 738–752. doi:10.1007/s10971-016-3968-5.

[18]   C. Sanchez, F. Ribot, B. Lebeau, Molecular design of hybrid organic-inorganic nanocomposites synthesized via sol-gel chemistry, J. Mater. Chem. 9 (1999) 35–44. doi:10.1039/A805538F.

[19]   Y. Hu, J.D. Mackenzie, Rubber-like elasticity of organically modified silicates, J. Mater. Sci. 27 (1992) 4415–4420. doi:10.1007/BF00541574.

[20]   J.L. Gurav, I.K. Jung, H.H. Park, E.S. Kang, D.Y. Nadargi, Silica aerogel: synthesis and applications, J. Nanomater. 2010 (2010) 409310. doi:10.1155/2010/409310.

[21]   A. Soleimani Dorcheh, M.H. Abbasi, Silica aerogel; synthesis, properties and characterization, J. Mater. Process. Technol. 199 (2008) 10–26. doi:10.1016/j.jmatprotec.2007.10.060.

[22]   D. Li, C. Zhang, Q. Li, C. Liu, M. Arıcı, Y. Wu, Thermal performance evaluation

of glass window combining silica aerogels and phase change materials for cold climate of China, Appl. Therm. Eng. 165 (2020) 114547. doi:10.1016/j.applthermaleng.2019.114547.

[23] S. Karami, S. Motahari, M. Pishvaei, N. Eskandari, Improvement of thermal properties of pigmented acrylic resin using silica aerogel, J. Appl. Polym. Sci. 135 (2018) 45640. doi:10.1002/app.45640.

[24] D. Li, V. Rohani, F. Fabry, A. Parakkulam Ramaswamy, M. Sennour, L. Fulcheri, Direct conversion of CO_2 and CH_4 into liquid chemicals by plasma-catalysis, Appl. Catal. B Environ. 261 (2020) 118228. doi:10.1016/j.apcatb.2019.118228.

[25] X.-D. Gao, Y.D. Huang, T.T. Zhang, Y.Q. Wu, X.M. Li, Amphiphilic SiO_2 hybrid aerogel: an effective absorbent for emulsified wastewater, J. Mater. Chem. A. 5 (2017) 12856–12862. doi:10.1039/C7TA02196H.

[26] A. Najafidoust, M. Haghighi, E. Abbasi Asl, H. Bananifard, Sono-solvothermal design of nanostructured flowerlike BiOI photocatalyst over silica-aerogel with enhanced solar-light-driven property for degradation of organic dyes, Sep. Purif. Technol. 221 (2019) 101–113. doi:10.1016/j.seppur.2019.03.075.

[27] J.E. Amonette, J. Matyáš, Functionalized silica aerogels for gas-phase purification, sensing, and catalysis: a review, Micropor. Mesopor. Mater. 250 (2017) 100–119. doi:10.1016/j.micromeso.2017.04.055.

[28] N. Ganonyan, N. Benmelech, G. Bar, R. Gvishi, D. Avnir, Entrapment of enzymes in silica aerogels, Mater. Today. 33 (2020) 24–35. doi:10.1016/j.mattod.2019.09.021.

[29] N. Bheekhun, A.R. Abu Talib, M.R. Hassan, Aerogels in aerospace: an overview, Adv. Mater. Sci. Eng. 2013 (2013) 406065. doi:10.1155/2013/406065.

[30] A. Percot, E.L. Zins, A. Al Araji, A.T. Ngo, J. Vergne, M. Tabata, A. Yamagishi, M.C. Maurel, Detection of biological bricks in space. The case of adenine in silica aerogel, Life 9 (2019) 82. doi:10.3390/life9040082.

[31] L.W. Hrubesh, Aerogel applications, J. Non. Cryst. Solids. 225 (1998) 335–342. doi:10.1016/S0022-3093(98)00135-5.

[32] M.A. Hasan, R. Sangashetty, A.C.M. Esther, S.B. Patil, B.N. Sherikar, A. Dey, Prospect of thermal insulation by silica aerogel: a brief review, J. Inst. Eng. Ser. D. 98 (2017) 297–304. doi:10.1007/s40033-017-0136-1.

[33] G. Jia, Z. Li, P. Liu, Q. Jing, Applications of aerogel in cement-based thermal insulation materials: an overview, Mag. Concr. Res. 70 (2018) 822–837. doi:10.1680/jmacr.17.00234.

[34] S. Karamikamkar, H.E. Naguib, C.B. Park, Advances in precursor system for silica-based aerogel production toward improved mechanical properties, customized morphology, and multifunctionality: a review, Adv. Colloid Interface Sci. 276 (2020) 102101. doi:10.1016/j.cis.2020.102101.

[35] C.J. Brinker, G.W. Scherer, Hydrolysis and condensation II: Silicates, in: C.J. Brinker, G.W.Scherer (Eds.), Sol-gel science: the physics and chemistry of sol-gel processing. Academic Press, San Diego, 1990: pp. 96–233. https://doi.org/10.1016/B978-0-08-057103-4.50008-8

[36] C.J. Brinker, G.W. Scherer, Hydrolysis and condensation I: Nonsilicates, in: C.J. Brinker, G.W.Scherer (Eds.), Sol-gel science: the physics and chemistry of sol-gel processing.Academic Press, San Diego, 1990: pp. 20–95. https://doi.org/10.1016/B978-0-08-057103-4.50007-6

[37] S. Rezaei, A.M. Zolali, A. Jalali, C.B. Park, Novel and simple design of nanostructured, super-insulative and flexible hybrid silica aerogel with a new macromolecular polyether-based precursor, J. Colloid Interface Sci. 561 (2020) 890–901. https://doi.org/10.1016/j.jcis.2019.11.072.

[38] M. de F. Júlio, L.M. Ilharco, Hydrophobic granular silica-based aerogels obtained from ambient pressure monoliths, Materialia. 9 (2020) 100527. https://doi.org/10.1016/j.mtla.2019.100527.

[39] H. Choi, V.G. Parale, T. Kim, Y.S. Choi, J. Tae, H.-H. Park, Structural and mechanical properties of hybrid silica aerogel formed using triethoxy(1-phenylethenyl)silane, Microporous Mesoporous Mater. 298 (2020) 110092. https://doi.org/10.1016/j.micromeso.2020.110092.

[40] L. Wang, G. Song, X. Qiao, G. Xiong, Y. Liu, J. Zhang, R. Guo, G. Chen, Z. Zhou, Q. Li, Facile Fabrication of flexible, robust, and superhydrophobic hybrid aerogel, Langmuir. 35 (2019) 8692–8698. https://doi.org/10.1021/acs.langmuir.9b00521.

[41] G. Horvat, M. Pantić, Ž. Knez, Z. Novak, Preparation and characterization of polysaccharide - silica hybrid aerogels, Sci. Rep. 9 (2019) 16492. https://doi.org/10.1038/s41598-019-52974-0.

[42] Z. Shariatinia, A. Esmaeilzadeh, Hybrid silica aerogel nanocomposite adsorbents designed for Cd(II) removal from aqueous solution, Water Environ. Res. 91 (2019) 1624–1637. https://doi.org/10.1002/wer.1162.

[43] P. Herman, I. Fábián, J. Kalmár, Mesoporous Silica–gelatin aerogels for the selective adsorption of aqueous Hg(II), ACS Appl. Nano Mater. 3 (2020) 195–206. https://doi.org/10.1021/acsanm.9b01903.

Materials Research Forum LLC
https://doi.org/10.21741/9781644900994-3

[44] S. Karamikamkar, A. Abidli, E. Behzadfar, S. Rezaei, H.E. Naguib, C.B. Park, The effect of graphene-nanoplatelets on gelation and structural integrity of a polyvinyltrimethoxysilane-based aerogel, RSC Adv. 9 (2019) 11503–11520. https://doi.org/10.1039/C9RA00994A.

[45] X. Zhao, Y. Zhu, Y. Wang, Z. Li, Y. Sun, S. Zhao, X. Wu, D. Cao, Hydrophobic, blocky silica-reduced graphene oxide hybrid sponges as highly efficient and recyclable sorbents, Appl. Surf. Sci. 486 (2019) 303–311. https://doi.org/10.1016/j.apsusc.2019.05.017.

[46] E. Tiryaki, Y. Başaran Elalmış, B. Karakuzu İkizler, S. Yücel, Novel organic/inorganic hybrid nanoparticles as enzyme-triggered drug delivery systems: dextran and dextran aldehyde coated silica aerogels, J. Drug Deliv. Sci. Technol. 56 (2020) 101517. https://doi.org/10.1016/j.jddst.2020.101517.

[47] N. Leventis, A. Sadekar, N. Chandrasekaran, C. Sotiriou-Leventis, Click synthesis of monolithic silicon carbide aerogels from polyacrylonitrile-coated 3D silica networks, Chem. Mater. 22 (2010) 2790–2803. https://doi.org/10.1021/cm903662a.

[48] G. Churu, B. Zupančič, D. Mohite, C. Wisner, H. Luo, I. Emri, C. Sotiriou-Leventis, N. Leventis, H. Lu, Synthesis and mechanical characterization of mechanically strong, polyurea-crosslinked, ordered mesoporous silica aerogels, J. Sol-Gel Sci. Technol. 75 (2015) 98–123. https://doi.org/10.1007/s10971-015-3681-9.

[49] A. Bang, C. Buback, C. Sotiriou-Leventis, N. Leventis, Flexible aerogels from hyperbranched polyurethanes: probing the role of molecular rigidity with poly(urethane acrylates) versus poly(urethane norbornenes), Chem. Mater. 26 (2014) 6979–6993. https://doi.org/10.1021/cm5031443.

[50] P. Paraskevopoulou, D. Chriti, G. Raptopoulos, C.G. Anyfantis, Synthetic polymer aerogels in particulate form, Materials . 12 (2019). https://doi.org/10.3390/ma12091543.

[51] G.D.Sorarù, E.Zera, R.Campostrini, Aerogels from preceramic polymers, in: L.Klein, M.Aparicio, A.Jitianu (Eds.), Handbook of sol-gel science and technology. Springer, Cham, 2018, pp. 1013-1037.

[52] H.K. Jung, B.M. Jung, U.H. Choi, Synthesis and characterization of silica aerogel-polymer hybrid materials, Mol. Cryst. Liq. Cryst. 687 (2019) 97–104. https://doi.org/10.1080/15421406.2019.1651058.

[53] D. Lasrado, S. Ahankari, K. Kar, Nanocellulose-based polymer composites for energy applications-A review, J. Appl. Polym. Sci. n/a (2020) 48959.

https://doi.org/10.1002/app.48959.

[54] H. Kargarzadeh, J. Huang, N. Lin, I. Ahmad, M. Mariano, A. Dufresne, S. Thomas, A. Gałęski, Recent developments in nanocellulose-based biodegradable polymers, thermoplastic polymers, and porous nanocomposites, Prog. Polym. Sci. 87 (2018) 197–227. https://doi.org/10.1016/j.progpolymsci.2018.07.008.

[55] L.Y. Long, Y.X. Weng, Y.Z. Wang, Cellulose aerogels: Synthesis, applications, and prospects, Polym. 10 (2018). https://doi.org/10.3390/polym10060623.

[56] M. Sánchez-Soto, L. Wang, T. Abt, L.G. De La Cruz, D.A. Schiraldi, Thermal, electrical, insulation and fire resistance properties of polysaccharide and protein-based aerogels, in: S.Thomas, L. A. Pothan, R. Mavelil-Sam (Eds.), Biobased aerogels: Polysaccharide and protein-based Materials, The Royal Society of Chemistry, 2018: pp. 158–176. https://doi.org/10.1039/9781782629979-00158.

[57] K. Ganesan, T. Budtova, L. Ratke, P. Gurikov, V. Baudron, I. Preibisch, P. Niemeyer, I. Smirnova, B. Milow, Review on the production of polysaccharide aerogel particles, Materials 11 (2018). https://doi.org/10.3390/ma11112144.

[58] M.A. Worsley, T.F. Baumann, Carbon Aerogels, in: L. Klein, M. Aparicio, A. Jitianu (Eds.), Handbook of sol-gel science and technology: Processing, characterization and applications. Springer International Publishing, Cham, 2018: pp. 3339–3374. https://doi.org/10.1007/978-3-319-32101-1_90.

[59] S. Long, H. Wang, K. He, C. Zhou, G. Zeng, Y. Lu, M. Cheng, B. Song, Y. Yang, Z. Wang, X. Luo, Q. Xie, 3D graphene aerogel based photocatalysts: Synthesized, properties, and applications, Coll. Surf. A Physicochem. Eng. Asp. 594 (2020) 124666. https://doi.org/10.1016/j.colsurfa.2020.124666.

[60] J.-H. Lee, S.J. Park, Recent advances in preparations and applications of carbon aerogels: A review, Carbon N. Y. 163 (2020) 1–18. https://doi.org/10.1016/j.carbon.2020.02.073.

[61] A. Lamy-Mendes, R.F. Silva, L. Durães, Advances in carbon nanostructure–silica aerogel composites: a review, J. Mater. Chem. A. 6 (2018) 1340–1369. https://doi.org/10.1039/C7TA08959G.

[62] M.A.B. Meador, E.F. Fabrizio, F. Ilhan, A. Dass, G. Zhang, P. Vassilaras, J.C. Johnston, N. Leventis, Cross-linking amine-modified silica aerogels with epoxies: mechanically strong lightweight porous materials, Chem. Mater. 17 (2005) 1085–1098. https://doi.org/10.1021/cm048063u.

[63] L.A. Capadona, M.A.B. Meador, A. Alunni, E.F. Fabrizio, P. Vassilaras, N. Leventis, Flexible, low-density polymer crosslinked silica aerogels, Polymer. 47

(2006) 5754–5761. https://doi.org/10.1016/j.polymer.2006.05.073.

[64] D.J. Boday, P.Y. Keng, B. Muriithi, J. Pyun, D.A. Loy, Mechanically reinforced silica aerogel nanocomposites via surface initiated atom transfer radical polymerizations, J. Mater. Chem. 20 (2010) 6863–6865. https://doi.org/10.1039/C0JM01448F.

[65] S. Rezaei, A. Jalali, A.M. Zolali, M. Alshrah, S. Karamikamkar, C.B. Park, Robust, ultra-insulative and transparent polyethylene-based hybrid silica aerogel with a novel non-particulate structure, J. Colloid Interface Sci. 548 (2019) 206–216. https://doi.org/10.1016/j.jcis.2019.04.028.

[66] M.E. Li, S.X. Wang, L.X. Han, W.J. Yuan, J.-B. Cheng, A.N. Zhang, H.-B. Zhao, Y.-Z. Wang, Hierarchically porous SiO_2/polyurethane foam composites towards excellent thermal insulating, flame-retardant and smoke-suppressant performances, J. Hazard. Mater. 375 (2019) 61–69. https://doi.org/10.1016/j.jhazmat.2019.04.065.

[67] S. Ookuma, K. Igarashi, M. Hara, K. Aso, H. Yoshidome, H. Nakayama, K. Suzuki, K. Nakajima, Porous ion-exchanged fine cellulose particles, method for production thereof, and affinity carrier, US5196527A, 4 May 1988.

[68] R. Mavelil-Sam, L.A. Pothan, S. Thomas,Polysaccharide and protein based aerogels: An introductory outlook, in: S. Thomas, L. A. Pothan, R. Mavelil-Sam (Eds.), Biobased aerogels: Polysaccharide and Protein-based Materials, The Royal Society of Chemistry, 2018: pp. 1–8. https://doi.org/10.1039/9781782629979-00001.

[69] R. Wang, X.-G. Ren, Z. Yan, L.J. Jiang, W.E.I. Sha, G.C. Shan, Graphene based functional devices: a short review, Front. Phys. 14 (2018) 13603. https://doi.org/10.1007/s11467-018-0859-y.

[70] E. Barrios, D. Fox, Y.Y. Li Sip, R. Catarata, J.E. Calderon, N. Azim, S. Afrin, Z. Zhang, L. Zhai, Nanomaterials in advanced, high-performance aerogel composites: a review, Polymers. 11 (2019) 726. https://doi.org/10.3390/polym11040726.

[71] W.S. Hummers, R.E. Offeman, Preparation of graphitic oxide, J. Am. Chem. Soc. 80 (1958) 1339. https://doi.org/10.1021/ja01539a017.

[72] W. Wang, J. Motuzas, X.S. Zhao, J.C. Diniz da Costa, 2D/3D assemblies of amine-functionalized graphene silica (templated) aerogel for enhanced CO_2 sorption, ACS Appl. Mater. Interfaces. 11 (2019) 30391–30400. https://doi.org/10.1021/acsami.9b07192.

Aerogels I: Preparation, Properties and Applications
Materials Research Foundations **84** (2020) 109-132

Materials Research Forum LLC
https://doi.org/10.21741/9781644900994-4

Chapter 4

Silica Aerogel

Nidhi Joshi[1], Ravi Kumar Pujala[2]*

[1]Department of Materials Engineering, Indian Institute of Science, Bangalore, India

[2]Soft and Active Matter Group, Department of Physics, Indian Institute of Science Education and Research (IISER) Tirupati, Andhra Pradesh 517507, India

*pujalaravikumar@iisertirupati.ac.in

Abstract

Over the last decade, silica aerogels have drawn enormous attention in science and technology due to their porous structure, low density, lightweight, exceptional thermal insulation property, high surface area, low refractive index, and excellent optical properties over conventional materials. Numerous advanced synthetic techniques have been applied to fabricate silica aerogels for their usage in various applications which extend from chemical sensors to thermal insulation system to drug delivery. This chapter presents the synthesis, properties, as well as review of the literature on emerging developments in the area of application of silica aerogels and future prospects.

Keywords

Aerogels, Silica, Porous, Insulation System, Biomedical

Contents

1.   Introduction

Silica aerogels are the unique lightweight materials having unusual properties, viz; low density (0.003-0.5 g/cm^3), high porosity (80-99.8%), specific surface area (500-1200 m^2/g), low index of refraction (~ 1.05), and very high thermal insulation value (0.005 W/mK) [1,2]. These porous materials result from wet gels, where the liquid content has been extracted through various drying methods, for example, supercritical drying, fluidized bed drying, ambient pressure, or atmospheric pressure drying, etc. so as to avoid pore collapse and leaving behind an intact nanostructure. The unique properties of silica aerogels are attributed to their remarkable network structures formed through clustered silica nanoparticles. The practical interest in the silica aerogels originates from their potential in the fields of catalysis, waveguides, drug delivery, thermal insulators, storage devices, supercapacitors, radiation detectors, sensor material, etc. [3-9]. Owing to the low value of thermal conductivity and density, they have been considered beneficial for various space/aerospace applications [10,11]. Since these aerogels are light weighted, fragile, and have poor mechanical strength, they are found to be attractive materials for the thermal insulation for windows and cerenkov radiators [6,12,13]. They have been perpetually used as a potential candidate in waste removal, biomedical, gas sensing, and air purification applications [14-17]. The unique characteristics of silica aerogels are attributed to the amount of mesoporosity that provides the bulk vacate space filled with air. They have bead-like structures forming the pearl-necklace like skeletal network; known as secondary particles that comprise of primary particles of silica as shown in Fig. 1. The coalescence of primary silica particles links each other to form secondary particles that are connected through siloxane bond (Si-O-Si) and result in the formation of three-dimensional (3D) network with pearl-necklace morphology [18-20]. It is anticipated that owing to their dendritic microstructure the particles fuse to form clusters and leads to the development of a 3D porous structure with pore size in the range of 5-100 nm.

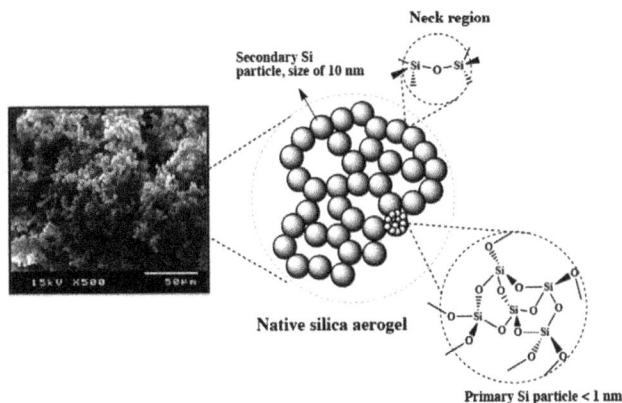

Fig. 1 *SEM images of silica aerogels along with its chemical structure. Reproduced with permission [29] Copyright 2019, Elsevier.*

Although the silica aerogels are sufficiently strong, their mechanical properties are not suitable and hence limit their use in some practical applications. Henceforth, their application possibility is strengthened by the introduction of reinforced micro/nanoscopic materials as a second phase such as carbon nanotubes, fibers, or nanoparticles, which enhance their remarkable properties [21-23]. For example, these silica aerogels form composites with water-soluble polymers, which provides thermo-responsive hydrogels exhibiting thermal and mechanical properties and reflects the features of the incorporated polymer [24,25]. Importantly, this kind of tuning may combine the properties and result in the formation of porous and enormously lightweight materials having extraordinary mechanical strength as well as toughness. With enormous development in the field, these modified silica aerogels present more fascinating properties and structural characteristics in contrast to native ones. The functionalization of their surfaces has been proven to be more promising in various other applications such as absorption of toxic substances and liquid transport on the nanoscale in chemical and biotechnological fields [22,25-28].

This chapter addresses the synthesis methodology, physicochemical properties, and recent developments in the area of applications of silica aerogels. First, the synthesis techniques to produce silica aerogels and their modification are discussed; second, the physical properties resulting from the morphology of the aerogels will be examined, and finally, the applications of silica aerogels in current scientific research and industrial developments will be presented in detail.

2.　Synthesis methodology

The transition from liquid (sol) to solid (gel) phase, known as sol-gel processing is one of the methodologies used to synthesize various materials having a uniform, and small particle sizes, with different structures. The prompt growth of sol-gel methodology during the last few decades has managed rapid development in the synthesis of porous materials. The most widely used methods for the synthesis of silica aerogels are sol-gel polymerization of silicon alkoxides, which can be accomplished in three steps, starting from the condensation of colloidal (silica) particles through sol-gel processing, aging of the gel and lastly supercritical drying process [29-31]. The schematic depicting the methodology for these aerogel through a sol-gel is shown in Fig. 2 [11].

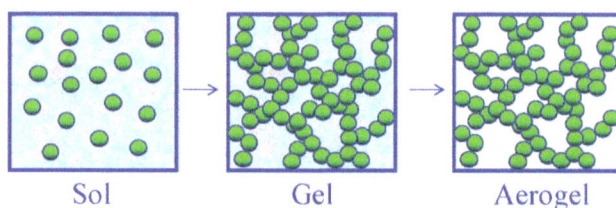

Sol　　　　　Gel　　　　　Aerogel

Fig. 2 *Scheme representing the fabrication of aerogels through a sol-gel process. Reproduced with permission [11] Copyright 2011, American Chemical Society.*

2.1　Bare silica aerogels

During the initial step of wet gel preparation, the source solution or precursor is dispersed with the catalyst (acid, or base) to initiate gelation in the system via hydrolysis and polycondensation reactions. Initially, it starts with the aggregation of primary particles to form secondary particles which eventually link together in a pearl necklace-type structure. The resulting siloxane bridges (Si-O-Si) provide backbone to the formation of silica matrix and are found to be porous in nature with solvent filled pores. The precursors range from alkoxides to acylates to amines and many others. This step of the methodology is carried out with an inorganic or organic precursor such as tetraethoxysilane (TEOS), tetramethoxysilane (TMOS), polyethoxydisiloxane (PEDS), or methyltriethoxysilane (MTES) which are initially involved as monomers [29,32-35]. The necessitate condition for precursors is that it should be soluble in the reaction media and reactive enough to participate in the reaction.

A variety of reaction media which are needed to homogenize the precursor mixture ranges from alcohols, dioxane, tetrahydrofuran, methylene chloride to ionic liquids. Additionally, various

Materials Research Forum LLC
https://doi.org/10.21741/9781644900994-4

precursors are not reactive enough towards gelling due to reduced partial positive charge from silicon, and thus, there is the paramount of a catalyst in the reaction media in order to reduce the gelation time from several days to hours [34]. Two approaches have been utilized; one of them involves a catalyst, basic or acidic, while the other consists of an acid catalyst facilitating hydrolysis reaction, leads to polymeric silica followed by a base-catalyzed set of condensation reactions and consequently leads to colloidal silica. Acid catalysis involves the entangled linear/randomly branched chains in silica sols with low cross-link density and in contrast, base catalysis form a network of uniform silica particles in the sol with high cross-link density. Varying different processing parameters such as the activity of the silicon alkoxides, solution pH, temperature, reaction media, additives, and nature of the catalyst provide materials with various microstructures and surface chemistry.

Once the wet gel is formed by hydrolysis and condensation reaction which is fragile in nature, the aging step involves further condensation and dissolution followed by reprecipitation of the sol particles, and thus enhances mechanical strength to the gel network as depicted in Fig. 3 [29]. The resultant morphology leads to a porous solid structure where the solvent is confined, and the final properties are strongly affected by the developed structure in the aging step. Importantly, there exist two simultaneous mechanisms during aging at different rates; one of which is a growth of necks between secondary silica particles and other is the dissolution of smaller ones followed by reprecipitation onto large particles [36]. These two mechanisms which occur simultaneously with different rates enhance the siloxane bonds and henceforth reinforce the network. This process is found to be crucial in order to prevent cracks in the formed gels.

The next salient step in sol-gel processing is drying of wet gels in a precisely controlled manner by the removal of liquid. During drying, the formed wet gel nanostructures couldn't withstand the capillary forces which occur via surface tension present at the liquid-vapor interface; thus, this presents a challenge to remove trapped solvent from the matrix for the silica aerogels. The supercritical drying prevents the formation of surface tension, which restricts the gel to collapse. The process consists of elevating the temperature as well as pressure beyond a critical point wherein the trapped liquid undergoes the transition to supercritical fluid so that the surface tension gets ceased, and finally, dry nitrogen is introduced to detach the liquid molecules from the formed wet gel. This drying methodology yields aerogels with high porosities, pore-volume, and surface area in contrast to other drying methods. The fluids suitable for this methodology are carbon dioxide, Freon, nitrous oxide, etc. however, carbon dioxide is mostly preferable amongst them all.

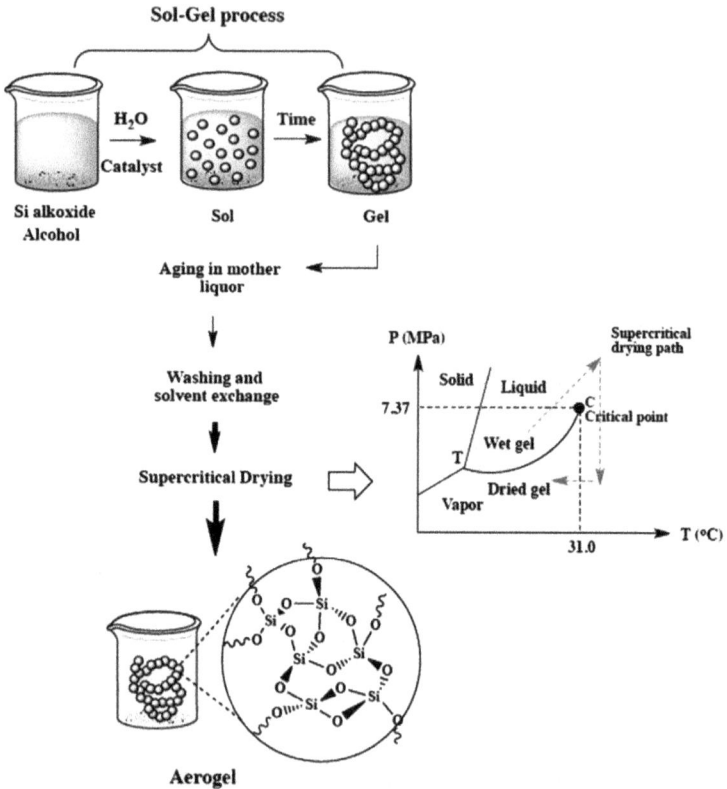

Fig. 3 *The methodology adopted for the synthesis of silica aerogels through the sol-gel process.*
Reproduced with permission [29] Copyright 2019, Elsevier.

Surprisingly, the supercritical drying method has definite restrictions in terms of safety, cost-efficiency, processing time, and other factors. Hence, the other method of drying known as ambient pressure drying introduced by Brinker [37], creates the surface of the gel with low energies simply by modifying with appropriate additives/agents in order to reduce surface tension. In such a modification process, the present hydroxyl (OH) group in the gel surface is chemically substituted through hydrophobic group and thus reduces the shrinkage that occurs in the subsequent drying process. To circumvent the above-mentioned issues, the *in-situ* attachment of the alkyl group to the siloxane backbone, as well as low degree of polymerization of silica,

Materials Research Forum LLC
https://doi.org/10.21741/9781644900994-4

provides the formed network to withstand capillary forces and compressive stress during the ambient pressure drying. The occurrence of capillary forces and compressive stress during the drying step provides the formed wet-gel to shrink and crack, and so there is a pre-requisite to balance these capillary forces before the drying procedure.

2.2 Modified silica aerogels

The bare silica aerogels are sensitive, fragile, hygroscopic at a very low value of stresses and thus restricts their application in industrial and commercial use, so it is a pre-requisite to introduce various organic polymers or fiber networks in silica precursors to form aerogels of high mechanical strength and stiffness. There are various applications where these aerogels need to be hydrophobic, and so there is a requirement to assess the hydrophobic behavior of the silica aerogels, which could be implemented by replacing hydrophilic hydroxy groups through surface modifications. The substances which are capable of modifying the wetting properties of the surface include methyltrimethoxysilane (MTMS), dimethyldichlorosilane (DMDC), hexamethyldisilazane (HMDZ), etc., through which substitution of hydrogen (H) from Si-OH by non-polar alkyl groups takes place [39-42]. The different types of precursors and/or additives used for the modification of silica aerogels are summarized in Table 1 [43-55].

A number of approaches have been used to substitute the terminal hydroxyl groups which could be merely incorporation of fluorinated chains covalently attached to the surface, or alteration of the silica network surface with other organic groups, or multi-step solvent exchange method [38]. The physicochemical properties and structure of silica aerogel in the modified aerogels/aerogels composites could be maintained simply by tuning reaction conditions such as temperature, solvent, etc. and the precursor ratio; possibly facilitating other properties such as high mechanical strength, dense network structure, and hydrophobicity. Hence, they are found to show improved thermal insulation and mechanical properties in contrast to bare silica aerogels. One of the intriguing and recent research work related to modification is the formation of aerogel-geopolymer composites which are known to provide the composite with favorable properties shown in Fig. 4 [24]. The modification of silica aerogels can be done by impregnating the silica aerogel in the prepared geopolymer leading to the formation of the resultant composite which can be used for insulating materials.

Table 1 *Some of the precursors and additives used in modified silica aerogels [43-55].*

Silica aerogel	Modified Silica aerogel		References
Precursor (Abbrev.)	Precursor (Abbrev.)	Additives for modified silica aerogel	
Tetramethoxysilane (TMOS)	Vinyltrimethoxysilane (VTMS)	Carbon (CNTs, MWCNTs, SWCNTS, GO)	[43-46]
Polyethoxydisloxane (PEDS)	Methyltriethoxysilane (MTES)	Polysaccharides (Cellulose, Pectin Carageenan, Chitosan)	[47-49]
Triethoxysilane (TrEOS)	Hexamethyldisilazane (HMDZ)	Fibers (Carbon, Ceramic, Nonwoven, Aramid)	[50-52]
Tetraethoxysilane (TEOS)	Trimethyldisiloxane (HMDSO)	Polymers (Epoxide, Polystyrene, Polyurethane, Polyacrylonitrile)	[53-55]
Sodium metasilicate (SS)	Trimethylchlorosilane (TMCS)		
Methyltrimethoxysilane (MTMS)	Aminopropyltriethoxysilane (APTES)		

There are three primary methods used for the alteration of silica aerogels known as the co-precursor method, vapor phase treatment, and surface derivatization method. Among these, the vapor phase treatment comprises heating the silica aerogel in the presence of methanol (M-OH) vapor for a specific time period in order to transit Si-OH to Si-OCH$_3$. The co-precursor method involves the addition of surface amending agent into silica solution before gelation while in the surface derivatization method, the gelation is done, followed by the placement of organic or inorganic network into a mixture consisting of surface modifying agent and the solvent. In the co-precursor method, the precursor is exchanged with organosilane and further proceeds by a supercritical drying step. It is to be noted that the amount of co-precursor used in this method affects the dynamics of the sol-gel process. The surface derivatization method modifies the surface of the formed wet gels before drying using a silylating agent. The hydrophobic behavior of silica aerogels aids the material to be transparent and thermally insulating and so are beneficial in various applications fields such as the Cherenkov detector, transparent window system, and many other applications [25,56-58]. The applications of these hydrophobic silica aerogels are illustrated in Table 2 [56-71].

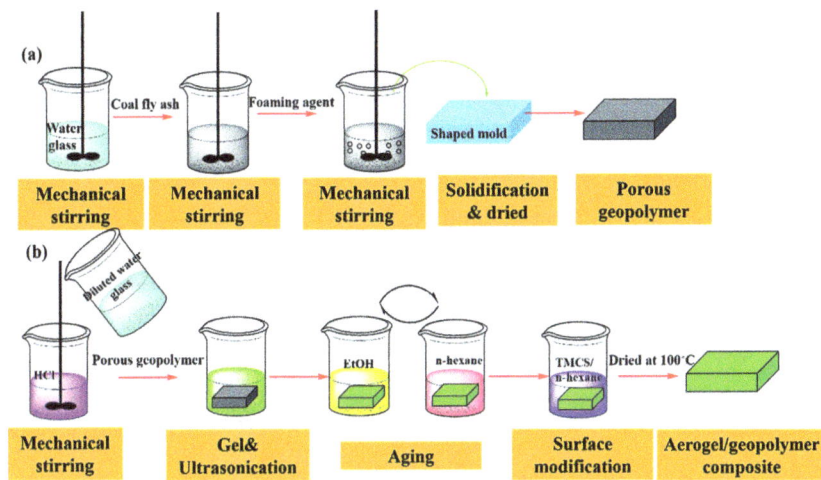

Fig. 4 *The scheme depicts the chemical reaction representations of the modification methodology to form silica composite aerogels. Reproduced with permission [24] Copyright 2018, The Royal Society of Chemistry.*

Table 2 *Some of the application of hydrophobic silica aerogels [56-71].*

Property	Application	References
Surface area	Oil spill clean-up Removal of pollutants	[59,60]
Porous structure	Drug Delivery Tissue Engineering	[61,62]
Refractive Index	Cherenkov Detector	[56,63,64]
Thermal conductivity	Themal insulation	[58,65-67]
Superhydrophobic surfaces	Self-cleaning windshield Anti-corrosion material	[68,69]
Transparency	Transparent windows	[57,70,71]

3.   Physico-chemical properties and applications

Silica aerogels exhibit distinctive features that make them fascinating and appealing in science as well as technology and because of the high porosity, pore accessibility, and surface area. They exhibit high porosity thus are rather brittle material. These aerogels produced by the methodology, as mentioned earlier, are hydrophilic and may contain up to 90 wt% air. As they are mainly composed of air, the 3D network in these aerogels comprises of porous type of morphology and shows remarkable properties. The versatile silica aerogels are characterized through different parameters which are gelling time, the shrinkage time during aging, apparent density, global shrinkage, and specific surface area. These structural characteristics have been studied using different techniques where the pattern characteristics could be correlated to the various structural changes at different length scales. These aerogels are not similar to Rayleigh scatterers as they show wavelength-independent components of scattering. In these cases, the scattering sources can be seen as occurring from inhomogeneities in the pores of the gel. The evolution of dynamical properties during the sol -gel transition, including the pore size, have been carried out on silica aerogels using light scattering [72,73].

The type of precursor is known to strongly influence the physical properties of these aerogels such as bulk density, distribution in pore size, surface area, amount of porosity, thermal conductivity, as well as microstructure as discussed in the above section. It has been found that different types of precursor in the synthesis provides aerogels with different surface area and variant pore size distribution which depends on the size of SiO_2 particles and pores in the network. Due to the pearl necklace-type fractal network that constitutes voids between the silica particles chains, the resultant silica aerogels have a low bulk density with approximately half the density of silica and hence they are fragile.

In the paradigm of the surface modified or hydrophobic silica aerogels, the general agreement is that the porous bare silica material formed shows brittle and elastic behavior; howsoever, their mechanical properties are less due to porosity and low connectivity in the network. Their weak mechanical properties are due to the large pore volume which could be reduced by greater content of monomer or through sintering or drying procedure, which leads to the collapse of the pores. Most importantly, the stress corrosion effect can lead to failure of silica aerogels, an impact that is favored by the OH content of the aerogels. The incorporation of cross-linker or polymer in the silica network shows substantial improvement in the mechanical properties through very insignificant changes in physical properties such as a slight increase in density etc. This introduction of cross-linker/polymer to the hydroxyl group on the silica gel surface will be an active reinforcing process for durable silica aerogels whose objective is to enhance the mechanical strength so that it could be able to adapt for the specific applications. These modified silica aerogels owe their low wavelength to the tortuosity of the robust network, the high porosity, and pore sizes beneath the mean free path length of air (see Fig. 4) [24].

Another intriguing class in hydrophobic silica aerogels is granular silica aerogels that associate the low value of thermal conductivity of the network with the high diffusion transmittance in certain spectral range and thus used in applications of building envelopes, cleanup of spillage, solvent removal, etc. [61-64,74-77]. These are of three types that are relatively regular spheres, semi-translucent and highly translucent irregularly fractured pieces. These are mesoporous that allows a decrease in the thermal conductivity of the gaseous phase because of the Knudsen effect, but their brittleness hinders various applications thus, a new type of aerogels known as bio-aerogels has been developed that can accomplish as conventional silica aerogels as well as synthetic organic ones. They are prepared by polymer dissolution, gelation through coagulation in a non-solvent phase, with further supercritical drying in the presence of carbon dioxide. Besides granular silica aerogels, different types of silica aerogels are prepared using transition metal alkoxides, organic, etc. Out of these, the widely studied aerogels are mixed oxide aerogels using alumina silicates because of the improved temperature stability compared to silica aerogels. The other approach towards the synthesis of aerogels is through doping, where the alcohol-soluble compound has been added to the sol before gelation. Thus, silica aerogels are doped with phosphors/florescent chromophore to enhance the output of radio-luminescent light sources [65,66,77-79].

The structure of silica aerogels is composed of small and spherical SiO_2 clusters of the size 3-4 nm that are linked to each other and result in the formation of chains, which further form a spatial grid with air-filled pores of 30-40 nm. They exhibit high porosity and thus are rather brittle material. The aerogels produced by the methodology, as mentioned earlier, are hydrophilic and may contain up to 10 wt% water. In contrast to hydrophilic aerogels, the hydrophobic aerogels are prepared via a two-step synthesis process, where the first one is identical to the one for hydrophilic, howsoever the hydrolysis and condensation are performed incompletely. The silicone oil obtained was purified via alcohol, followed by mixing with alcohol-free solvent, and so the polymerization occurs in the presence of alkali catalyst. The consecutive steps comprise of the same mentioned for hydrophilic that are aging and drying. The other approach for synthesizing hydrophobic aerogels is to replace the hydrophilic group with hydrophobic ones, which can be done by substituting hydrogen in OH with $Si(CH_3)_3$. Though aerogels are hydrophilic, they become wet with atmospheric moisture and water and thus deteriorate with time. The surface group Si-OH is known to be the primary source of hydrophilicity to enhance condensation reaction in aerogels, and so the replacement of this group with Si-R (R = alkyl) result in hydrophobic aerogels. Mostly, the hydrophobic aerogels synthesis is done using a co-precursor such as methyltrimethoxysilane (MTMS).

Silica aerogels have fascinating properties that make them a potential candidate for application in numerous fields as shown in Fig. 5 [80]. Some of their applications are described in the next section.

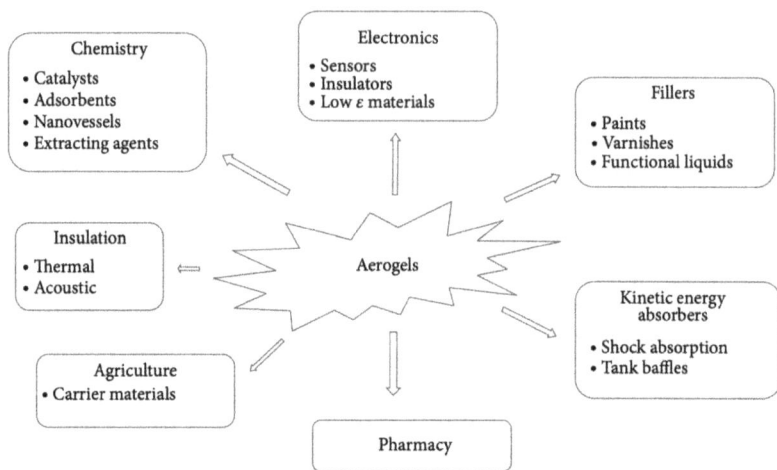

Fig. 5 *Applications of aerogels in different fields. Reproduced with permission [80]*
Copyright 2014, Journal of Materials.

3.1 Thermal insulating application

The critical property of the silica aerogels of poor thermal conductivity (0.020 W/m.K) could be exploited in various applications for example architectural purpose, heat and cold storage appliances, transport vehicles, and vessels. The extraordinary small pore size in the silica aerogels provides remarkable physical, thermal, and acoustical properties, specifically in thermal insulation. The low content of silica and the presence of dead ends in their three-dimensional network results in low intrinsic thermal conductivity as it provides a tortuous path for the transport of heat. These have been used as a transparent thermal superinsulation in buildings, containers, refrigerators, and refrigerated vehicle due to better environmental compatibility. The translucent aerogels insulation has been applied to huge parts in the buildings for daylighting as they exhibit better heat conductance compared to window panes. In some applications, the surface modification of silica aerogels is needed as the uptake of water led the materials to deteriorate and thus limit its application and so, the hydrophobic silica aerogels are found to be more beneficial in various applications [64-66]. The tailoring of silica aerogels with the addition of additives and/or polymer dampens the brittleness and provides durable as well as flexible material that can be used in clothing, blankets, sleeping bags, apparel, and/or thermal bridges. The aerogels should be highly hydrophobized as their contact with water lead to a thrashing of the aerogel structure, and so, used with a vacuum that prevents water exclusion while the vacuum decreases the thermal conductivity. In addition to this, the aerospace domains use the silica aerogels for the thermal insulation of the Mars Rover, space suits, collection of aerosol particles, designing of tank baffles, and in protection of space mirrors [10].

3.2 Optical property application

One of the best uses of aerogels is to fabricate ultrapure, high-density silica glass below the silica melting temperature through sintering. An efficient radio-luminescence light source can be achieved by incorporating radioactive tritium and a phosphor in silica aerogels. In addition to this, translucent aerogels are used for solar covers and collectors while transparent aerogels are used for solar windows [9,12]. The presence of pores in the silica aerogels acts as scattering centers and depending on these centers, the scattering efficiency is traced. These aerogels shows maximum transparency in the infrared region of the spectrum and so can be used for desired applications such as super-insulating windows. The heterogeneities in the aerogel network lead to Rayleigh scattering which is responsible for the material coloration in different modes and thus is needed for optical applications. Their low value of refractive index led the potential to be used in the cerenkov counters, where the elementary particles travel with a velocity in a low-density medium in a radiator which is silica aerogel. In these counters, these are placed intermediate as compressed gases and cryogenic liquids in the radiators based on their refractive index value.

3.3 Electronic application

The aerogels are known to be suitable dielectrics for ultrafast integrated circuits (ICs) because of low dielectric constants, good thermal stability, and high porosity and could be exploited for microwave electronics as well as high voltage insulators. These aerogels show different types of behavior varying from superconducting to thermoelectric to piezoelectric. The silica nanowires can be assembled to produce functional micro-photonics devices that leads to the fabrication of essential devices such as linear waveguides, branch couplers, and waveguides bend. Due to their minimal size than prevailing comparable devices with less optical loss, these aerogels can be used in optical applications such as communications and sensing. The high surface area and three-dimensional network in silica aerogels amplify the electrifiable interface and facilitate the electronic conduction and hence they can be potential candidates for supercapacitor [5]. In contrast to the normal capacitor, the silica aerogel capacitor exhibit low impedance and absorb high peak current because of high surface area ($\sim$500 m^2/g).

3.4   Acoustic insulation applications

Silica aerogels are the outstanding materials for acoustic devices where the sound propagation velocity is much less and so can be used for building the corresponding acoustic delay lines [4]. The minimum acoustic impedance usually characterizes the highly porous aerogels compared to solid and liquid substances. The high porosity in these aerogels is related to its low static density which traps the air in its nanorange pores providing a higher acoustic density. In addition to this, the low cross-section interconnecting the silica particles leads to low rigidity, and hence this results in low acoustic impedance and can be used as a subwavelength flexural in fragile acoustic metamaterials for transmission lines. These aerogels can also be used as acoustic antireflecting $\lambda/4$ – layer for piezoelectric materials. Besides this, another possible application of aerogels is echoless terminations in the audio-frequency band, where granular aerogels are most probably used. The low longitudinal sound velocity and acoustic impedances are found to be responsible for them to be used as acoustic insulation applications.

3.5   Biomedical applications

The potential of silica aerogels is practically exploited in the medical field too due to their biocompatibility and could be fabricated in different shapes and sizes, varying from beads to disks, and further monoliths. These aerogels hold excellent ability for their applicability to be used in the biomedical field, for example, drug delivery, biosensing, tissue engineering, etc. The natural polymers such as pectin, cellulose, etc. can be converted into aerogels, which further can be used as scaffolding materials, blood vessels, artificial cartilage, etc. which are environment-friendly and can withstand the toxicity and yield. These can be used as drug delivery matrices to enhance the therapeutic effect with modified drug release profiles and govern the delivery of active biomolecules. The incorporation of drugs into silica/functionalized silica aerogels can be achieved by two approaches, either the addition of drugs during the sol-gel procedure using means of post-processing [15]. The incorporated drug should meet particular criteria such as solubility, stability, etc. to form these aerogels to be used as drug delivery. Surprisingly, the hydrophobic silica aerogels are most commonly used for enzyme encapsulation and drug delivery where the hydrophobicity of the resultant aerogels could control the absorption and release of the drugs [25, 61,62]. Their high porosity can trace these silica aerogels as dry powder carriers for pulmonary delivery in the form of particles. The enhanced dissolution of inadequately water-soluble drugs, good air flowability, and high stability of drugs lead the biopolymer-based aerogels for these applications.

3.6   Environmental applications

The aerogels can be used as a filter for absorbing long-lived nuclear waste contained in liquid radioactive wastes and purify air consisting of dangerous aerosol particles of nanometer dimensions due to the presence of large surface area. Also, they are used in the purification of gases that emerges from the exhaust pipe of a car [17]. Silica aerogels reinforced with a metal oxide such as titanium oxide has a large binding capacity for gasoline, which get absorbed and decomposed with the formation of carbon dioxide.

3.7 Others applications

3.7.1 Space and detector

These light weighted silica aerogels have been used to collect contaminants in order to shield space mirrors from volatile organics. These aerogels to be used in space type applications require the material property to fulfill desirable thermal features of the outer space. Among the various applications of the aerogels, the foremost is its usage as a source of cerenkov radiation [6]. The silica aerogels used in cerenkov counters for the detection of relativistic particles attributes to a low index of refraction that allows determining the momentum of relativistic particles in a range that does not include compressed gases or by liquids. Due to their high transparency, their use as a radiator for threshold as well as differential counter is carried out, which could help to determine the particle velocity of the Cherenkov radiation. The stacks of tiles made of aerogels sintered to different densities and index of refraction have been used in Cerenkov threshold detectors.

3.7.2 Oil spill clean-up

Oil spills are formed by occasional accidental episodes of supertankers, war, oil drilling, and naturally occurring events. There is a necessity for absorbent materials for cleaning-up of oil spills that requires to undergo the transition from liquid (sol) to semi-solid through which the removal of oil spills becomes easier. For a material to be suitable absorbent, it should be hydrophobic, large uptake capacity, reusability, and biodegradability. Since silica aerogels acquire low surface energy, so it absorbs liquid which possesses surface energy less than that of aerogels. The hydrophobic nature of these aerogels can be retained up to higher temperatures (~350-400 °C) and so can be used for the application of absorption and desorption of accidental spillage of oils and organic type liquids [59,60].

3.7.3 Aerospace

The fragile nature of the aerogels is advantageous in space applications such as capturing the cosmic dust of hypervelocity particles from the comet tail [11]. The structural characteristics of aerogels are such that it slows the particles which are traveling at extreme velocities followed by capturing them so that these can be examined on reaching the earth. The thermal insulation requirement for the various aerospace missions requires a low value of thermal conductivity as well as low density, which are fulfilled by silica aerogels and thus are used in these applications.

These insulation properties of aerogels protect the electronic components for several months/years and hence its widespread use in aerospace. The vehicle and launch pad operations need durable and lightweight insulation and so the aerogels out-perform multilayer insulation system in mild vacuum and ambient pressures. The typical methodology to enhance the mechanical properties of silica aerogels is the polymer reinforcement that leads to a build-up of conformal coatings on the particles. Mostly, the epoxy reinforced aerogels show twice the order of magnitude in strength compared to bare silica aerogels.

Conclusions and future prospects

Silica aerogels are the class of highly porous materials with appealing properties such as lightweight, high specific surface area, tunable surface chemistry, as well as tunable properties, which can overcome the limitation of various technologies. They have well-known characteristics, for example low thermal conductivity, low density, high surface area and very high porosity. Their favorable characteristics make them advantageous in biomedical, process engineering, material engineering, environmental, and many more. The chapter presents an overview of the sol-gel synthesis for silica aerogels including a short introduction to hydrophobic silica aerogels. As illustrated here, their physico-chemical properties could be perceived through the proper selection of precursor type, and optimization of reaction conditions during sol-gel processing. In the field of thermal insulation, they are being utilized as a promising material for the construction of a building. The high surface area could provide a platform for these materials to be used in the biomedical field. Also, due to high electrical conductivity, these could be employed in the development of integrated circuits. Moreover, the studies have shown that the hydrophobicity supplemented with the silica aerogels enhance their properties and widened the application areas of current interest. The incorporation of the contrasting agent to modify from hydrophilic to hydrophobic led them to be adequate to use in different fields such as biomedical, electronics, spacecraft, and others. In conclusion, we anticipate that with research on silica aerogels composites, they will be used for variant commercial applications too.

There is a scope for additional work on the modified silica aerogels which could be applicable in numerous applications, specifically on the superhydrophobic silica aerogel. In recent times, the other fascinating current area of research is the usage of smart silica aerogels in wastewater treatment and corrosion protection. A detailed study of these is still required and taking their physico-chemical properties and morphology into consideration, they can be used in wide areas of research and technology.

References

[1] J. Fricke, Aerogels–highly tenuous solids with fascinating properties, J. Non-Cryst. Solids 100 (1988) 169–173. https://doi.org/10.1016/0022-3093(88)90014-2

[2] C.E. Carraher Jr., General topics: Silica aerogels–properties and uses, Polym. News 30 (2005) 386–388. https://doi.org/10.1080/00323910500402961

[3] D.A. Ward, E.I. Ko, Preparing catalytic materials by the sol-gel method, Ind. Eng. Chem. Res. 34 (1995) 421–433. https://doi.org/10.1021/ie00041a001

[4] V. Gibiat, O. Lefeurre, T. Woignier, J. Pelous, J. Phalippou, Acoustic properties and potential applications of silica aerogels, J. Non-Cryst. Solids 186 (1995) 244–255. https://doi.org/10.1016/0022-3093(95)00049-6

[5] X. Du, C. Wang, T. Li, M. Chen, Studies on the performance of silica aerogel electrodes for the application of supercapacitor, Ionics 15 (2009) 561–565. https://doi.org/10.1007/s11581-009-0315-7

[6] J.P. Cunha, F. Neves, M.I. Lopes, On the reconstruction of cherenkov rings from aerogels radiators, Nucl. Instrum. Meth. A 452 (2000) 401–421. https://doi.org/10.1016/S0168-9002(00)00452-6

[7] L.C. Alexa, G.M. Huber, G.J. Lolos, F. Farzanpay, F. Garibaldi, M. Jodice, A. Leone, R. Perrino, Z. Papandreou, D.L. Humphrey, P. Ulmer, R. Deleo, Empirical tests and model of a silica aerogel cherenkov detector for CEBAF, Nucl. Instrum. Meth. A 365 (1995) 299–307. https://doi.org/10.1016/0168-9002(95)00515-3

[8] C.T. Wang, C.L. Wu, I.C. Chen, Y.H. Huang, Humidity sensors based on silica nanoparticle aerogel thin films, Sensor. Actuator. B Chem. 107 (2005) 402–410. https://doi.org/10.1016/j.snb.2004.10.034

[9] M. Rubin, C.M. Lampert, Transparent silica aerogels for window insulation, Sol. Energy Mater. 7 (1983) 393–400. https://doi.org/10.1016/0165-1633(83)90012-6

[10] S.M. Jones, Aerogel: space exploration applications, J. Sol-Gel Sci. Technol. 40 (2006) 351–357. https://doi.org/10.1007/s10971-006-7762-7

[11] J.P. Randall, M.A.B. Meador, S.C. Jana, Tailoring mechanical properties of aerogels for aerospace applications, ACS Appl. Mater. Interfaces, 3 (2011) 613–626. https://doi.org/10.1021/am200007n

[12] V. Wittwer, Development of aerogel windows, J. Non-Cryst. Solids 145 (1992) 233–236. https://doi.org/10.1016/S0022-3093(05)80462-4

[13] C. Buratti, E. Moretti, Glazing systems with silica aerogel for energy savings in buildings, Appl. Energy 98 (2012) 396–403. https://doi.org/10.1016/j.apenergy.2012.03.062

[14] H. Maleki, Recent advances in aerogels for environmental remediation applications: A review, Chem. Eng. J. 300 (2016) 98–118. https://doi.org/10.1016/j.cej.2016.04.098

[15] I. Smirnova, J. Mamic, W. Arlt, Adsorption of drugs on silica aerogels, Langmuir, 19 (2003) 8521–8525. https://doi.org/10.1021/la0345587

[16] P.A.S. Jorge, P. Caldas, C.C. Rosa, A.G. Oliva, J.L. Santos, Optical fiber probes for fluorescene based oxygen sensing, Sensor. Actuator. B Chem. 103 (2004) 290–299. https://doi.org/10.1016/j.snb.2004.04.086

[17] J.E. Amonette, J. Matyas, Functionalized silica aerogels for gas-phase purification, sensing, and catalysis: A review, Micropor. Mesopor. Mat. 250 (2017) 100–119. https://doi.org/10.1016/j.micromeso.2017.04.055

[18] T. Woignier, J. Phalippou, Mechanical strength of silica aerogels, J. Non-Cryst. Solids 100 (1988) 404–408. https://doi.org/10.1016/0022-3093(88)90054-3

[19] M.A. Aegerter, N. Leventis, M.M. Koebel, Aerogels handbook, Springer Science & Business Media, New York, 2011.

[20] G. Zhang, A. Dass, A.M. Rawashdeh, J. Thomas, J.A. Counsil, C.S. Leventis, E.F. Fabrizio, F. Ilhan, P. Vassilaras, D.A. Scheiman, L. Mccorkle, A. Palczer, J.C. Johnston, M.A. Meador, N. Leventis, Isocyanate-crosslinked silica aerogel monoliths: preparation and characterization, J. Non-Cryst. Solids 350 (2004) 152–164. https://doi.org/10.1016/j.jnoncrysol.2004.06.041

[21] M.A.B. Meador, S.L. Vivod, L. Mccorkle, D. Quade, R.M. Sullivan, L.J. Ghosn, N. Clark, L.A. Capadona, Reinforcing polymer cross-linked aerogels with carbon nanofibers, J. Mater. Chem. 18 (2008) 1843–1852. https://doi.org/10.1039/B800602D

[22] S. Kabiri, D.N.H. Tran, S. Azari, D. Losic, Graphene-diatom silica aerogels for efficient removal of mercury ions from water, ACS Appl. Mater. Interfaces 22 (2015) 11815–11823. https://doi.org/10.1021/acsami.5b01159

[23] H. Nakamura, Y. Matsui, Silica gel nanotubes obtained by the sol-gel method, J. Am. Chem. Soc. 117 (1995) 2651–2652. https://doi.org/10.1021/ja00114a031

[24] Y. Huang, L. Gong, Y. Pan, C. Li, T. Zhou, X. Cheng, Facile construction of the aerogel/geopolymer composite with ultra-low thermal conductivity and high mechanical performance, RSC Adv. 8 (2018) 2350–2356. https://doi.org/10.1039/C7RA12041A

[25] I. Smirnova, S. Suttiruengwong, W. Arlt, Feasibility study of hydrophilic and hydrophobic silica aerogels as drug delivery systems, J. Non-Cryst. Solids 350 (2004) 54-60. https://doi.org/10.1016/j.jnoncrysol.2004.06.031

[26] S. Standeker, Z. Novak, Z. Knez, Adsorption of toxic organic compounds from water with hydrophobic silica aerogels, J. Colloid Interface Sci. 310 (2007) 362–368. https://doi.org/10.1016/j.jcis.2007.02.021

[27] D. Wang, E. Mclaughlin, R. Pfeffer, Y.S. Lin, Adsorption of oils from pure liquid and oil-water emulsion on hydrophobic silica aerogels, Sep. Purif. Technol. 99 (2012) 28–35. https://doi.org/10.1016/j.seppur.2012.08.001

[28] A.V. Rao, M.M. Kulkarni, S.D. Bhagat, Transport of liquids using superhydrophobic aerogels, J. Colloid Interface Sci. 285 (2005) 413–418. https://doi.org/10.1016/j.jcis.2004.11.033

[29] H. Maleki, L. Duraes, A. Portugal, An overview on silica aerogels synthesis and different mechanical reinforcing strategies, J. Non-Cryst. Solids 385 (2014) 55–74. https://doi.org/10.1016/j.jnoncrysol.2013.10.017

[30] M. Stolarski, J. Walendziewski, M. Steininger, B. Pniak, Synthesis and characteristic of silica aerogels, Appl. Catal. A Gen. 177 (1999) 139–148. https://doi.org/10.1016/S0926-860X(98)00296-8

[31] A.S. Dorcheh, M.H. Abbasi, Silica aerogel: synthesis, properties and characterization, J. Mater. Process. Technol. 199 (2008) 10–26. https://doi.org/10.1016/j.jmatprotec.2007.10.060

[32] Z. Deng, J. Wang, J. Wei, J. Shen, B. Zhou, L. Chen, Physical properties of silica aerogels prepared with polyethoxydisiloxanes, J. Sol-Gel Sci. Technol. 19 (2000) 677–680. https://doi.org/10.1023/A:1008754504788

[33] A. Jitianu, A. Britchi, C. Deleanu, V. Badescu, M. Zaharescu, Comparative study of the sol-gel processes starting with different substituted Si-alkoxides, J. Non-Cryst. Solids 319 (2003) 263–279. https://doi.org/10.1016/S0022-3093(03)00007-3

[34] M.A. Einarsrud, E. Nilsen, A. Rigacci, G.M. Pajonk, S. Buathier, D. Valette, M. Durant, B. Chevalier, P. Nitz, F. Ehrburger-Dolle, Strengthening of silica gels and aerogels by washing and aging processes, J. Non-Cryst. Solids 285 (2001) 1–7. https://doi.org/10.1016/S0022-3093(01)00423-9

[35] A.V. Rao, G.M. Pajonk, S.D. Bhagat, Comparative studies on the surface chemical modification of silica aerogels based on various organosilane compounds of the type

RnSiX4-n, J. Non-Cryst. Solids 350 (2004) 216–223.
https://doi.org/10.1016/j.jnoncrysol.2004.06.034

[36] R.A. Strom, Y. Masmoudi, A. Rigacci, G. Petermann, L. Gullberg, B. Chevalier, M.A. Einarsrud, Strengthening and aging of wet silica gels for up-scaling of aerogel preparation, J. Sol-Gel Sci. Technol. 41 (2007) 291-298. https://doi.org/10.1007/s10971-006-1505-7

[37] C.J. Brinker, G.W. Scherer, Sol-gel science: The physics and chemistry of sol-gel processing, Academic press, Boston, 1990.

[38] A.V. Rao, P.B. Wagh, Preparation and characterization of hydrophobic silica aerogels, Mater. Chem. Phys. 53 (1998) 13–18. https://doi.org/10.1016/S0254-0584(97)02047-6

[39] A.V. Rao, M. Kulkarni, G.M. Pajonk, D.P. Amalnerkar, T. Seth, Synthesis and characterization of hydrophobic silica aerogels using trimethylethoxysilane as a co-precursor, J. Sol-Gel Sci. Technol. 27 (2003) 103–109. https://doi.org/10.1023/A:1023765030983

[40] A.V. Rao, M.M. Kulkarni, D.P. Amalnerkar, T. Seth, Surface chemical modification of silica aerogels using various alkyl-alkoxy/chloro silanes, Appl. Surf. Sci. 206 (2003) 262–270. https://doi.org/10.1016/S0169-4332(02)01232-1

[41] S.D. Bhagat, A.V. Rao, Surface chemical modification of TEOS based silica aerogels synthesized by two step (acid-base) sol-gel process, Appl. Surf. Sci. 252 (2006) 4289–4297.

https://doi.org/10.1016/j.apsusc.2005.07.006

[42] Y. Duan, S.C. Jana, B. Lama, M.P. Espe, Hydrophobic silica aerogels by silylation, J. Sol Gel Sci. Technol. 437 (2016) 26–33. https://doi.org/10.1016/j.jnoncrysol.2016.01.016

[43] A.L. Mendes, R.F. Silva, L. Duraes, Advances in carbon nanostructure-silica aerogel composites: A review, J. Mater. Chem. A. 6 (2018) 1340–1369. https://doi.org/10.1039/C7TA08959G

[44] K. Chen, Z. Bao, A. Du, X. Zhu, J. Shen, G. Wu, Z. Zhang, B. Zhou, One-pot synthesis, characterization and properties of acid-catalyzed resorcinol/formaldehyde cross-linked silica aerogels and their conversion to hierarchical porous carbon monoliths, J. Sol-Gel Sci. Technol. 62 (2012) 294–303. https://doi.org/10.1007/s10971-012-2722-x

[45] J. Huang, H. Liu, S. Chen, C. Ding, Hierarchical porous MWCNTs-silica aerogel synthesis for high-efficiency oily water treatment, J. Environ. Chem. Eng. 4 (2016) 3274–3282. https://doi.org/10.1016/j.jece.2016.06.039

[46] S. Dervin, Y. Lang, T. Perova, S.H. Hinder, S.C. Pillai, Graphene oxide reinforced high surface area silica aerogels, J. Non-Cryst. Solids 465 (2017) 31–38. https://doi.org/10.1016/j.jnoncrysol.2017.03.030

[47] J.C.H. Wong, H. Kaymak, P. Tingaut, S. Brunner, M.M. Koebel, Mechanical and thermal properties of nanofibrillated cellulose reinforced silica aerogel composites, Micropor. Mesopor. Mat. 217 (2015) 150–158. https://doi.org/10.1016/j.micromeso.2015.06.025

[48] J. Feng, D. Le, S.T. Nguyen, V.T. Nien, D. Jewell, H.M. Duong, Silica-cellulose hybrid aerogels for thermal and acoustic insulation applications, Colloids Surf. 506 (2016) 298–305. https://doi.org/10.1016/j.colsurfa.2016.06.052

[49] M.R. Ayers, A.J. Hunt, Synthesis and properties of chitosan-silica hybrid aerogels, J. Non-Cryst. Solids 285 (2001) 123–127. https://doi.org/10.1016/S0022-3093(01)00442-2

[50] A. Karout, P. Buisson, A. Perrard, A.C. Pierre, Shaping and mechanical reinforcement of silica aerogel biocatalysts with ceramic fiber felts, J. Sol-Gel Sci. Techn. 36 (2005) 163–171. https://doi.org/10.1007/s10971-005-5288-z

[51] Z.T. Mazraeh-shahi, A.M. Shoushtari, A.M. Bahramian, M. Abdouss, Synthesis, structure and thermal protective behavior of silica aerogel/PET nonwoven fiber composite, Fibers Polym. 15 (2014) 2154–2159. https://doi.org/10.1007/s12221-014-2154-z

[52] Z. Li, L. Gong, X. Cheng, S. He, C. Li, H. Zhang, Flexible silica aerogel composites strengthened with aramid fibers and their thermal behavior, Mater. Des. 99 (2016) 349–355. https://doi.org/10.1016/j.matdes.2016.03.063

[53] M.A.B. Meador, A.S. Weber, A. Hindi, M. Naumenko, L. McCorkle, D. Quade, S.L. Vivod, G.L. Gould, S. White, K. Deshpande, Structure-property relationships in porous 3D nanostructures: Epoxy-cross-linked silica aerogels produced using ethanol as the solvent, ACS Appl. Mater. Interfaces 4 (2009) 894–906. https://doi.org/10.1021/am900014z

[54] K. Chang, Y. Wang, K. Peng, H. Tsai, J. Chen, C. Huang, K. Ho, W. Lien, Preparation of silica aerogel/polyurethane composites for the application of thermal insulation, J. Polym. Res. 21 (2014) 338–343. https://doi.org/10.1007/s10965-013-0338-7

Materials Research Forum LLC
https://doi.org/10.21741/9781644900994-4

[55] K. Khezri, H. Mahdavi, Polystyrene-silica aerogel nanocomposites by in situ simultaneous reverse and normal initiation technique for ATRP, Micropor. Mesopor. Mat. 228 (2016) 132–140. https://doi.org/10.1016/j.micromeso.2016.03.022

[56] A.K. Gougas, D. Ilie, S. Ilie, V. Pojidaev, Behavior of hydrophobic aerogel used as a Cherenkov medium, Nucl. Instrum. Meth. A 421 (1999) 249–255. https://doi.org/10.1016/S0168-9002(98)01192-9

[57] A.V. Rao, G. Pajonk, D. Haranath, Synthesis of hydrophobic aerogels for transparent window insulation applications, Mater. Sci. Technol. 17 (2001) 343–348. https://doi.org/10.1179/026708301773002572

[58] D. Ge, L. Yang, Y. Li, J. Zhao, Hydrophobic and thermal insulation properties of silica aerogel/epoxy composite, J. Non-Cryst. Solids 355 (2009) 2610–2615. https://doi.org/10.1016/j.jnoncrysol.2009.09.017

[59] M.H, Sorour, H.A. Hani, G.A. Al-Bazedi, A.M. El-Rafei, Hydrophobic silica aerogels for oil spills clean-up, synthesis, characterization and preliminary performance evaluation, J. Porous Mater. 23 (2016) 1410–1409. https://doi.org/10.1007/s10934-016-0200-5

[60] L.W. Hrubesh, P.R. Coronado, J.H. Satcher Jr., Solvent removal from water with hydrophobic aerogels, J. Non-Cryst. Solids 285 (2001) 328–332. https://doi.org/10.1016/S0022-3093(01)00475-6

[61] U. Guenther, I. Smirnova, R.H.H. Neubert, Hydrophobic silica aerogels as dermal drug delivery systems- Dithranol as a model drug, Eur. J. Pharm. Biopharm. 69 (2008) 935–942. https://doi.org/10.1016/j.ejpb.2008.02.003

[62] J. Stergar, U. Maver, Review of aerogel-based materials in biomedical applications, J. Sol-Gel Sci. Technol. 77 (2016) 738–752. https://doi.org/10.1007/s10971-016-3968-5

[63] M. Ishino, J. Chiba, H. Enyo, H. Funahashi, A. Ichikawa, M. Ieiri, H. Kanda, A. Masaike, S. Mihara, T. Miyashita, T. Murakami, A. Nakamura, M. Naruki, R. Muto, K. Ozawa, H.D. Sato, M. Sekimoto, T. Tabaru, H. Yokogawa, Mass production of hydrophobic silica aerogel and readout optics of Cherenkov light, Nucl. Instrum. Meth. A 457 (2001) 581–587. https://doi.org/10.1016/S0168-9002(00)00786-5

[64] M. Tabata, I. Adachi, H. Kawai, T. Sumiyoshi, H. Yokogawa, Hydrophobic silica aerogel production at KEK, Nucl. Instrum. Meth. A 668 (2012) 64–70. https://doi.org/10.1016/j.nima.2011.12.017

[65] J.E. Fesmire, Aerogel insulation systems for space launch applications, Cryogenics 46 (2006) 111–117. https://doi.org/10.1016/j.cryogenics.2005.11.007

[66] K. Maghsoudi, S. Motahari, Mechanical, thermal, and hydrophobic properties of silica aerogelepoxy composites, Appl. Polym. Sci. 135 (2017) 45706–45710. https://doi.org/10.1002/app.45706

[67] P.B. Wagh, S.V. Ingale, Comparison of some physic-chemical properties of hydrophilic and hydrophobic silica aerogels, Ceram. Int. 28 (2002) 43-50. https://doi.org/10.1016/S0272-8842(01)00056-6

[68] X. Li, D. Reinhoudt, M. Crego-Calama, What do we need for a superhydrophobic surface?Areview on the recent progress in the preparation of superhydrophobic surfaces, Chem. Soc. Rev. 36 (2009) 1350–1368. https://doi.org/10.1002/chin.200744242

[69] A.V. Rao, M.M. Kulkarni, S.D. Bhagat, Transport of liquids using hydrophobic aerogels. J. Colloid Interface Sci. 285 (2005) 413–418. https://doi.org/10.1016/j.jcis.2004.11.033

[70] A.A. Pisal, A.V. Rao, Development of hydrophobic and optically transparent monolithic silica aerogels for window panel applications. J. Porous Mater. 24 (2017) 685–695. https://doi.org/10.1007/s10934-016-0305-x

[71] K.H. Lee, S.Y. Kim, K.P. Yoo, Low density, hydrophobic aerogels, J. Non-Cryst. Solids 186 (1995) 18–22. https://doi.org/10.1016/0022-3093(95)00066-6

[72] A.J. Hunt, Light scattering for aerogel characterization, J. Non-Cryst. Solids 225 (1998) 303–306. https://doi.org/10.1016/S0022-3093(98)00048-9

[73] A. Hasmy, E. Anglaret, M. Foret, J. Pelous, R. Jullien, Small-angle neutron scattering investigation of long-range correlations in silica aerogels: simulations and experiments, Phys. Rev. B 50 (1994) 6006–6009. https://doi.org/10.1103/PhysRevB.50.6006

[74] G. Qin, Y. Yao, W. Wei, T. Zhang, Preparation of hydrophobic granular silica aerogels and adsorption of phenol from water, Appl. Surf. Sci. 280 (2013) 806–811.https://doi.org/10.1016/j.apsusc.2013.05.066

[75] M. Reim, G. Reichenauer, W. Korner, J. Manara, M. Arduini-Schuster, A. Beck, J. Fricke, Silica-aerogel granulate-Structural, optical and thermal properties, J. Non-Cryst. Solids 350 (2004) 358–363. https://doi.org/10.1016/j.jnoncrysol.2004.06.048

[76] M. Reim, W. Korner, J. Manara, S. Korder, M. Arduini-Schuster, F.J. Potter, Silica aerogel granulate material for thermal insulation and daylighting, Sol. Energy 79 (2005) 131–139. https://doi.org/10.1016/j.solener.2004.08.032

[77] L.W. Hrubesh, P.R. Coronado, J.H. Satcher Jr, Solvent removal from water with hydrophobic aerogels, J. Non-Cryst. Solids 285 (2001) 328–332. https://doi.org/10.1016/S0022-3093(01)00475-6

[78] Y. Li, Y. Liu, M. Wang, X. Xu, T. Lu, C.Q. Sun, L. Pan, Phosphorus-doped 3D carbon nanofiber aerogels derived from bacterial-cellulose for highly-efficient capacitive deionization, Carbon 130 (2018) 377-383. https://doi.org/10.1016/j.carbon.2018.01.035

[79] S.A. Glauser, H. Lee, Luminescent studies of fluorescent chromophore-doped silica aerogels for flat panel display applications, MRS Online Proceedings Library A 471, 1997. https://doi.org/10.1557/PROC-471-331.

[80] P.C. Thapliyal, K. Singh, Aerogels as promising thermal insulating materials: An overview, Journal of materials (2014) 1–10. https://doi.org/10.1155/2014/127049.

Aerogels I: Preparation, Properties and Applications
Materials Research Foundations **84** (2020) 133-154

Materials Research Forum LLC
https://doi.org/10.21741/9781644900994-5

Chapter 5

Carbon Aerogels

Subhajit Kundu[1], Debarati Mitra[1], Mahuya Das[2,*]

[1]Department of Chemical Technology, University of Calcutta, Kolkata-700009, India

[2]Regent Education and Research Foundation, Bara Kathalia, Barrackpore, Kolkata- 700121, India

*d_mahuya@yahoo.com

Abstract

Aerogel is a porous solid material derived from gel in which the liquid is replaced by gas. The wide application of aerogel in the field of adsorbent is due to its macro dimension. It can be removed very easily from an aqueous reaction medium in comparison to the nanostructured powdered materials. Currently, carbon-based aerogel, having a 3D micro and mesoporous network structure interconnected by nanosized primary particles, are getting much attention as adsorbents for their large surface area and higher adsorption capacity. Actually, carbon aerogels have some unique backbone density and connectivity properties along with very small pore size, high porosity and high surface area which facilitates its wide range of applications including electrical fields, hydrogen storage and adsorption, catalyst support, thermal insulation, optical fields, sensor, broadband non-reflective materials etc.

Keywords

Aerogel, Classification, Porous, Carbon Aerogel, Adsorption

Contents

1.   Introduction

The aerogels were first synthesized by Samuel Kistler in 1932 where he and his co-workers concluded that this special kind of material retains porosity and shape on replacing the liquid by gas [1]. They investigated that this type of material has very high specific pore volume. Later in 1950, Monsanto et al. [2] followed Kistlerls process to synthesize silica based aerogels. During 1960's Teichner and his co-workers [3] gelled trimethylsiloxane (TMOS) solution by applying water in excess amount using acid or base catalyst. The synthesized material was also characterized both structurally as well as morphologically. Organic aerogels were first introduced in the period of late 1980s by Pekala et al. [4] at Lawrence Livermore National Laboratory (LLNL) where they synthesized an aerogel by polycondensation of resorcinol with formaldehyde. This type of aerogel later was known as Resorcinol-formaldehyde (RF) aerogel. Probably this was the first initiation to synthesize carbon aerogel. Here it is worth mentioning that there is a distinct difference between the term 'aerogel' and 'xerogel'. According to IUPAC definition, xerogels are "formed by the removal of all swelling agents from a gel" whereas the aerogels are defined as "aerogels are a gel composed of a microporous solid in which the dispersed phase is gas". Pekala et al. [5] investigated on their synthesized carbon aerogel and found that RF aerogel has great stiffness, no ductility and both the compressive strength and modulus are very high. Overall it can be said that carbon aerogels have a 3D micro and mesoporous network structure interconnected by nanosized primary particles. Based on the above technology some researchers developed carbon-carbon cryogel type composites. The comparison of density between silica and carbon aerogel was done by Sun et al. [6] and Worsley et al. [7]. Though there lots of adsorbents available, the aerogel, especially carbon based aerogel has the uniqueness to be an ideal adsorbent, they have microstructure which is highly hydrophobic 3D network structure. Adsorption of pollutants from waste aqueous solutions is highly appreciable using these special types of materials due to their high sorption capacity along with high surface area. Furthermore very low density along with flexible surface phenomenon and the macro size of these aerogels makes them highly suitable towards adsorption. Previously different types of aerogels were used to control the water pollution by removing harmful contaminants like metal oxides, different carbon wastes, polymers etc. from waste water. Nowadays carbon based aerogels are being used mainly due to high surface area and high efficiency of adsorption. There are various carbon aerogels such as flexible aerogels, carbon nano tube aerogels, nano diamond aerogels or even many metal doped aerogels. The electrical conductivity and mechanicals properties of activated carbon aerogel were investigated by Baumann et al. [8] whereas the optical, electronic and thermal properties of diamond aerogels were analyzed by Pauzauskie et al. [9]. Different organic source for carbon aerogel has been presented in the Table 1 [10-15].

Table 1   Different organic sources for carbon aerogel [10-15]

Reagent	Density (g/cm^3)	Specific surface area (m^2/g)	Pore size (nm)	Ref.
Melamine-Formaldehyde (MF)	0.10-0.80	875-1025	~50	[10]
Phenol-Formaldehyde (PF)	0.10-0.25	350-600	~10	[11]
Phenol-Melamine (PM)	0.53-0.71	600-800	~10	[12]
Cresol-Formaldehyde (CF)	0.15-0.4	800	~30	[13]
Resorcinol-Formaldehyde (RF)	0.03-0.60	400-1000	~50	[14]
Polyvinyl Chloride (PVC)	0.08-0.52	300-700	~2-20	[15]

Actually, carbon aerogels have some unique backbone density, backbone connectivity properties along with very small pore size (10-30nm), high porosity (>80%) high surface area (400-900 m^2/g) which facilitates its wide range of applications including electrical fields, hydrogen storage and adsorption, catalyst supports, materials for thermal insulation, optical fields, sensor, broadband non-reflective materials, etc.

2.   Types of carbon aerogels

2.1   Low flexible-carbon aerogel

In the low-flexible-carbon aerogel there are lots of elongated pores. The dimensions of the pores are generally 100-110 μm long and 15-20 μm thick [16]. The structure of this type of aerogel is greatly dependent on the external compression on the aerogel [17,18]. The compression up to 15% of the initial shape shows the characteristic structural change, where above 15%, the two pores in the nearest vicinity touch with each other. When the load is removed, the re-opening of the pores occurs [19].

2.2   Super flexible-carbon aerogel

The flexibility analysis of this type of carbon aerogel is done by the same procedure as that of low flexible aerogel. Here some large oval shaped and small round shaped pores are observed. Up to 10% of compression the change in structure is observed [20]. Above 30% of compression the shapes of the oval pores become changed. By subsequent release of load gradually the pore walls gradually returns to its initial position and also the pores regain their shape i.e. reappearance of the tiny pore at the corner and regaining of shape of the round pore. The developed background network disappears without rejoiningthe broken chain however maintaining a very close proximity among the ends of the chains [21].

2.3   Carbon nano tube aerogels

Many materials including some polymers and silica although can be fabricated on cabonaceous materials pure carbon nano tube aerogels preparation is quite troublesome. Very few works have been done with very significant results till now. Ya-Li Li et al. [22] have synthesized an endless continuous fiber carbon nano tube based aerogel with an appearance of elastic smoke. The structure of these types carbon aerogel is largely dependent on the source of carbon and the alcohols and ketones used in the process. To get a mixed type i.e. thin continuous as well as thick fiber based aerogel the source of carbon can be petroleum as it is a mixture of different hydrocarbons. Another work has been done by Shen et al. [23] where they used the carbon nano tube powder in the dispersed phase of water and then they mixed sodium dodecyl benezenesulfonate as surfactant to synthesize the aerogel maintain the optimum ratio of water, carbon powder and surfactant under ultrasonic waves.

2.4   Graphene nano aerogel

Graphene oxide has unique property of uniform aqueous dispersion. Also it has the oxygen atoms in the basal planes and edges which facilitates its high covalent character. Due to these characteristics grapheme oxide can be used to prepare grapheme nano aerogel. The general method of the synthesis is the hydrothermal method using reducing agents. Xu et al. [24] found that due to conjugated structure this aerogel has very high conductivity as well as mechanical properties. Cheng et al. [25] followed the same route of preparation only by changing the temperature as $1500°C$ to improve the mechanical property and conductivity of the aerogel. Moreover reduction of grapheme oxide can also be done by chemical regents such as vitamin C, mercaptoacetic acid, mercapto ethanol, hydrazine etc. But these methods give grapheme nanoerogel with very low porosity.

Polymers are commonly used as building material to more strong 3D gel network in the aerogel framework. Carbon nanotube aerogel mixed with sodium carboxymethylcellulose as binder gives a low density, ultra-low thermal diffusivity along with very high mechanical strength and conductivity. Due to the chain structure of the composite scanning electron microscopy (SEM) image shows the formation of spindle stripes. Zhang et al. [26] have used carbon nano tube with poly (amido amine) in presence of Fe_3O_4nano powder to get a 3D hydrogel structure of the carbon nano tube/ Fe_3O_4

composite. Graphene oxide/graphite nano tube composites can also be made using different reducing agents in the grapheme oxide dispersion medium. Gao et al. [27] have treated $ZrOCl_2.8H_2O$ in graphene oxide with dimethyl formaldehyde solution followed by subsequent drying under supercritical condition in presence of CO_2 and carbonization in argon environment to get graphenenano ZrO_2 composite aerogel. There is another work where Liu et al. [28] have prepared graphene nanopolyaniline Co_3O_4 ternary hybrid aerogel composite.

2.5   Nano-diamond aerogel

The nano diamond aerogel has unique property of tunable optical index of refraction having the range from 1 to 25. It also has very high surface area with low density. It is applicable in the electron gun emission field, as anti-reflective coatings etc. Pauzauskiea et al. [29] have used laser-heated diamond anvil cell in the neon environment to prepare nano diamond aerogel. Neon was used to fill the void volume in the aerogel. Transmission electron microscopy (TEM) image shows the surface morphology of the aerogel. The sol-gel process can be applied using resorcinol and formaldehyde as precursor through an acid-catalyzed conversion process in polar aprotic solvent like acetonitrile. Roldan et al. [30] used nano diamond on graphene nano aerogel so that the restacking of nano graphene and the agglomeration of nano diamond can be restricted.

2.6   Ni-doped carbon aerogel

Ni-doped carbon aerogel is usually prepared by carbonization at $500°C$ with the release of gases by the decomposition of the organic matrix leading to an enhancement in the BET surface area along with the development in the microporosity. An increase in the pyrolysis to $1800°C$ leads to reduction in this parameter. Heat treatment up to $500°C$ results in decrease of macropore volume due to the shrinkage of aerogel, but above $500°C$ temperature, decrease of mesoporosity was observed with an increase of macropore volume [31].

2.7   Pt, Pd, Ag and Ru-doped carbon aerogel

Carbon aerogel doped by Pt-, Pd- and Ag particle using resorcinol, formaldehyde and the coordination compounds of Pr, Pd and Au as the precursor followed by pyrolysis at $1000°C$ in N_2 flow and steam activation at $900°C$. These particles are actually encapsulated by the carbon matrix. Among all these doped carbon aerogel the steam activated Pt-aerogel exhibited larger macropore as well as mesopore volume. Pt doped carbon aerogel can be fabricated by adding an ultrasonically prepared aqueous dispersion of commercially available Pt black, having particle size in the range of 200– 500nm, to the sol. Again this aerogel can be prepared by dipping of the carbon aerogel in the solution of hexachloro platinic acid and chloroform followed by pyrolysis and reduction of the sample with H_2. Supercritical deposition method is also used for the synthesis of Pt-doped carbon aerogelsin which monolithic carbon aerogels has been impregnated in dimethyl (1,5-cyclooctadiene)platinum (II) solution in supercritical CO_2. Ru-containing carbon aerogels has also been prepared by impregnation. In this case the carbon aerogel is impregnated with ruthenium acetate solution followed by sublimation leading to high metal loading with good dispersion. Heating this sample above $500°C$ facilitate migration of the Ru-particles diffused to the outer surface of the carbon aerogel forming a thick coating [32].

2.8   Ce, Zr-based carbon aerogel

Ce-, Zr-doped carbon aerogels have been fabricated using the solution of those metal nitrates in an resorcinol-formaldehyde mixture at different pHs followed by carbonization at $1050°C$. The metal salts alters the initial pH of the mixture and hence the mechanism of the process. The surface area of the product developed ranges from 80 to 800 m^2/g with a microporous volume in the range of 0.03 to 0.20 cm^3/g and the micropore size ranging from 0.65 to 0.90 nm. It has been reported that along with initial pH, one of the important factor to control the chemistry of the synthesis process, the nature of the doped metal also affects the sol–gel chemistry. For example, when carbon aerogels doped with Zr develops microporosity in the aerogel, whereas doping only with Ce generates very low microporosity in the product [33].

3.   General characteristics and properties

3.1   Bulk density and porosity

Bulk density or relative porosity is an indication of the handling properties of the carbon aerogel material. The smaller the sol particles the lower will be the density which facilitates good mechanical properties [34]. The porosity can be high enough up to 99%, but for specific applications like for double layer capacitor (DLC) porosity of the material can be controlled. The bulk density is generally around 50-500kg/m^3. As these kinds of materials have very high surface area sample should be well degassed before determining the density.

3.2   Backbone density

Backbone density or helium density is another important property of carbon aerogel. The amount of helium required to fill the accessible pore volume per mass of the aerogel is considered as the backbone density. The backbone density depends on the degree of activation as in the non-activated state helium gas may be trapped by carbon micropores in the aerogel. The bulk density (backbone) is related with the specific micropore volume as follows:

$$\frac{1}{\rho_{bbd}} = V_{mp} + \frac{1}{\rho_{npc}}$$

where, ρ_{bbd} is the bulk density of the carbon backbone, V_{mp} is the specific micropore volume and ρ_{npc} is the density of the non-porous carbon [34]. In un-activated sate generally the value of ρ_{bbd} is 1.4 g/cm^3 and ρ_{npc} is 2.2 g/cm^3.

3.3 Backbone connectivity

The backbone connectivity plays a very important role which affects many physical properties of the aerogel.

i. The backbone connectivity influences the intrinsic properties of the aerogel like conductivity. With the increase of the microcrystalline size the conductivity increases.

ii. Another important property which is related to the backbone connectivity is the tortuosity. It the average ratio of the distance along the backbone phase between the two sides of a sample and the minimum distance between the starting and the end of the path. This affects the transport properties of carbon aerogel is dependent on tortuosity.

iii. The backbone connection is also affected by the mass fraction of the aerogel. Generally the higher the density the higher will be the thermal conductivity.

iv. Another important parameter is the neck size among the backbone particles. For thermal and electronic transport, there are found some weak spots which in turn lowers down the cross-sectional are of the aerogel [35, 36].

3.4 Pore connectivity

Apart from the backbone connectivity, aerogel pore connectivity analysis is also very important.

i. The pore connectivity influences the diffuse property of the aerogel which can be analyzed by Pulsed Gradient Spin Echo (PGSR) NMR spectroscopy, gas permeability, etc. This analysis also provides the value of porous phase tortuosity.

ii. The pore connectivity also highlights the viscous transport which is aaffected by both the porous phase tortuosity and the effective pore diameter [37].

3.5   Pore size

Pore size is another valuable parameter. It can be determined by fluorescence tomography, small angle X-ray scattering (SAXS), mercury-porosimetry, nitrogen adsorption analysis, etc. When density is less than 200 kg/m³ and moduli of compression is less than 200MPa, the pore size will be less than 50 nm. But when these factors are ignored mesporous size distribution can be obtained [38]. The mean pore size can also be calculated using the given formulae:

$$\text{Mean pore size} = \frac{4(V_{p,t} - V_{mp})}{S_{ex}} \text{ and } V_{p,t} - V_{mp} = \frac{1}{\rho_{bulk}} - \frac{1}{\rho_{bbd}}$$

where, $V_{p,t}$ is the total pore volume, S_{ex} is the external specific surface area and ρ_{bulk} is the bulk density [39].

3.6   Thermal properties

The total thermal conductivity for monolithic carbon aerogels is given by the equation,

$$\lambda_{total} = \lambda_g + \lambda_s + \lambda_e + \lambda_r$$

neglecting coupling between the different heat transfer. Here λ_g is the thermal conductivity of gas, λ_s is for solid, λ_e is occurred by the electrons and λ_r is for radiation. The infrared-optical extinction of the carbon skeleton is mass-dependent and very high i.e. 1000 m²/kg. This also renders very low total thermal conductivity. The λ_r is calculated by obtaining the index of radiation. The λ_g can be found from Kundsen equation. The value of λ_e can be found using Wiedemann-Franz law and Lorenz number. The low value of λ_e indicates the total thermal conductivity is basically due to the fact of heat transfer by the phonons. In comparison to organic and silica aerogels, carbon aerogel has lower thermal resistance which is an advantage of carbon aerogel in thermal insulation at very high temperature as it suppress the radiative transport [40-43].

3.7   Electrical properties

The electrical conductivity occurred at the drift of the delocalized charge carriers from one conducting zone to another zone by hopping or tunneling. The electrical conductivity can be calculated using the relation:

Materials Research Forum LLC
https://doi.org/10.21741/9781644900994-5

$$\sigma_{el} = k. \, \rho^t$$

where, t is the scaling exponent, ρ is the density and k is a constant depends on the particle interconnectivity. Generally the electrical conductivity decreases at ambient temperature which is further drastically decreased below 100K of temperature. The carbon aerogel having density between 100 to 820 kg/m^3 exhibits electrical conductivity in between 1 Scm^{-1} to 55 S cm^{-1} at 280K and a scaling component value of 1.7. The electrical conductivity value of Resorcinol-formaldehyde derived aerogel is directly proportional to the increase in pyrolysis temperature showing no conducting network up to 500°C whereas a substantially high value is observed above 700°C. The reason is probably the removal of oxygen containing groups from the polymeric network at around 500°C leaving behind only the hydrogen atoms as the main element with carbon. This stage of pyrolysis leads to the formation of aromatic chain ribbons in the condensed state which have isolated oxygen containing groups resulting inhibition of the mobility of charge carrier due to binding of charge carriers at the edge planes with huge number of hydrogen atoms. Above 500°C the linkage between carbon and hydrogen breaks down leading to rearrangement in the micrographic area. This affects the formation of large pores which in turn increases the micropore volume and particle porosity [43-47].

3.8 Electrochemical properties

Carbon aerogels are finding their application as electrodes of electrochemical double layer capacitors owing to the controllable pore-size distribution in their micro network, low electrical resistivity, high specific surface area, etc. Carbon aerogels has become highly demanding materials in the field of electrochemical double layer capacitors which is used for storage of energy and supercapacitors having high power densities. The gel type structure reduces the energy of activation required for transportation of between carbon particles in comparison to compacted carbon powder electrode. The capacitance of resorcinol-formaldehyde and phenol-formaldehyde based carbon aerogel reaches a maximum at about 800 kg/m^3 density and decreases with higher densities. But the capacitance is not affected with the size of particle in the skeleton. The variation of capacitance and density is due to increase in the quantity of material per unit volume which in turn results in the increase of surface area per unit volume at higher density. Another factor is the surface area/mass. The later one is predominant by 35-45% above a fraction of solid materials [48-53].

3.9 Mechanical properties

Structurally, aerogels are less efficient than that of foams. The mechanical property of carbon aerogel is dependent on the uniaxial modulus which is directly proportional to the density of the material by the following relation:

$$\text{uniaxial modulus} \propto \rho^{\alpha}$$

where, ρ is the density and α is the scaling exponent. Generally the value of α is greater than 2. It was considered that there is lots of dangling mass which is responsible in the contribution of mass nut not to the mechanical strength. Using pyrolysis technique Pekala et al. [56] found the value of scaling exponent as 2.75 whereas Gross and Fricke followed the same methodology and found the value as 3 to 3.5. They compare the scaling exponent value of carbon aerogel with silica aerogel and organic aerogel and concluded that the value of α is less in the former case. In the carbon aerogels fabricated from strongly diluted resorcinol-formaldehyde sols, a density variation between 300 and 600 kg/m^3 can be introduced through shrinkage upon different cure treatments [54-56].

3.10 Gas-transport properties

If we consider a zero-order approximation level, the voids of the carbon aerogels can be represented by independent parallel tubes capillary model. The permeability is related by the following equation:

$$P = \frac{\alpha \cdot p_{av}}{\mu} + \beta \cdot \left(\frac{T}{M_g}\right)^{0.5}$$

where, assuming small pressure gradients p_{av} is the average gas pressure for the gas transport, μ is the viscosity of the fluid, T is the temperature and M_g is the gas molecular mass. The parameters α and β depends on the tortuosity. The first expression represents the transport properties and the second expression indicates the molecular diffusion. Here Knudsen number, the ratio of the mean free path of a molecule to the mean distance between the pore walls, plays a key role [57-61].

3.11 Optical properties

Investigation of infrared-optical as well as optical properties of carbon aerogels revealed that the carbon aerogel synthesized from supercritically dried resorcinol-formaldehyde aerogel, with a ratio of around 300 having a density of 117-120 kg/m^3, can be used as diffused reflectors with reflectance comparable or superior to non-reflective Martin

Black. No reflectance band was observed between wavelength 2.5 pm to 14.5 pm for the carbon aerogel sample. Light scattering study showed that the behavior of carbon aerogel was quasi-Lambertian type with asymmetric scattering. In fact the scattering was of forwarded scattering as well as backward scattering [62].

4. Applications

4.1 Electrochemical field

Electrochemical double-layer capacitors (EDLCs) are that kind of supercapacitors which have less energy storage capability but high power. Nowadays these materials are used in battery storage, even electrical vehicles. Due to high surface area, momolithic structure, high conductance, carbon aerogel materials are used in these field of electrochemical applications. The self-discharge property of electrochemical double layer capacitor is influenced by the stability of the charges at the interface between electrode and electrolyte [63].

Nickel-cadmium based batteries are better over the lithium ion based batteries for the requirement of high-energy densities. For electrical vehicles the nickel-cadmium based batteries are used. These metals doped carbon aerogels have higher charge capacity [64-65].

4.2 Hydrogen storage

In the quest to find new technologies, application of hydrogen as energy and also the storage of hydrogen is utmost important. Statistics given by the US department of energy 50% of the total cost is invested for the storage of hydrogen. Their primary objective was the economy, safety, and either the technology is practically possible or not. [66-68].

4.3 Catalyst support

Sol–gel polymerization forming emulsion of phenol, melamine, and formaldehyde has been used to fabricate spherical carbon aerogels (SCAs) having controlled mesopore size and particle size. Investigation of the rate of adsorption and the capacity of biomolecules having varying molecular dimensions, containing L-phenylalanine (Phe), achymotrypsin (Chy), vitamin B12 (VB), and bovine serum albumin (BSA), onto SCAs.Size of the mesopore can easily be kept within the range of 5 to 10 nm simply by maintaining the catalyst concentration in the initial reaction medium whereas size of the spherical particle can be adjusted within 50–500 mm by altering the stirring speed. The SCAs synthesized by this way possess large pore volume i.e. $>1cm^3/g$, high specific surface area which is >600 m^2/g and hardness ten times larger than that of commercially available spherical activated carbon particles. The adsorption rate of VB increases with the increase of mesopore size and the decrease of particle size. For small molecule like Phe, the key factor for determining the capacity to adsorb is specific surface area whereas for larger molecules like VB, Chy, and BSA it is the pore size of SCAs, which must be effectively larger than the biomolecular size [69-72].

4.4   Thermal insulation

Aerogels usually exhibit a high thermal resistance by virtue of their nanostructure and porosity. The gaseous heat conduction through the material having pores 1–100 nm wide, can be reduced under ambient pressure and can be suppressed completely by complete evacuation. The carbon skeleton is also capable of a very efficient reduction of the radiative heat transfer along with very low solid thermal conductivity due to the high porosity [73-76].

4.5   Adsorbent for waste water treatment

Carbon aerogel microstructures possess exceptionally high surface area, well defined 3D structures, hydrophobicity, flexible surface chemistry, low density, high porosity, etc. making themselves effective and efficient sorbents in removal of contaminants from aqueous solutions. In comparison to the powdered nanostructured materials the carbon aerogel materials are advantageous because due to its larger dimension they can easily be removed from an aqueous solution. They are very efficient in removal of various kinds of pollutants from wastewater e.g. carbons, polymers, metal oxides and hence forth [77-80].

4.6   Photocatalyst for waste water treatment

Carbon-based aerogels exhibited high visible light photocatalytic activity reducing the band gap energy and slowing down e/h+ pair recombination rate leading to higher recovery and reusability of the catalyst.The unique physicochemical, electrical, and optical properties of one dimensional carbon nanotube and two dimensional graphene nanostructure make them suitable to develop the semiconductor aerogel composite photocatalysts applicable for the adsoption of various types of pollutants from waste water [81-83].

4.7   Sensor application

Graphene nano (GN) aerogels with its low electrical and thermal resistivity, very high surface area, unique mechanical elasticity and large carrier mobility are highly applicable in the field of electronics, composites, sensors, hydrogen storage etc. Every Graphene nano (GN) sheets are also carrier of those aforementioned remarkable properties, and also suitable for fabrication of GN-based sensors with respect to the properties like electrical conductance, etc. There are examples of GN based sensor for the detection of NO, NO_2 (to tens of parts per billion) and NH_3. Normally the sensitivity generates by the movement of p-type electrons to the holes concentration followed by withdrawal of grapheme nano materials to the acceptors of NO_2. Adsorption of NO_2 leads to the increase in conductance where as adsorption of NH_3 decreased conductance while detecting as the concentration of charge carriers on GN has been reduced. The sensor based on GN foam are much more sensitive than commercially available polymer sensors, owing to the effective charge carrier movement among the highly porous foam, few-layer GN component, and the adsorbed gas. Very good catalytic performance was shown by the GN-based sensor used for NO reduction [84-87].

Conclusions

Carbon aerogels possesses porosity, high surface area, low density and flexible surface chemistry. Formaldehyde based resins like Novolac, resol, UF, etc. are main source for carbon aerogel preparation. There are different types of carbon aerogel depending upon the flexibility and the constituent carbon particle. Furthermore different types of transition metal particle like Pt, Pd, etc. doping in this aerogel developed another variety of carbon aerogel. Carbon aerogel has been characterized by different types of properties like backbone density, backbone connectivity, pore size, thermal, electrical, electrochemical properties. Carbon aerogels having some unique backbone density, backbone connectivity properties along with very small pore size, high porosity and high surface area are very much suitable in wide range of applications including electrical fields, hydrogen storage and adsorption, catalyst supports, materials for thermal insulation, optical fields, sensor, broadband non-reflective materials, etc.

References

[1] S.S. Kistler, Coherent expanded aerogels, J .Phys. Chem. 36 (1932) 52–64. https://doi.org/10.1021/j150331a003.

[2] C.J. Brinker, G.W. Scherer, Sol-gel science: The physics and chemistry of sol-gel processing, Academic Press, New-York, 1990.

[3] A.E. Gash, T.M. Tillotson, Jr. J.H. Satcher, L.W. Hrubesh, R.L. Simpson, New sol-gel synthetic route to transition and main-group metal oxide aerogels using inorganic salt precursors. J. Non Cryst. Solids. 285 (2001) 22–28. https://doi.org/10.1016/S0022-3093(01)00427-6.

[4] J. Eid, A.C. Pierre, G. Baret, Preparation and characterization of transparent Eu doped Y2O3 aerogel monoliths for application in luminescence, J. Non Cryst. Solids. 351 (2005) 218–227. hal-00012515.

[4] R.W. Pekala, F. M. Kong, Resorcinol-formaldehyde aerogels and their carbonized derivatives, Abstr. Pap. Am. Chem. Soc. 197 (1989) 113-115. DOE contract no: W-7405-ENG-48.

[5] R.W. Pekala., F.M. Kong, New organic aetogels based upon a phenolic-furfural reaction, J. Non. Cyst. Solids, 188 (1995) 34-40. https://doi.org/10.1016/0022-3093(95)00027-5.

[6] H.W. Sun, Z. Xu, C. Gao, Multifunctional, ultra-flyweight, synergistically assembled carbon aerogels, Adv. Mater. 25 (2013) 2554-2560. https://doi.org/10.1002/adma.201204576.

[7] M.A. Worsley, T.Y. Olson, J.R.I. Lee, T.M. Willey, M.B. Nielson, S.K. Roberts, P.J. Pauzauskie, J. Biener, Jr. J.H. Satcher, T.F. Baumman, High surface area, sp2-cross-linked 3-D graphene monoliths, J. Phys. Chem. Lett. 2 (2011b) 921-925. https://doi.org/10.1021/jz200223x.

[8] T.F. Baumann, M.A. Worsley, Carbon aerogels, in: L. Klein et al. (Eds.), Handbook of sol-gel science and technology, Springer, New York, 2016, pp. 1-27

[9] P. J. Pauzauskie, M.J. Crane, M.B. Lim, X. Zhou, Synthesis and characterization of a nanocrystalline diamond aerogel, Proc. Natl. Acad. Sci. U.S.A., 108 (2011) 8550-8553. doi:10.1038/micronano.2017.32.

[10] R.W. Pekala, Organic aerogels from the polycondensation of resorcinol with formaldehyde, J. Mater Sci. 24 (1989) 3221–3227. https://doi.org/10.1007/BF01139044.

[11] R.W. Pekala, C.T. Alviso, X. Lu, J. Gross, J. Fricke, New organic aerogels based upon a phenolic-furfural reaction, J Non-Cryst Solids, 188 (1995) 34–40. https://doi.org/10.1016/0022-3093(95)00027-5.

[12] R. Zhang, Y. Lu, L. Zhan, X. Liang, G. Wu, L. Ling, Monolithic carbon aerogels from sol–gel polymerization of phenolic resoles and methylolated melamine, Carbon 41 (2002) 1660–1663. https://doi.org/10.1016/S0008-6223(03)00112-X.

[13] W. Li, G. Reichenauer, J. Fricke, Carbon aerogels derived from cresol–resorcinol–formaldehyde for supercapacitors, Carbon 40 (2002) 2955–2959. https://doi.org/10.1016/S0008-6223(02)00243-9.

[14] C. T. Alviso, R. W. Pekala, J. Gross, X. Lu, R. Caps, J. Fricke, Resorcinol–formaldehyde and carbon aerogel microspheres, Micropor. Macropor. Mater 431 (1996) 521–525. https://doi.org/10.1557/PROC-431-521.

[15] J. Yamashita, T. Ojima, M. Shioya, J. Hatori, Y. Yamada, Organic and carbon aerogels derived from poly (vinyl chloride), Carbon 41 (2003) 285–294. https://doi.org/10.1016/S0008-6223(02)00289-0.

[16] R.W. Pekala, C.T. Alviso, J.D. LeMay, Organic aerogels: Microstructural dependence of mechanical properties in compression, J. Noncryst. Solids 125 (1990) 67–75. https://doi.org/10.1016/0022-3093(90)90324-F.

[17] M. Schwan, L. Ratke, Flexibilisation of resorcinol-formaldehyde aerogels, J. Mater. Chem. A 1 (2013) 13462–13468. https://doi.org/10.1039/C3TA13172F.

[18] M. Schwan, R. Tannert, L. Ratke, New soft and spongy resorcinol–formaldehyde aerogels, J. Supercrit. Fluids 107 (2016) 201–208. https://doi.org/10.1016/j.supflu.2015.09.010.

[19] G. Zhou, F. Li, H.M. Cheng, Progress in flexible lithium batteries and future prospects, Energy Environ. Sci. (2014) 1307–1338. https://doi.org/10.1039/C3EE43182G.

[20] C. Wang, G.G. Wallace, Flexible electrodes and electrolytes for energy storage, Electrochem. Acta 175 (2015) 87–95. https://doi.org/10.1016/j.electacta.2015.04.067.

[21] H. Sun, Z. Xu, C. Gao, Multifunctional, ultra-weight, synergistically assembled carbon aerogels, Adv. Mater. 25 (2013) 2554–2560. https://doi.org/10.1002/adma.201204576.

[22] Y-L. Li, I.A. Kinloch, A.H. Windle, Direct spinning of carbon nanotube fibers from chemical vapor deposition synthesis, Science, 304 (2004) 276-278. https://doi.org/10.1002/adma.201204576.

[23] Y. Shen, A. Du, XL. Wu, X-G. Li, J. Shen, B. Zhou, Low-cost carbon nanotube aerogels with varying and controllable density, J. Sol Gel. Sci. Technol. 79 (2016) 76-82. https://doi.org/10.1002/adma.201204576.

[24] Y. Xu, K. Sheng, C. Li, G. Shi, Self-assembled graphene hydrogel via a one-step hydrothermal process , ACS Nano, 4 (2010) 4324-4330. https://doi.org/10.1021/nn101187z.

[25] Y. Cheng, S. Zhou, P. Hu, G. Zhao, Y. Li, X. Zhang, W. Han, Enhanced mechanical, thermal and electrical properties of graphene aerogels via supercritical ethanol drying and

high-temperature thermal reduction, Sci. Reports 7:1439 (2017) 1-11.
https://doi.org/10.1038/s41598-017-01601-x.

[26] M. Zhang, S. Fang, A.A. Zakhidov, S.B. Less, A.E. Aliev, C.D. Williams, K.R.
Atkinson, R.H. Baughman, Strong, transparent, multifunctional, carbon nanotube sheets,
Science 309 (2005) 1215-1219. doi:10.1126/science.1115311.

[27] Y-D. Gao, Q.Q. Kong, Z. Liu, X-M. Li, C-M. Chen, R. Cai, Graphene oxide
aerogels constructed using large or small graphene oxide with different electrical,
mechanical and adsorbent properties, RSC Adv. 6 (2016) 9851-9856.
https://doi.org/10.1039/C5RA26922A.

[28] P. Liu, Y. Huang, Synthesis of reduced graphene oxide-conducting polymers-Co3O4
composites and theor excellent microwave absorption properties, RSC Adv. 3 (2013)
19033-19039. https://doi.org/10.1039/C3RA43073A.

[29] P.J. Pauzauskie, J.C. Crowhurst, M.A. Worsley, T.A. Laurence, A.L.D. Kilcoyne, Y.
Wang, T.M. Willey, K.S.Visbeck, S.C. Fakra, W.J. Evans, J.M. Zang, Jr.J.H. Satcher,
Synthesis and characterization of a nanocrystalline diamond aerogel, Proc. Natl. Acad.
Sci. U.S.A. doi:10.1073/pnas.1010600108.

[30] L. Roldan, A.M. Benito, E. Garc-a-Bordeje, Self-assembled graphene aerogel and
nanodiamond hybrids as high performance catalysts in oxidative propane
dehydrogenation, J. Mater Chem. A 3 (2015) 24379-24388.
https://doi.org/10.1039/C5TA07404E.

[31] S. Manandhar, P.B. Roder, J.L. Hanson, M. Lim, B.E. Smith, A. Mann, P.J.
Pauzauskie, Rapid sol–gel synthesis of nanodiamond aerogel, J. Mater Res. 29 (2014)
2905-2911. https://doi.org/10.1557/jmr.2014.336.

[32] M. Schwan, L. Ratke, Flexibilisation of resorcinol–formaldehyde aerogels,
J. Mater. Chem. A 1 (2013) 13462. https://doi.org/10.1016/j.cap.2017.03.003.

[33] R. Tannert, M. Schwan, Z. Rege, M. Eggeler, J.C.D. Silva, M. Bartsch, B. Milow,
M. Itskov, L. Ratke, The three-dimensional structure of flexible resorcinol-formaldehyde
aerogels investigated by means of holotomography, J. Sol-Gel Sci. Technol. 84 (2017)
391–399. https://doi.org/10.1007/s10971-017-4363-6.

[34] G. Reichenauer, Structural characterization of aerogels, in: M.A. Aegerter,
N. Leventis, M.M. Koebel (Eds.), Aerogels handbook, Springer, New York, 2011,
pp. 449–498.

[35] J.U. Keller, M.U. Goebel, T. Seeger, Oscillometric-gravimetric measurements of pure gas adsorption equilibria without the non-adsorption of helium hypothesis, Adsorption 23 (2017) 753–766. https://doi.org/10.1007/s10450-017-9893-2.

[36] G. Reichenauer, C. Stumpf, J. Fricke, Characterization of SiO2, Rf and carbon aerogels by dynamic gas-expansion, J. Non-Cryst. Solids 186 (1995) 334–341. https://doi.org/10.1016/0022-3093(95)00057-7.

[37] H. Lu, H. Luo, N. Leventis, Mechanical characterization of aerogels, in: M.A. Aegerter, N. Leventis, M.M. Koebel (Eds.), Aeroogels handbook, Springer, New York, 2011, pp. 499–535.

[38] J. Gross, J. Fricke, Ultrasonic velocity-measurements in silica, carbon and organic aerogels, J. Non-Cryst. Solids 145 (1992) 217–222. https://doi.org/10.1016/S0022-3093(05)80459-4.

[39] W. Behr, A. Haose, G. Reichenauer, J. Fricke, Self and transport diffusion of fluids in SiO2 aerogels studied by NMR pulsed gradient spin echo and NMR imaging, J. Non-Cryst. Solids 225 (1998) 91–95. https://doi.org/10.1016/S0022-3093(98)00012-X.

[40] W. Behr, V.C. Behr, G. Reichenauer, Self diffusion coefficients of organic solvents and their binary mixtures with CO2 in silica alcogels at pressures up to 6 MPa derived by NMR pulsed gradient spin echo, J. Supercrit. Fluids 106 (2015) 50–56. https://doi.org/10.1016/j.supflu.2015.05.024.

[41] G. Reichenauer, J. Fricke, Gas transport in sol-gel derived porous carbon aerogels, Symposium FF-Dynamics in small confining systems III 464 (1996) 345. . https://doi.org/10.1557/PROC-464-345.

[42] G. Reichenauer, Aerogels, in: A. Seidel (Ed.), Kirk-Othmer encyclopedia of chemical technology, Wiley, Hoboken, 2008.

[43] M. Schwan, L. Ratke, Flexibilisation of resorcinol–formaldehyde aerogels, J. Mater. Chem. A 1 (2013) 13462-13468. https://doi.org/10.1016/j.carbon.2020.02.073.

[44] R. Saliger, V. Bock, R. Petricevic, T. Tillotson, S. Gris, J. Fricke, Carbon aerogels from dilute catalysis of resorcinol with formaldehyde, J. Non-Cryst. Solids 221 (1997) 144–150. https://doi.org/10.1016/S0022-3093(97)00411-0.

[45] M. Thommes, K. Kaneko, A.V. Neimark, J.P. Olivier, F. Rodriguez-Reinoso, J. Rouquerol, K.S.W. Sing, Physisorption of gases, with special reference to the evaluation of surface area and pore size distribution (IUPAC technical report), Pure Appl. Chem. 87 (2015) 1051–1069. https://doi.org/10.1515/pac-2014-1117.

[46] X. Lu, O. Nilsson,, J. Fricke, R.W. Pelala, Thermal and electrical conductivity of monolithic carbon aerogels, J. Appl. Phys. 73 (1993) 581-584. https://doi.org/10.1063/1.353367.

[47] V. Bock, O. Nilsson, J. Blumm, J. Fricke, Thermal properties of carbon aerogels, J. Non-Cryst. Solids, 185 (1995) 233-239. https://doi.org/10.1016/0022-3093(95)00020-8.

[48] Y.S. Touloukian, R.W. Powell, C.Y. Ho, P.G. Klemens, Thermo- physical Properties of Matter -the TPRC data series, Thermal conductivity-nonmetallic solids, Data book, 2 (1971)..

[49] A.W.P. Fung, G.A.M. Reynolds, H.Wang, M.S. Dresslehaus, G. Dresselhaus, R.W. Pekala, Relationship between particle size and magnetoresistance in carbon aerogels prepared under different catalyst conditions, J. Non- Cryst. Solids 186 (1995) 200-208. https://doi.org/10.1016/0022-3093(95)00056-9.

[50] G.A.M. Reynolds A.W.P. Fung, H. Wang, M.S. Dresslehaus, R.W. Pekala, Morphological effects on the transport and magnetic properties of polymeric and colloidal carbon aerogels, Phys. Rev. 50 (1994) 18590-18600. doi:10.1103/physrevb.50.18590.

[51] G.M. Jenkins, K.Kawamu, Polymeric carbons - carbon fiber, glass and char, 1st ed. Cambridge University Press, Cambridge, UK (1976).

[52] R.W. Pekala, J. Fricke, in J. E. Mark (Ed.), Encyclopedia of materials: organic-inorganic hybrid materials, Elsevier, Amsterdam, 2000.

[53] R.W. Pekala, S.T. Mayer, J.F. Poco, J.L. Kaschmitter, Structure and performance of carbon aerogel electrodes, Mat. Res. SOC. Symp. Proc. MRS Warrendale, 349(1994) 79-87. https://doi.org/10.1557/PROC-349-79.

[54] H. Probstle, C. Schmitt, J. Fricke, Button cell supercapacitors with monolithic carbon aerogels, Power Sources 205 (2002) 189-194. https://doi.org/10.1016/S0378-7753(01)00938-7.

[55] Z. Yang, J. Li, X. Xu, S. Pang, C. Hu, P. Guo, S. Tang, H.-M. Cheng, Synthesis of monolithic carbon aerogels with high mechanical strength via ambient pressure drying without solvent exchange, J. Mater. Sc. Technol. 50 (2020) 66-74.https://doi.org/10.1016/j.jmst.2020.02.013.

[56] J. Gross, G.W. Scherer, C.T.Alviso, R.W. Pekala, Elastic properties of crosslinked Resorcinol-Formaldehyde gels and aerogels, J. Non-Cryst. Solids 211 (1997) 132-142. https://doi.org/10.1016/S0022-3093(96)00621-7.

[57] J. Gross, J. Fricke, Ultrasonic velocity measurements in silica, carbon and organic aerogels, J. Non-Cryst. Solids, 145 (1992) 217-222. https://doi.org/10.1016/S0022-3093(05)80459-4.

[58] X. Lu, M.C. Arduini-Schuster, J. Kuhn, O. Nilsson, J. Fricke, R.W. Pekala, Thermal conductivity of monolithic organic aerogels, Science 255 (1992) 971-972. https://doi.org/10.1126/science.255.5047.971.

[59] G. Wei, L. Wang, L. Chen, X. Du, C. Xu, X. Zhang, Analysis of gas molecule mean free path and gaseous thermal conductivity in confined nanoporous structures, Int. J. Thermophys. 36 (2015) 2953-2966.https://doi.org/10.1007/s10765-015-1942-z.

[60] A. Emmerling, J. Fricke, Scaling properties and structure of aerogels, J. Sol-Gel Sci. Technol. 8 (1997) 781–788. https://doi.org/10.1023/A:1018381923413.

[61] K. Swimm, V. Reichenauer, S. Vidi, H.-P. Ebert, Gas pressure dependence of the heat transport in porous solids with pores smaller than 10 μm, Int. J. Thermophys. 30 (2009) 1329–1342. https://doi.org/10.1007/s10765-009-0617-z.

[62] H.-P. Ebert, Thermal properties of aerogels, in: M. A. Aegerter, N. Leventis, M.M. Koebel (Eds.), Aerogels handbook, Springer, New York, 2011, pp. 537–564.

[63] F. Hemberger, S. Weis, G. Reichenaur, H.-P. Ebert, Thermal transport properties of functionally graded carbon aerogels, Int. J. Thermophys. 30 (2009) 1357–1371. https://doi.org/10.1007/s10765-009-0616-0.

[64] S.R. Meier, M.L. Korwin, C.I. Merzbacher, Carbon aerogel: a new nonreflective material for the infrared, Applied Optics 39 (2000) 3940-3944. https://doi.org/10.1364/AO.39.003940.

[65] R. Saliger, U. Fischet, C. Herta, J. Fricke, High surface area carbon aerogels for supercapacitors, J Non-Cryst Solids 225 (1998) 81-85. https://doi.org/10.1016/S0022-3093(98)00104-5.

[66] J.M. Miller, B. Dum, T.D. Tran, R.W. Pekala, Deposition of ruthenium nanoparticles on carbon aerogels for high energy density supercapacitor electrode,. J Electrochem. Soc. 144 (1997) 309-311. https://doi.org/10.1149/1.1838142.

[67] J.L. Kaschmitter, S.T. Mayer, R.W. Pekala, Process for producing carbon foams for energy storage devices, US Patent 5789338 assigned to Regents of the University of California (1998).

[68] S.T. Mayer, R.W. Pekala, J.L. Kashmitter, The aerocapacitor: an electrochemical double-layer energy storage device, J Electrochem. Soc. 140 (1993) 446-451. https://doi.org/10.1149/1.2221066.

[69] A. Lherbier, X. Blasé, Y.M. Niquet, F. Trizon, S. Roche, Charge transport in chemically doped 2D graphene, Phys. Rev. Lett. 101 (2008) 036808-036811. https://doi.org/10.1103/PhysRevLett.101.036808.

[70] S. Meng, E. Kaxiras, Z.Y. Zhang, Metal-diboride nanotubes as high-capacity hydrogen storage media, NanoLett. 7 (2007) 663-697. https://doi.org/10.1021/nl062692g.

[71] F.J. Maldonado-Hodar, C. Moreno-Castilla, J. Rivera-Utrilla, M.A. Ferro-Garcia, Metal-carbon aerogels as catalysts and catalyst supports, Stud. Surf. Sci. Catal. 130 (2000) 1007-1012. https://doi.org/10.1016/S0167-2991(00)80330-4.

[72] J. Kima, J.W. Gratea, P. Wang, Nanostructures for enzyme stabilization, Chem. Eng. Sci. 61 (2006) 1017-1026. https://doi.org/10.1016/j.ces.2005.05.067.

[73] N. Job, A. Thery, R. Pirard, J. Marien, L. Kocon, J. N. Rouzand, F. Beguin, J. P. Pirard, Carbon aerogels, cryogels and xerogels: Influence of the drying method on the textural properties of porous carbon materials, Carbon 43 (2005) 2481-2494. https://doi.org/10.1016/j.carbon.2005.04.031.

[74] D. Kalpana, K.S. Omkumar, S.S. Kumar, N.G. Renganathan, A novel high power symmetric ZnO/carbon aerogel composite electrode for electrochemical supercapacitor,Electrochem. Acta 52 (2006) 1309-1315. https://doi.org/10.1016/j.electacta.2006.07.032.

[75] S.Q. Zhang, C.G. Huang, Z.Y. Zhou, Z. Li, Investigation of the microwave absorbing properties of carbon aerogels, Mater. Sci. Eng. B 90 (2002) 38-41. https://doi.org/10.1016/S0921-5107(01)00750-4.

[76] J. Fricke (Ed.) Proceedings of the first international symposium on aerogels (ISA 1), Springer-Verlag, Berlin, Proc. Phys. 6 (1986) 167-173. https://doi.org/10.1007/978-3-642-93313-4

[77] J. Fricke, Proceedings of the third international symposium on aerogels (ISA 3), J. Non-Cryst. Solids 145 (1992) 141-145. https://doi.org/10.1016/S0022-3093(05)80444-2.

[78] V.P. Mahida, M.P. Patel, A novel approach for the synthesis of hydrogel nanoparticles and a removal study of reactive dyes from induatrial effluent, RSC Adv. 6 (2016) 21577-21589. https://doi.org/10.1039/C5RA19441E.

[79] H. Maleki, Recent advances in aerogels for environmental remediation applications: a review, Chem. Eng. J.300 (2016) 98-118. https://doi.org/10.1016/j.cej.2016.04.098.

[80] H. Wang, Y. Gong, Y. Wang, Cellulose-based hydrophobic carbon aerogels as versatile and superior adsorbents for sewage treatment, RSC Adv.4 (2014) 45753-45759. https://doi.org/10.1039/C4RA08446B.

[81] J. Li, L. Zhang, H. Liu, A novel carbon aerogel prepared for adsorption of copper (II) ion in water, J. Porous Mater.24 (2017) 1575-1580. https://doi.org/10.1007/s10934-017-0397-y.

[82] G. Gorgolis, C. Galiotis, Graphene aerogels: a review, 2D Mater 4 (2017) 032001. https://doi.org/10.1088/2053-1583/aa7883.

[83] X. Li, J. Yu, S. Wageh, A.A. Al-Ghamdi, J. Xie, Graphene in photocatalysis: a review, Small 12 (2016) 6640-6696. doi:10.1002/smll.201600382.

[84] M. Nawaz, W. Miran, J. Jang, D.S. Lee, One-step hydrothermal synthesis of porous 3D reduced graphene oxide/TiO2 aerogel for carbamazepine photo-degradation in aqueous solution, Appl. Catal. B Environ. 203 (2017) 85-95. https://doi.org/10.1016/j.apcatb.2016.10.007.

[85] M.A. Worsley, P.J. Pauzauskie, T.Y. Olson, J. Biener, Jr. J.H. Satcher, T.F. Baumann, Synthesis of graphene aerogel with high electrical conductivity, J. Am. Chem. Soc. 132 (2010) 14067-14069. https://doi.org/10.1021/ja1072299.

[86] J.P. Randall, M.A. B. Meador, S.C. Jana, Tailoring mechanical properties of aerogels for aerospace applications, ACS Appl. Mater. Interfaces 3 (2011) 613-626. https://doi.org/10.1021/am200007n.

[87] D. Chen, H. Feng, J. Li, Graphene oxide: preparation, functionalization, and electrochemical applications, Chem. Rev.112 (2012) 6027-6053. https://doi.org/10.1021/cr300115g.

Materials Research Forum LLC
https://doi.org/10.21741/9781644900994-6

Chapter 6

Magnetic Aerogels

Praveen Kumar Yadav[1], Jyoti Raghav[2], Sapa Raghav[3], Pallavi Jain[*4]

[1]Academy of Science and Innovative Research (AcSIR), Chemical and Food BND Group, IRM (BND), CSIR-National Physical Laboratory, Dr K.S. Krishnan Marg, New Delhi-110012, India

[2]School of Engineering and Applied Science, Bennett University, Greater Noida, Uttar Pradesh, India.

[3]Department of Chemistry, Banasthali Vidyapith, Banasthali, Tonk 304022, Rajasthan, India

[4]Department of Chemistry, SRM Institute of Science & Technology, Delhi-NCR Campus, Modinagar-210204, India

* palli24@gmail.com, pallavij@srmist.edu.in

Abstract

Aerogels are exceptional solid-state materials having interconnected 3D solid networks with numerous air-filled pores, large specific surface area (SSA), ultra-low density, etc. The hybridization of aerogels with inorganic metals produces a unique composite with exceptionally improved properties such as flexibility, compressibility, controllable magnetization, electrical conductivities, etc. and these hybrid composites are called the magnetic aerogels (MAg). The development of MAg has increased the applicability of aerogels and provides more opportunity for its application in various new fields as well. So, the present chapter focuses more on the development of various types of MAgs along with their applications.

Keywords

Aerogels, Specific Surface Area (SSA), Flexibility, Magnetization, Magnetic Aerogels

Contents

1.   Introduction

Aerogels are very unique material with excellent physical characteristics. Its formation is also very interesting where air replaces the liquid solvent present in the gel without disturbing the network structure or the volume of the gel body. These materials offer porosity up to 99.9%, extremely high SSA along with the ultralow density to the materials. Additionally, aerogels also provide some excellent properties like exceptional acoustic and optical properties, low dielectric permittivity, as well as extraordinarily low thermal conductivity, due to their exclusive topological porous structure. Due to these properties of aerogels, it has 15 entries in Guinness World Records for material properties.

Being the lightest known solid material, current aerogel production is very high and hence, they are produced at industrial scale. Largely, their unique characteristics like high-porosity (>90% v/v), low-density (typically <0.2 g cm^{-3}) and nanostructured solids are attained due to drying of almost all kind of wet gels without affecting their volume and hence, the gels are produced as monoliths or particles of various shapes and sizes. Drying process is very important factor in aerogel formation which varies with the type and mechanical stability of the gels. Generally, a simple evaporative drying and freeze drying both could collapse the nanostructure during the solvent exchange or conversion in pores. But the surface modification in gels may allow these drying methods without any destruction in sample. The gel can be prepared by both organic and inorganic precursors or their combinations in a suitable solvent and could produce stable 3D network also. Additionally, the starting material and its processing after gel formation are the two major concerns in case of aerogel production.

From definition point of view, aerogels have been defined with different definitions and there is no consensus among the researchers regarding the definition of aerogel. But according to IUPAC an aerogel is "a gel comprised of a microporous solid in which the dispersed phase is a gas". Though, this definition is also very inconvenient for the aerogel community due to the inclusion of microporous glass and zeolites. But the two recent proposed definitions i.e. firstly, "An aerogel is an open, non fluid colloidal network or polymer network that is expanded throughout its whole volume by a gas and is formed by the removal of all swelling agents from a gel without substantial volume reduction or network compaction" and secondly, "Aerogels are solids that feature very low density and high SSA and consist of a coherent open porous network of loosely packed, bonded particles or fibers" are combinedly more satisfactory and comprehensive [1-5].

Presently different types of aerogels are synthesized with the hybridization of inorganic and organic materials in order to improve the properties like flexibility, compressibility, controllable magnetization, electrical conductivities etc. of the aerogel for various applications. Hence, the aerogels produced by the hybridization with magnetic inorganic elements are called the MAg. These hybrid aerogels showed the upper hand over the non-hybrid or normal aerogels due to enhanced super paramagnetic property and their additional mechanical strength. Unlike, the ferrogels and solvent-swollen gels, the MAgs have flexibility, high porosity lightweight and large SSA, and with these additional properties it could be applied in many fields like in microfluidics devices as electronic actuators and many more. The most recent application of MAg is its utilization in the adsorptive removal of pyrethroid insecticides such as cypermethrin, fenpropathrin, and lambda-cyhalothrin from tea beverages and juices.

So, the different types of MAg have been prepared and are as follows

I.   Cellulose magnetic aerogels

II. Magnetic graphene aerogel

III. Carbon magnetic aerogel

IV. Magnetic silica aerogels

V.   Magnetic pectin aerogel

2.   Cellulose magnetic aerogels

Cellulose aerogels (CAg) are the third generation and advanced aerogels which are developed with the help of the synthetic polymer and inorganic. Additionally, these aerogels have absorption properties due to the presence of natural polymers which are abundant and renewable. Cellulose MAgs offer very unique characteristics like high strength, very low density, and ductility in comparison to the inorganic or polymeric aerogels, individually. The variation in structure and other properties of the aerogel materials can be seen with the variation of preparation methods and cellulose sources.

Hence, the research for functional cellulose aerogels is most recent and young research topic for the researchers, in these days. So, a number of studies have been reported on the modifications of cellulose aerogels to obtained electrical [6], improved mechanical [7], magnetic or superhydrophobic [8] properties [9]. Therefore, it has been observed that the new combinations of properties can give a new insight from the application point of view. Here, we have discussed some of the advancements done in the cellulose aerogels to obtain MAgs.

A highly flexible magnetic cellulose aerogels has been developed by Liu et al. [10] using a new and facile method. Here, nanostructured cellulose nanofibrils which formed ductile or tough networks have been used as template for the development of material with advanced functional properties. The cellulose MAg preparation followed the two-step procedure, firstly, cellulose hydrogel films was prepared using LiOH/urea solvent, after that, NPs of $CoFe_2O_4$ were synthesized in the porous structured cellulose scaffolds followed by the freeze-drying resulted in $CoFe_2O_4$/cellulose MAgs. The porosity of the composite aerogel was reported up to 52% where the internal SSA reported around 300–320 m^2/g with the density range of 0.25–0.39 g/cm^3 of the sample. The other properties of the magnetic cellulose aerogel like mechanical strength, super paramagnetic properties, lightweight, flexibility, high porosity with large SSA, etc. supported the expectation of application in new and many fields.

Wang et al. [11] reported a MAg which was synthesized by using $MnFe_2O_4$ and carboxylated cellulose and it was utilized as an adsorbent for removal study of Cu (II). Here, cellulose hydrogel offers porous structure, in which the NPs of $MnFe_2O_4$ were developed by using co-precipitation method. These magnetic NPs encouraged the adsorption property of material as well as their recyclability. The adsorption equilibrium time (120 min) and the relationship between adsorption capacity and adsorption temperature of material were also studied. The maximum adsorption capacity at the initial concentration (150 mg/L) was reported as 73.70 mg/g, at RT. The material showed the pseudo secondary reaction kinetic along with the Langmuir isotherm adsorption model for the adsorption of Cu (II). The XPS analysis showed that the Cu (II) adsorption was mostly due to the active adsorption by the carboxyl and hydroxyl groups. Additionally, the good reusability of material was also claimed.

A new carbon aerogel doped with nitrogen-doped was synthesized using sodium carboxymethyl cellulose. The synthesis method included carbonization, freeze-drying, sol-gel, and KOH activation processes. Whereas, collagen and ferric trichloride were utilized as nitrogen source and cross-linking agent, respectively. The morphology of carbon aerogels reported as three-dimensional and porous with high SSAs and exceptional magnetic properties. Its application as electrode, the CA-N 0.5, exhibited a specific capacitance of 185.3 F/g at the current density of 0.5 A/g in a 6 M KOH electrolyte. The exceptional cycling stability was also obtained due to the specific capacitance retention reported which was 90.2% after 5000 charge/discharge cycles. Additionally, it has also displayed outstanding adsorption capacities of 230.4 and 238.2 mg/g for MB and MG, respectively. Finally, it can be said about the N-doped carbon aerogels that this material can be used as pollutant adsorbent in aqueous solutions as well as electrodes in supercapacitors [12].

Regarding the selective and instant adsorptive removal of oil from surface water, a hydrophobic and highly porous magnetic CAg with high porosity, was developed by Chin et al. [13]. It is reported that, within 10 min, the composite material can absorb the oil around 28 times of its own weight and it could be recovered very fast from the water surface. A 3D macroscopic superhydrophobic magnetic porous carbon aerogel (3DSMPC) was prepared by using popcorn carbonization for selective adsorption of oil where biomass used as precursor. The hygroscopic popcorn was carbonized into popcorn carbon with an additional property such as chemical durability as well as the magnetic property developed during carbonization process. Later on, the super hydrophobicity obtained in the material by the hydrolysis with octyl trichlorosilane of the popcorn carbon. Finally, this material was used for the adsorption of the organic solvents (dichloroethane, cyclohexane, benzene), edible oil (colza oil, soya-bean oil, corn oil), spill oil (kerosene, gasoline, diesel oil, engine oil,), and so on. Other properties of 3DSMPC are its excellent super hydrophobicity with static water contact angle of 151.6°, excellent magnetic saturation (3.3 emu g^{-1}), low density (0.095 g cm^{-3}), and extraordinary selective adsorption of edible oil, organic solvent and fuel oil from water. Additionally, the adsorbent material also showed exceptional reusability because the adsorbed organic pollutants could be easily recovered by simple distillation [14].

A new three-dimensional magnetic bacterial cellulose nanofiber/GO polymer aerogel (MBCNF/GOPA) was synthesized and used for the dye (malachite green (MG)) removal study. The composite has Fe_3O_4 NPs, polyvinyl alcohol (PVA), bacterial cellulose nanofibers (BCNFs), and GO nano-sheets. The effect of various parameters such as (temperature, contact time, pH of dye solution, initial dye concentration, and adsorbent dosage) on adsorption was also studied. This composite material showed high adsorption capacity (270.27 mg/g) at large pH range. Also, the experimental data followed the Langmuir isotherm model, however, the adsorption kinetics followed the PSO model. The morphology of the material showed large SSA, light weight, super paramagnetic behavior at 25 °C, 3D interconnected porous structure, etc. The adsorbent efficiency was reported as 93% and the material can be retrieved from contaminated water with the help of a small magnet. The adsorbent (MBCNF/GOPA) has also shown good reusability in water purification [15].

In spite of these cellulose based MAg synthesis, a new category of cellulose based aerogel i.e. magnetic nanocellulose aerogels are also developed. This aerogel showed some unique properties such as very low density (9.2 mg cm^{-3}), high adsorption capacity (68.06 mg/g), good hydrophobic and magnetic properties, and offer a very efficient adsorbent for oil spills treatment from water [16].

3. Magnetic graphene aerogel

Revolutionary progress of information technology has boomed the more advanced, portable, multifunctional, fashionable and wearable electronic devices in the market. These advancements have enhanced the quality of life and conveniences of human being. Apart from the benefits, the electronic devices cause some serious concerns by generating electromagnetic waves around us, which is also called electromagnetic pollution. So, the development of high-performance microwave absorption materials (MAMs) such as magnetic particles, various carbon materials and their composites, conducting polymers, etc. is one of the most important tools to mitigate the adverse effect of electromagnetic waves [17, 18].

In this regard, different type of MAgs have been developed to attenuate the effect of electromagnetic waves and the developed high-performance magnetic graphene aerogel (GA@Ni) nanocomposite is one of these. The developed material has shown improved microwave absorption and super paramagnetic properties with low density. The synthesis method includes two-step procedure, the first step is hydrothermal reaction and the next step is in situ pyrolysis. The microwave absorption (MA) capability of the material is due to their exceptional impedance matching and synergistic effect. Experimental results reveal that (GA@Ni) can be used as superior microwave absorption materials and the two-step method will open an avenue for the fabrication of synergistic microwave absorbers [19].

Quana et al. [20] developed a graphene-based aerogel i.e. magnetic Fe_3O_4 NPs ornamented graphene aerogel (Fe_3O_4/GAg) using different Fe_3O_4: GO mass ratios of 0.5:1, 1:1, and 2:1, in the presence of ethylenediamine as a reducing agent. The mass ratio of 1:1 was found to be best for the synthesis of Fe_3O_4/GAg (FGAg2). The FGAg2 was used for the adsorption study of bisphenol A (BPA) for which the initial BPA concentration, contact time and pH studies were carried out. The adsorption data were followed by the Langmuir isotherm model, PSO kinetic models and intraparticle

Materials Research Forum LLC
https://doi.org/10.21741/9781644900994-6

diffusion. The maximum adsorption capacity of FGA2 for BPA was reported as 253.80 mg/g.

In contrast to GO and its precursors, the magnetic properties of GO/partially reduced GO aerogels have not been broadly examined. The intrinsic magnetism in carbon-based materials is the bone of contention, possibly due to the presence of transition metal impurities. Hence, the purity of material is examined by at least three methods such as EDS, elemental analysis, EPR at helium temperatures and then the material is used for magnetic studies.

The inside foam structure was confirmed by computer tomography which also confirmed various stacking inside and at the surface, as well as some structural defects. Electrical measurement exhibited a linear relationship between current and voltage as well as its temperature dependency. The paramagnetism and antiferromagnetic interactions led to the weak ferromagnetic behavior observed by the magnetic susceptibility measurements. EPR study authorizes two component paramagnetism of Curie and Pauli type [21]. Another material, graphene aerogel/Fe_3O_4/polystyrene composites with reticulated graphene structure was synthesized by solvothermal method. The porous NPs of Fe_3O_4 acted as a cross linkers between graphene plates. Incorporation of polystyrene and Fe_3O_4 led to the formation of a porous substructure on the reticulated graphene surfaces which successfully improves the hydrophobicity of the material. The composite has shown very low density (0.005 g cm^{-3}) corresponding to a volume porosity of 99.7%. Hence, the composite offered the intake capacity for crude oil was 40 times to its own mass, after 10 water-oil separation cycles. Additionally, the material has also shown very good recovery and reusability due the easy collection of the material with the help of magnets [22].

Graphene based aerogels motivated for the design and development of highly efficient synergistic microwave absorption materials owing to their adjustable dielectric properties and low density, and the porosity of the material offer plentiful active sites for the support of functional magnetic components. Keeping these properties in mind, a super-hydrophobic magnetic graphene aerogel was developed following the facile chemical reduction-assembly method as well as *in-situ* thermal decomposition method. The developed material has revealed the improved microwave absorption performance due to the enhanced synergistic effect and attenuation characteristic. The composite with 4.25 wt.% of functional fillers revealed minimum reflection loss value (-51.6 dB) at 14.6 GHz with maximum effective absorption bandwidth (*EAB*, below -10 dB) of 6.5 GHz and absorber thickness of 2.4 mm. In comparison to other reported graphene-based absorbers, such kind of high-performance synergistic microwave absorbers with low filler loading, lightweight and super-hydrophobic properties may flourish research interest for future practical application [23].

4.   Carbon magnetic aerogel

The carbon-based MAg has shown some unique characteristics such as large SSA, low density, high conductivity, magnetism, good microwave adsorbent, low cost and good recyclability. These materials are utilized for different application purposes like dye and oil spill removal from contaminated water, microwave absorption, energy storage, bioelctrocatalysis, biosensors, etc.

Hence, in this regard, a new carbon-based MAg i.e. conductive Ni/carbon aerogel has been developed successfully using an autocatalytic reduction method. The composite morphology showed nano-porous structure with low densities, appropriate electrical conductivities, high SSA and controllable magnetization. These properties of aerogel composite make it a very good microwave absorbent having strong and controllable EM absorption with ultrathin thickness. The filler loading of 10 wt% in the wax, causes a minimum RL of -57 dB at 13.3 GHz with the thickness of 2mm. The synergistic effect of the weak magnetic loss, good impedance match and medium dielectric loss, led to the high microwave absorption performance of the material [24].

Ye et al. [25] synthesized a Mn and Fe oxides (Mag-FMBO) loaded carbon aerogels based on konjac glucomannan (KGM) and their performance for dyes adsorptive removal study. The material was characterized by various methods, including vibrating sample magnetometer (VSM), X-ray diffraction (XRD), Fourier transformation infrared spectroscopy (FTIR) and Brunner Emmett Teller (BET) surface area analysis. The Mag-Carbon aerogel material was utilized as adsorbents for the removal study of dyes [cationic MB and anionic MO]. The adsorption equilibrium data followed Langmuir isotherm model for both MB and MO dyes. The maxi-mum uptake capacity of magnetic carbon aerogel for MB and MO was noticed as 9.37 and 7.42 mg/g at 303 K, respectively. The adsorption processes followed the PSO model for both MB and MO dyes. Additionally, the composite material exhibited good reusability and easily regenerated by ethanol from the dyes (MO or MB) and the material was also separated from aqueous solution by using the magnet.

Multifunctional magnetic superhydrophobic carbonaceous (MSC) aerogel with hierarchically porous structures is developed by the carbonization method with the help of disposable cotton balls. The MSC aerogel showed ultrahigh porosity (>99.9%) and ultralow density (<2 mg/cm^3), and super oleophilicity (θ water=160.5°, θ oil=0°, a rolling contact angle of 1.8°) and robust super hydrophobicity as well. This MAg was used as adsorbent with good absorption capacity (61-113 g/g) for various oily liquids. The most important and beneficial property of the MSC aerogel is its compressibility and fire resistance property. It can be recycled easily using combustion, squeezing and distillation depending on the type of pollutant. Moreover, this material can be utilized in supercapacitors also. All these characteristics make MSC a comprehensive and versatile material for various applications in energy storage as well as oil spill cleanup [26]. Another carbon-based MAg is magnetic iron oxide/carbonaceous aerogel composites (Fe3O4/CA) which is spongy in nature, and synthesized by hydrothermal method using crude biomass, followed by freeze drying process. The isotherm parameter study showed Freundlich-isotherm model as the best fit for the experimental results and for the adsorption kinetics PSO reaction kinetic model was the best fit model. The high adsorption capacity (185.89 mg/g), high-performance and recyclability of the Fe_3O_4/CA adsorbent is owing to the presence of a, micro and mesoporous structure, high surface area, and surface hydroxyl group on the adsorbent. The present material i.e. Fe_3O_4/CA nanocomposites aerogel is low cost, efficient adsorbent and recyclable material which can be utilized for the removal of the toxic dyes as well as in multiple applications for the field of energy storage [27].

With little modification in carbo aerogel, another aerogel i.e. magnetic ferro ferric oxide carbon aerogel (Fe_3O_4-CA) was developed. Afterward, a film of the composite material was synthesized by mixing the ionic liquid (IL) with the Fe_3O_4-CA, which was analysed by Atomic force microscopy (AFM). This film was utilized as an electrochemical interface for accelerating electrochemistry of myoglobin (Mb) and glucose oxidase (GOx). The results showed a pair of well-defined quasi-reversible redox peaks which was due to the direct electron transfer of Mb and GOx on the film (Fe_3O_4-CA /IL) surface. Furthermore, the restrained Mb showed exceptional bioelectrocartalytic activity for H_2O_2 reduction. The biosensor demonstrated broad linear response ranging from 10 to 1450 µM for H_2O_2. The experimental results confirmed that Fe_3O_4-CA/IL films are a potential biocompatible interface due to their abundant mesoporous structures, exceptional electron transfer activities and large SSA [28].

5.   Magnetic silica aerogels

Aerogel is anebulous strong structure-controllable material having the porosity of 80.0% to 99.8%. The colloidal molecule and pore size 2 to 60 nm and 1 to 100 separately. Its SSA that it can achieve is approximately 1000 m^2/g and the density is 0.06 kg/m^3 [15]. The unique interface effect is seen in the aerogels due to its porous and small structure, which makes the superior performance of aerogel incomparable to the other materials. As a significant part of aerogels mainly carbon aerogels (CAg) are broadly utilized in different areas, for example, electrochemical, ecological science, and investigative chemistry. But the complete separation of CAg is difficult owing to its nano size, low thickness, and low adsorption capacity. Therefore, the research shifted towards the development of magnetic CAgs using the other members of the carbon family like Si and developed an attractive magnetic silica aerogel ($(Fe_3O_4@SiO_2)$. This aerogel has significant features of magnetism as well as aerogel basic features, which make it more attractive for many applications.

The $Fe_3O_4@SiO_2$ aerogels were effectively synthesized by two-advance reactant sol-gel strategy utilizing tetraethylorthosilicate (TEOS) precursor solution. The synthesized aerogel was utilized to adsorb 3 pyrethroid bug sprays from tea and juice drinks by utilizing solid phase extraction. For this process, a few impacting parameters which could affects magnetic solid-phase extraction which comprising contact time, pH, adsorbent dose, volume of elution i.e. solvent, ionic quality, and type of elution dissolvable type were deliberately researched and upgraded. Under ideal conditions, a superior linearity appeared for 3 pyrethroid bugs i.e. 0.4-4, 0.04-4, 0.08-8 ng/mL for fenpropathrin, lambda-cyhalothrin, and cypermethrin, respectively with 0.9985–0.9996 value of residual square, which shows linearity relation. The MSPE technique, was used to detect the trace of pyrethroid insecticides in tea beverages and in juices based on $Fe_3O_4@SiO_2$aerogels, having high affectability and productive [29].

Further Li et al [30] reported, a unique MAg of Fe, Ti and Si i.e. ternary $Fe_3O_4@TiO_2/SiO_2$ aerogel having great photocatalytic action was arranged effectively by consolidating sol-gel and straightforward hydrothermal techniques with $Fe_3O_4@TiO_2/SiO_2$ core-shell microspheres and using industrial fly ash as beginning material, SiO_2 aerogels were prepared. The magnetic and structural properties $Fe_3O_4@TiO_2/SiO_2$ aerogel composite were examined by XRD, FTIR, Transmission electron microscopy (TEM), BET, UV-vis/DRS and VSM). The outcomes indicated that the surface of silica was covered with scattered microspheres of $Fe_3O_4@TiO2$ core-shell. The synthesized aerogel had phenomenal magnetic properties at RT, high adsorption ability to natural poisons, and solid light ingestion in the noticeable area. The photocatalytic movement of $Fe_3O_4@TiO_2/SiO_2$ aerogel degrade for Rhodamine B under visible light, was deliberately examined by under different conditions to ensure higher rate of degradation such as pH, catalyst concentration, irradiation time, and starting substrate dose. The process followed the Langmuir-Hinshelwood model and the first order kinetics. The synthesized ternary aerogel shows higher degradation rate for RhB as compared to pure TiO_2 and $Fe_3O_4@TiO_2$, which implied that synthesized aerogel can fill in as a proficient for the removal of dangerous natural dyes in wastewater [30].

A streamlined methodology is projected to build up a magnetite hybrid silica multifunctional aerogel by exploiting the adaptability of the sol-gel procedure and adaptability of the crosslinking response. To take advantage of this, magnetite NPs at the concentration approximately 0.05 to 0.2 were mixed with the initial solution having the

Materials Research Forum LLC
https://doi.org/10.21741/9781644900994-6

silica as the starting material as well as the cross linker. The active silylation agent was used for chemical coating of magnetite surface to compatibilized the surface chemistry of NPs with silica chemistry for the powerful and feasible linkage of magnetite to the silica network. It is shown that with dopant concentration the properties of magnetite aerogels have been improved as compared to the undoped. With the addition of dopant at particular concentration, the compressive strength and the density (a little bit) of material increased whereas the active surface area of the material decreased. Significantly, a great arrangement of progress with respect to the thermal insulation execution for doped aerogels has been acquired because of the cloudy nature of the magnetite dopants at temperatures over 300 K. The synthesized aerogels show much better mechanical and physical properties with higher magnetism at RT, which make this hybrid aerogel utilized for so many applications, going from specific magnetic separation for natural cleaning and magnetic medication conveyance to catalysis, and so on [31].

Amirkhani et al. [32] synthesized magnetic silica aerogel by sol-gel process by utilizing Fe_3O_4 NPs and sodium silicate for surface modification. Alcoholic solution was utilized to prepared aerogel by the immersion of silica gel in nanoparticles. Structural and surface analysis were carried out by utilizing FESEM, TEM, and FTIR, magnetism and analyzed by VSM, and crystal structure by XRD. The structural analysis shows that the surface of aerogel was homogeneous, mesoporous, and with surface area of 520 m^2/g, and shows ferromagnetism. The aerogel was super hydrophobicity. The synthesized aerogel was utilized to immobilize Candida rugosa by scattered alcoholic solution of aerogel composite. The impact of the sonication abundancy and lipase immobilization time affects the yield. The studies of sonication for lipase immobilization were perusal by statically by utilizing reaction surface strategy (RSM). Correlation between the exhibition scattered and non-scattered backings uncovered the positive impact of scattering process on immobilization yield and protein movement. The accomplishment of immobilization was affirmed by confocal laser filtering magnifying lens (CLSM), demonstrating that chemical is very much circulated in the scattered attractive silica aerogel. A relative report between free and immobilized lipase on scattered help, was directed regarding temperature, pH, temperature stability and the kinetic parameters. Most extreme adsorption limit of lipase was assessed as 81.9 mg/g dependent on Langmuir isotherm.

6. Magnetic pectin aerogel

For biomedical applications, a novel material, containing magnetic NPs has been synthesized and investigated. The polymeric matrix in combination with the fillers having magnetic property make this composite versatile. Here, the utilization of gelatin aerogels as a biodegradable network containing maghemite NPs (c-Fe_2O_3 NPs) is examined. The c-Fe_2O_3 NPs-stacked gelatin aerogels are created in two unique morphologies (tube shaped stone monuments and microspheres) arranged by a mix of sol–gel and supercritical drying strategies. On account of the aerogel microspheres, the sol–gel strategy was replaced by the emulsion-gelation system. The acquired aerogel-based materials were assessed with respect to their physical appearance and steadiness and their textural and attractive properties. The molecule size dissemination and morphology of the aerogel microspheres were furthermore breaking down by laser scattering spectrometry, static picture investigation and filtering electron microscopy. Procedure parameters affected the appropriation of the c-iron oxide NPs inside the material. Though the c-iron oxide NPs were homogeneously disseminated all through the aerogel stone monuments, c-iron oxide NPs were generally stored on the external surface of the aerogel microspheres. The magnetic properties of maghemite were saved at last after general handling [33-37].

The combination of magnetic Fe_2O_4 NPs in aerogels has been accounted for as pure Fe_2O_4 aerogels, $CoFe_2O_4$–cellulose composite aerogels and iron oxide–silica composite aerogels. Garcia-Gonzalez [38] reported maghemite nanoparticles were fused in biocompatible and biodegradable gelatin gel frameworks as mass chambers and microspheres. To acquire attractive gelatin aerogels supercritical drying was used. The properties of the final materials were analysed as to their potential application as attractive focusing on sedate conveyance vehicles.

Conclusions

Aerogels are exceptional solid-state materials having interconnected 3D solid networks with numerous air-filled pores, large SSA, ultra-low density, etc. The hybridization of aerogels with inorganic metals produce a unique composite with exceptionally improved properties such as flexibility, compressibility, controllable magnetization, electrical conductivities, etc. and these hybrid composites are called MAgs. There are different types of MAgs like magnetic cellulose, pectin, carbon, silica and graphene aerogels which have been discussed in this chapter. Additionally, these MAgs, synthesized by different methods such as sol-gel, hydrothermal, autocatalytic reduction, chemical reduction, freeze drying, etc. showed wide applicability in various fields i.e. oil spill removal, inorganic toxic contaminant removal, dye removal, as microwave adsorbent, biosensor application, etc.

Acknowledgements

The authors are thankful to the Director NPL for his encouragement. One of the author Mr. Praveen Kumar Yadav is thankful to AcSIR and to UGC also for providing fellowship to carry out his Ph.D. work. Dr. Sapna is thankful to Banasthali Vidyapith and Ms. Jyoti is thankful to Bennett University. Dr Pallavi Jain thankful to SRM Institute of Science & Technology.

References

[1] S.S. Kistler, 1931. Coherent expanded aerogels and jellies, Nature 127 (1931) 741. https://doi.org/10.1038/127741a0

[2] S. Mulik, C. Sotiriou-Leventis, G. Churu, H. Lu, N. Leventis, Cross-linking 3D assemblies of nanoparticles into mechanically strong aerogels by surface-initiated free-radical polymerization, Chem. Mater. 20 (2008) 5035–46. https://doi.org/10.1021/cm800963h

[3] D.H. Everett, Manual of symbols and terminology for physicochemical quantities and units, Appendix II: definitions, terminology and symbols in colloid and surface chemistry, Pure Appl. Chem. 31 (2009) 577–638. https://doi.org/10.1351/pac197231040577

[4] N. Leventis, A. Sadekar, N. Chandrasekaran, C. Sotiriou-Leventis, Click synthesis of monolithic silicon carbide aerogels from polyacrylonitrile-coated 3D silica networks, Chem. Mater. 22 (2010) 2790–803. https://doi.org/10.1021/cm903662a

[5] L. Falk, A. Nikita, S. Christian, P. Antje P, R. Thomas, Bacterial cellulose aerogels: from lightweight dietary food to functional materials. Funct. Mater. Renew. Sources 1107 (2012) 57–74. https://https://doi.org/ 10.1021/bk-2012-1107.ch004

[6] O. Ikkala, R.H.A. Ras, N. Houbenov, J. Ruokolainen, M. Pääkkö, J. Laine, Solid state nanofibers based on self-assemblies: From cleaving from self-assemblies to multilebel hierarchical constructs, Faraday Discussions, 143 (2009) 95–107. https://doi.org/10.1039/B905204F

[7] A.J. Svagan, L.A. Berglund, P. Jensen, Cellulose nanocomposite biopolymer foam-hierarchical structure effects on energy absorption, ACS Appl. Mater. Interfaces, 3 (2011) 1411–1417. https://doi.org/10.1021/am200183u

[8] H. Jin, M. Kettunen, A. Laiho, H. Pynnönen, J. Paltakari, A. Marmur, et al. Superhydrophobic and superoleophobic nanocellulose aerogel membranes as

bioinspired cargo carriers on water and oil, Langmuir, 27 (2011) 1930–1934.
https://doi.org/10.1021/la103877r

[9] M. Pääkkö, J. Vapaavuori, R. Silvennoinen, H. Kosonen, M. Ankerdors, T.
Lindström, T., et al. Long and entangled native cellulose I nanofibers allow flexible
aerogels and hierarchically porous templates for functionalities. Soft Matter. 4 (2008)
2492–2499. https://doi.org/10.1039/B810371B

[10] S. Liu, Q. Yan, D. Tao, T. Yu, X. Liu, Highly flexible magnetic composite
aerogels prepared by using cellulose nanofibril networks as templates, Carbohy.
Polymers 89 (2012) 551– 557. https://doi.org/10.1016/j.carbpol.2012.03.046

[11] X. Wang, S. Jiang, S. Cui, Y. Tang, Z. Pei, H. Duan, Magnetic-controlled aerogels
from carboxylated cellulose and MnFe2O4 as a novel adsorbent for removal of Cu
(II), Cellulose 26 (2019) 5051–5063.https://doi.org/ 10.1007/s10570-019-02444-7

[12] M. Yu, Y. Han, J. Li, L. Wang, Magnetic N-doped carbon aerogel from sodium
carboxymethyl cellulose/collagen composite aerogel for dye adsorption and
electrochemical supercapacitor, Int. J. Biol. Macromol. 115 (2018) 185-193.
https://doi.org/10.1016/j.ijbiomac.2018.04.012

[13] S.F. Chin, A.N.B. Romainor, S.C. Pang, Fabrication of hydrophobic and magnetic
cellulose aerogel with high oil absorption capacity, Mater. Letters 115 (2014) 241–
243.d oi.org/10.1016/j.matlet.2013.10.061

[14] J. Dai, R. Zhang, W. Ge, A. Xie, Z. Chang, S. Tian, Z. Zhou, Y. Yan, 3D
macroscopic superhydrophobic magnetic porous carbon aerogel converted from
biorenewable popcorn for selective oil-water separation, Mater. Design 139 (2018)
122–131. https://doi.org/10.1016/j.matdes.2017.11.001

[15] P. Arabkhani, A. Asfaram, Development of a novel three-dimensional magnetic
polymer aerogel as an efficient adsorbent for malachite green removal, J. Hazrd.
Mater. 384 (2019) 121394. https://doi.org/10.1016/j.jhazmat.2019.121394

[16] H. Gu, X. Zhou, S. Lyu, D. Pan, M. Dong, S. Wu, T. Ding, X. Wei, I. Seok, S.
Wei, Z. Guo, Magnetic nanocellulose-magnetite aerogel for easy oil adsorption, J.
Colloid Interface Sci. 560 (2020) 849-856. https://doi.org/10.1016/j.jcis.2019.10.084

[17] B. Wen, M.S. Cao, M.M. Lu, W.Q. Cao, H.L. Shi, J. Liu, Reduced graphene
oxides: light-weight and high efficiency electromagnetic interference shielding at
elevated temperatures, Adv. Mater 26 (2014) 3484-3489.
https://doi.org/10.1016/j.jcis.2019.10.084.

[18]   P.B. Liu, Y. Huang, J. Yan, Y. Zhao, Magnetic graphene@PANI@porous TiO2 ternary composites for high-performance electromagnetic wave absorption, J. Mater. Chem. C 4 (2016) 6362-6370. https://doi.org/10.1039/C6TC01718E

[19]   D. Xu, S. Yang, P. Chen, Q. Yu, X. Xiong, J. Wang, Synthesis of magnetic graphene aerogels for microwave absorption by a in-situ pyrolysis, Carbon, 146 (2019) 301-312.https://doi.org/10.1016/j.carbon.2019.02.005

[20]   L.D. Quana, N.H. Dang, T.H. Tua, V.N.P. Linha, L.T.M. Thya, H.M. Nama, M.T. Phong, N.H. Hieua, Preparation of magnetic iron oxide/graphene aerogel nanocomposites for removal of bisphenol A from water, Synthetic Metals, 255 (2019) 116106. https://doi.org/10.1016/j.synthmet.2019.116106

[21]   K. Tadyszaka, K. Chybczyńska, P. Ławniczak, A. Zalewska, B. Cieniek, M. Gonet, M. Murias, Magnetic and electric properties of partially reduced graphene oxide aerogels, J. Magnetism Magnetic Mater. 492 (2019) 165656. https://doi.org/10.1016/j.jmmm.2019.165656

[22]   S. Zhou, W. Jiang, T. Wang, Y. Lu, Highly Hydrophobic, Compressible, magnetic polystyrene/Fe3O4/graphene aerogel composite for oil-water separation, Industrial Eng. Chem. Res. 54 (2015) 5460-5467. https://doi.org/10.1021/acs.iecr.5b00296

[23]   D. Xu, J. Liu, P. Chen, Q. Yu, J. Wang, S. Yang, X. Guo, In situ growth and pyrolysis synthesis of super-hydrophobic graphene aerogels embedded with ultrafine β-co nanocrystals for microwave absorption, 7 (2019) 3869-3880. https://doi.org/10.1039/C9TC00294D

[24]   H.B. Zhao, Z.B. Fu, X.Y. Liu, X.C. Zhou, H.B. Chen, M.L. Zhong, C.Y. Wang, Magnetic and conductive Ni/carbon aerogels toward high-performance microwave absorption, Industrial Eng. Chem. Res. 57 (2017) 202-211. https://doi.org/10.1021/acs.iecr.7b03612

[25]   S. Yea, W. Jina, Q. Huanga, Y. Hua, Y. Lia, B. Lia, KGM-based magnetic carbon aerogels matrix for the uptake of methylene blue and methyl orange, Inter. J. Biol. Macromolecul. 92 (2016) 1169–1174. https://doi.org/10.1016/j.ijbiomac.2016.07.106

[26]   Y. Lu, Z. Niu, W. Yuan, Multifunctional magnetic superhydrophobic carbonaceous aerogel with micro/nano-scale hierarchical structures for environmental remediation and energy storage, Appl. Surface Sci. 480 (2019) 851-860. https://doi.org/10.1016/j.apsusc.2019.03.060

[27]   P. Shanmugama, W. Wei, K. Qiana, Z. Jianga, J. Lua, J. Xiea, Efficient removal of erichrome black T with biomass-derived magnetic carbonaceous aerogel sponge, Mater. Sci. Eng. B 248 (2019) 114387. https://doi.org/10.1016/j.mseb.2019.114387

[28]   M. Li, S. Dong, N. Li, H. Tang, J. Zheng, Magnetic Fe3O4 Carbon aerogel and ionic liquid composite films as an electrochemical interface for accelerated electrochemistry of glucose oxidase and myoglobin, RSC Adv. 5 (2015) 14704-14711. https://doi.org/10.1039/C4RA13400A

[29]   T. Feng, X. Ye, Y. Zhao, Z. Zhao, S. Hou, N. Liang, L. Zhao, Magnetic silica aerogels with high efficiency for selective adsorption of pyrethroid insecticides in juice and tea beverages, New J Chem. 43 (2019) 5159-5166. https://doi.org/10.1039/C8NJ05962D

[30]   Z.D. Li, H.L. Wang, X.N. Wei, X.Y. Liu, Y.F. Yang, W.F. Jiang, Preparation and photocatalytic performance of magnetic Fe3O4@TiO2 core-shell microspheres supported by silica aerogels from industrial fly ash, J. Alloy Comp. 659 (2015) 240-247. https://doi.org/10.1016/j.jallcom.2015.10.297

[31]   H. Maleki, L. Durães, B.F.O. Costa, R.F. Santos, A. Portugal, Design of multifunctional magnetic hybrid silica aerogels with improved properties, Microporous Mesoporous Mater. 232 (2016) 227-237. https://doi.org/10.1016/j.micromeso.2016.06.025

[32]   L. Amirkhani, J. Moghaddas, H. Jafarizadeh-Malmiri, Candida rugosa lipase immobilization on magnetic silica aerogel nanodispersion, RSC Adv. 6 (2016) 12676-12687. https://doi.org/10.1039/C5RA24441B

[33]   J.W. Long, M.S. Logan, C.P. Rhodes, E.E. Carpenter, R.M. Stroud, D.R. Rolison, Nanocrystalline iron oxide aerogels as mesoporous magnetic architectures, J. Am. Chem. Soc., 126 (2004) 16879–16889. https://doi.org/10.1021/ja046044f

[34]   J.W. Long, M.S. Logan, C.P. Rhodes, E.E. Carpenter, R.M. Stroud, D.R. Rolison, Nanocrystalline iron oxide aerogels as mesoporous magnetic architectures, J. Am. Chem. Soc. 126 (2004) 16879–16889. https://doi.org/10.1021/ja046044f

[35]   E. Taboada, R.P. del Real, M. Gich, A. Roig and E. Molins, Faraday rotation measurements in maghemite-silica aerogels, J. Magn. Magn. Mater. 301 (2006) 175–180. https://doi.org/10.1016/j.jmmm.2005.06.019

[36]   M. Popovici, M. Gich, A. Roig, L. Casas, E. Molins, C. Savii, D. Becherescu, J. Sort, S. Surinach, J.S. Munoz, M.D. Baro, J. Nogues, Ultraporous single phase iron

oxide− silica nanostructured aerogels from ferrous precursors, Langmuir 20 (2004) 1425–1429. https://doi.org/10.1021/la035083m

[37] R.T. Olsson, M.A.S. Azizi Samir, G.S. Alvarez, L. Belova, V. Strom, L.A. Berglund, O. Ikkala, J. Nogue´s, U.W. Gedde, Making flexible magnetic aerogels and stiff magnetic nanopaper using cellulose nanofibrils as templates, Nat. Nanotechnol. 5 (2010) 584–588. https://doi.org/10.1038/nnano.2010.155

[38] C.A. Garcıa-Gonzalez, E. Carenza, M. Zeng, I. Smirnova, A. Roig, Design of biocompatible magnetic pectin aerogel monoliths and microspheres, RSC Adv. 2 (2012) 9816–9823. https://https://doi.org/ 10.1039/C2RA21500D

Aerogels I: Preparation, Properties and Applications
Materials Research Foundations **84** (2020) 172-200
Materials Research Forum LLC
https://doi.org/10.21741/9781644900994-7

Chapter 7

Properties of Aerogels

M.A. Velazco-Medel, L.A. Camacho-Cruz, L. Duarte-Peña, and E. Bucio*

Departamento de Química de Radiaciones y Radioquímica, Instituto de Ciencias Nucleares, Universidad Nacional Autónoma de México, Circuito Exterior, Ciudad Universitaria, México CDMX 04510, México

*ebucio@nucleares.unam.mx

Abstract

Since the development of aerogels in 1934 by Kistler, these substrates have been the focus of a lot of research due to their interesting properties. The peculiar nature of their structure has allowed for the discovery of many useful properties in many fields. Due to these properties, such materials may be used for thermal and acoustic insulation, microelectronics, optical arrangements, and even for developing medical devices. In the interest of describing these materials, this chapter outlines the most relevant properties of different aerogels (silica, clay, polymeric, carbonaceous, etc.) focusing on the description of their properties and their uses.

Keywords

Aerogels, Thermal Conductivity, Acoustic Insulation, Biocompatibility, Electrical Conductivity, Mechanical Properties, Optical Properties, Structure and Morphology

Contents

1.   Introduction

Since the discovery and development of aerogels as novel materials, these substrates have found uses in many fields. In some applications, these materials have fulfilled the requirements that no other solid material can. Aerogels have a plethora of peculiar properties because of the uniqueness of their structure. For example, aerogels have the lowest thermal conductivities of any solid substrate, very low refractive indexes and dielectric constants, very low Young's moduli, and sound speeds within the materials are as low as 100 m/s (while the speed of sound in air is 343 m/s). In additionto this, aerogels may be chemically modified to present other properties such as electrical conductivity or even be synthesized as being biocompatible materials that may be useful as drug delivery devices. Due to all these interesting properties, aerogels have been used to replace many materials in microelectronics, optical devices, thermal insulation implementations, acoustic equipment, high energy particle physics, and medicine.

Even when aerogels composed of different materials exhibit similar properties, it is important to consider that aerogels can be manufactured with different backbone materials. The most common aerogels are silica-based; however, in recent years, metal-oxide aerogels, polymeric aerogels, clay-based aerogels, polysaccharide aerogels, and carbonaceous aerogels have risen in importance, each having specific applications.

Considering the interesting behavior of aerogels and the great variety of properties they exhibit, in this chapter, some of the most relevant properties of aerogels will be briefly described and the uses derived from these properties will be presented.

2.   Structure

The microstructural properties of aerogels such as morphology, porosity, and pore size determine parameters in aerogels that are critical for finding the correct material for many applications. It is important to point out that there exist a great variety of aerogel microstructures determined by the identity of the backbone of the aerogel and the synthetic route used to produce it [1].

In general, aerogels present structures with high porosity and very high surface areas. The pores in aerogels may be of different sizes and according to this size, they are classified as mesopores (2 nm < diameter pore < 50 nm) and micropores (diameter pore < 2 nm), with a great variety of morphologies such as spheres, cylinders, rectangle, etc., as illustrated in Fig. 1 [2–4].

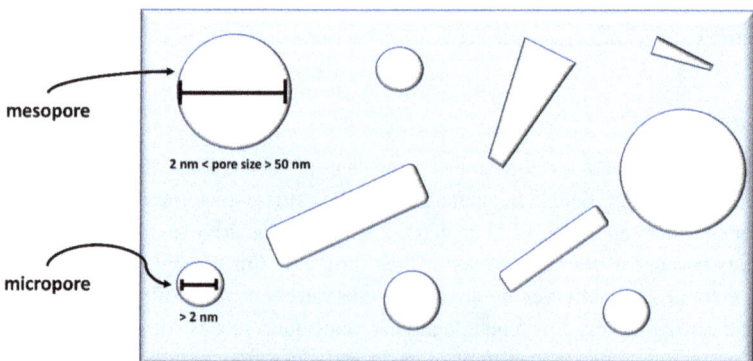

Fig. 1 Morphology and pore size in aerogels [3].

With this variety of morphologies; porosity, pore size, and connectivity between pore and backbone represent the best parameters to describe the structure of these materials. These terms are defined as follows [5]:

I.   Porosity: void fraction per unit volume. This parameter is directly related to the density of the material.

II.   Pore size: distribution of sizes of the pores. This parameter allows for the characterization of an aerogel and determines its possible applications.

III. Connectivity between pore and backbone: As its name implies it defines the amount of solid material that is around pores of the aerogel. This parameter is related to the permeability of the material.

The identity of the aerogel is an important parameter in its final structure. For example, silica aerogels present a structure of irregular spherical nanoparticles that tend to aggregate and condense to form clusters, which ultimately determine their final porosity. This behavior results in great range of porosities of silica aerogels [4,6]. Another example of how the composition of material determines the properties of the final aerogel is the case of starch aerogels fabricated with wheat, maize, and high amylose maize that show different pore density depending on the material (0.23, 0.24 and 0.10 g cm^{-3}respectively) [7]. Currently, there are many studies about biopolymer aerogels, mainly those that use polysaccharides such as cellulose, alginate, chitosan and starch as backbone materials. Table 1 shows some examples of the variation of density and surface area depending on the identity of the backbone material [1, 6-8].

Table 1 Parameters of aerogels depending on the backbone material [1, 6-8].

Aerogel	Density [g cm^{-3}]	Surface area [m^2 g^{-1}]	Reference
Silica	0.003 – 0.500	500 - 1200	[6]
Biopolymers (alginate, starch...)	0.09 – 8.00	14 - 400	[7,8]
Silica-biopolymers	0.12	500 - 900	[1]

The effect of drying method is also important in determining the final structure. Although this varies depending on the material, the general trend is that supercritical CO_2 drying leads to the highest surface area (>500 m^2 g^{-1}). Additionally, it is possible to obtain a wide range of densities, due to the production of different pore sizes, normally in the mesoporous range. In contrast, freeze drying usually produces aerogels of low or ultra-low densities (< 0.1 g cm^{-3}) and lower surface areas because of the formation of macropores [9].

To characterize these morphologies, there exist some useful methods and tools. Scanning electron microscopy (SEM) and X-ray tomography are some techniques that allow to analyze the morphology and the structural distribution of aerogels. For example, in the works by Glenn and Irving [13] it is reported that biopolymer aerogels show fibrillar blocks while silica aerogels exhibit archetypal pearl-necklace. These techniques might be useful, the preparation and manipulation of the sample (by coating or grinding)may affect the validity and the reproducibility of the results.

Pore size can be evaluated by many methods like nitrogen sorption, mercury intrusion, and small-angle X-ray scattering (SAXS); Nevertheless, the first two techniques are limited by the deformation and the collapse of the pores in very low-density aerogels. Besides, it has been shown that large specific pore volume aerogels presentunreliable results [2].

3.   Thermal properties

The thermal properties of aerogels (like almost all other properties) depend directly on the identity of the major component in the aerogel (backbone) [10]. One of the most important properties of aerogels is their low thermal conductivity (λ) which makes aerogels good thermal insulator material with high thermal resistance [11].

Aerogels behave differently in the way they transfer energy and heat than other solid materials because of their nanostructure and high porosities. Due to this, they usually follow the behavior of heat transfer of porous media. Additionally, heat transport within aerogels is a combination of three factors: solid thermal conductivity, conduction through gas, and radiative conductivity.

In aerogels, convection and conduction are poor, since they are mainly composed of gases with poor heat conduction and lack of mobility. Therefore, when the porosity of the material increases, the thermal conductivity of the aerogel decreases since more pores mean a lower amount of solid network where the heat may be transferred [12]. Even though gases are poor heat conductors, they may also transport thermal energy through the aerogel. The contribution of the gaseous phase to the thermal conductivity depends on the gas pressure and it can greatly modify the effective conductivity of the material.

Thermal properties of a wide variety of aerogels have been studied: silica, clay, metal oxide, carbon, polymeric, and other-aerogels. Generally,their thermal conductivity is almost zero, but it may differ depending on the identity of the backbone of the aerogel.

Aerogels I: Preparation, Properties and Applications   Materials Research Forum LLC
Materials Research Foundations **84** (2020) 172-200   https://doi.org/10.21741/9781644900994-7

3.1   Silica aerogels

Silica aerogels are light porous material (porosity >95%) with higher thermal insulation compared to most conventional thermal insulating materials such as polymer foams (e.g. polyurethane derivatives λ= 0.02-0.029 W K^{-1} m^{-1} and expanded polystyrene λ= 0.029-0.055 W K^{-1} m^{-1}), or inorganic wool and foam glass.

Silica aerogels (SiO_2) have total thermal conductivity within the range of 0.003-0.1 W K^{-1} m^{-1} [13,14]. This thermal conductivity value is significantly lower than that of air (λ= 0.026 W K^{-1} m^{-1}), even when a great amount of the material is composed of trapped gases. This has usually been explained with the Knudsen effect, which describes the diffusion of particles in a space that is similar in size to the free path of motion of these particles. When this scenario happens in an aerogel, gas particles easily collide with the backbone of the material, travelling a lot slower than in open space, thus transferring heat much less effectively through convection and diffusion [15]. Thermal conductivity is, however, very dependent on these trapped gases. For instance, at room temperature, some silica aerogels have thermal conductivity of 0.015 W K^{-1} m^{-1} while in air and atmospheric pressure; however, under vacuum the thermal conductivity decreases to 0.01 W K^{-1} m^{-1}. Interestingly, for silica aerogels, the total thermal conductivity has high infrared radiation contributions, meaning that most of the transferred heat is transmitted through IR radiation [16,17].

Composite-silica aerogels with low thermal conductivity have been prepared as alternatives to improve the mechanical properties and thermal stability of resulting materials [18,19]. For example, cellulose-silica aerogels have been produced with thermal conductivities of λ = 0.027 W K^{-1} m^{-1}[20–23], and polypropylene-silica aerogels have achieved conductivities of λ= 0.12 W K^{-1} m^{-1} [24].

3.2   Organic and polymeric aerogels

Thermal conductivities of organic and polymeric aerogels have also been determined and some of these materials have presented good thermal insulation. For example, some pure polysaccharide aerogels produced by gelation have shown low thermal conductivities [25] with values of λ for pectin derivatives being between 0.021 and 0.023 W K^{-1} m^{-1}, for alginate being 0.08 W K^{-1} m^{-1}, and for chitosan aerogels being 0.022-0.029 W K^{-1} m^{-1} [26]. However, it has been found that the low thermal conductivities depend directly on the chemical structure of the aerogel [27]. Additionally, polymer aerogels with good thermal insulation behavior do not present high thermal stability (polysaccharide aerogels for example decompose at ~230°C). This last parameter has been improved when obtaining polymeric composites with silica [28–30].

Other organic aerogels derived from resorcinol-formaldehyde aerogels (RFA) present low thermal conductivity ($\lambda = \sim 0.01$ W K^{-1} m^{-1}), thus materials manufactured with RFA are mainly used as thermal insulators, RFA is also combined with other fragments to improve its properties [11,31].

Cross-linked polymers are also used to prepare aerogels with high porosity, excellent mechanical properties, low densities, and thermal conductivities. Aerogels derived from polyimide with 90% porosity have presented λ around 0.014 W K^{-1} m^{-1} with better mechanical properties than silica aerogels, and with thermal decompositions above 500°C [32–34].

As another example, polyurethane (PU) aerogels have also demonstrated high thermal insulation with λ between 0.017 and 0.024 W K^{-1} m^{-1} [35].

3.3 Carbon aerogels

Aerogels made of carbon nanomaterials also have high thermal stabilities and temperature insulation since they present porosities around 99%. Additionally, they present decomposition temperatures above 800°C. These aerogels are produced from other organic aerogels and then treated by pyrolysis or by other chemical synthesis of the carbonaceous materials, this is the case of carbon aerogels from RFA or poly(vinyl alcohol) [36]. Other derivatives are produced directly from carbon nanomaterials [37,38]. Taking advantage of the high thermal stability, the thermal conductivity of some carbon aerogels has been measured at high temperatures (near 1500°C), obtaining values of 0.09 in vacuum and 0.12 at 0.1 MPa in an argon atmosphere [39]. The same research group from the previous experiments demonstrated that the thermal conductivity via the solid network of carbon aerogels strongly increases when pyrolysis or high annealing temperatures are applied during the process of synthesis [40].

Carbon aerogels are used to fabricate high thermostable materials for high temperature insulation because of their properties. Although their thermal conductivity is typically higher than that of silica aerogels ($\lambda = 0.4$-0.8 W K^{-1} m^{-1}) [41,42]. It is important to mention that carbon nanotubes and graphene are the nanomaterials which are most commonly employed to produce these carbon aerogels, and the choice of nanomaterials with different characteristics is crucial in determining the final properties of the material. [43–45].

4. Electrical properties

The electrical properties of aerogels, similarly to their other properties, are unique and worth discussing. With respect to this field, several studies have been conducted to evaluate the potential applications of aerogels as components in electronics. Although aerogels produced from different materials may share properties such as their thermal behavior or their porosity, electrical properties of aerogels vary greatly depending on the identity of the materials which may form them. In general, there exist two main categories when dealing with electrically relevant aerogels. First, some aerogels present low dielectric constants and very low electrical conductivities which make them useful for applications involving electrical insulation. Examples of these substrates are polymeric and silica-based aerogels. In contrast, when aerogels contain (or are composed of…) materials that are electrically conductive, the resulting aerogel may present good electrical conductivity with high capacitances due to their large surface areas. Examples of this last kind of materials are aerogels of metal oxides and aerogels which contain carbonaceous compounds such as graphene and carbon nanotubes. In the following sections, the properties of both types of aerogels will be contrasted and some characteristics of the base materials will be presented.

4.1   Aerogels with low conductivity

Aerogels composed of non-conductive materials such as silica or polymers are intrinsically bad conductors because of the non-conductivity of their backbone materials. Even when resistivity values for aforementioned materials are high, aerogels tend to be even less conductive than their solid counterparts; for instance, aerogels from silica may present resistivities up to $\sim 10^{13}$ Ω m which are higher in comparison with silicon dioxide which has typical resistivity values of $\sim 10^{10}$ Ω m. This effect is very common in insulating hydrogels and is caused mainly by two properties of the materials. First, the density of the solid frame of the aerogel is extremely low; therefore, conduction paths are poor. The second cause is the fact that the porous nature of the material allows for air and gases to be trapped within its structure; consequently, pores of the aerogel are good insulators and prevent the movement of charge carriers. Both effects may be seen in Fig. 2 [46].

Although the high resistivity of the materials may seem as a disadvantage for aerogel uses in microelectronics, many applications of aerogels focus on their ability to display insulator behavior and even function as excellent dielectric materials. Additionally, aerogels may be combined with other substrates to enhance the already useful properties of these materials [47].

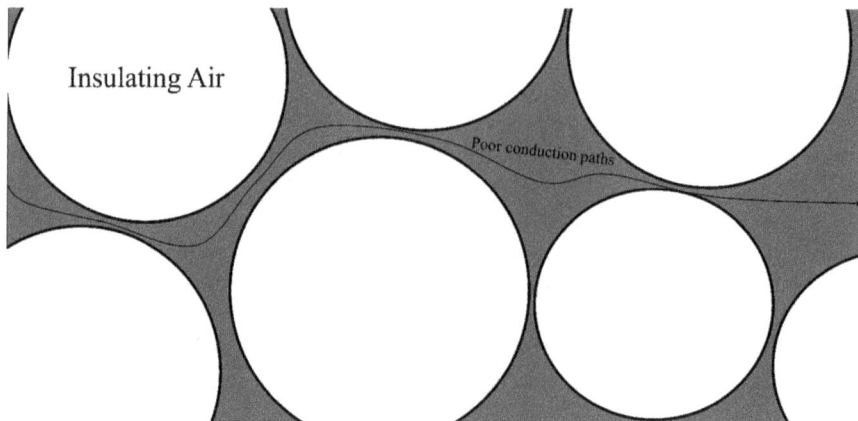

Fig. 2 Poor conductivity of aerogel materials due to porous nature and low density of solid frame [46].

4.2 Low dielectric constant materials

The study of the dielectric properties of aerogels started in the 1970s. In this decade, the study of the behavior of mostly silica-based aerogels was described as being similar to those of gases, in which the most notorious similarity to these systems is the linear dependence of the dielectric constant (the measured real value of this constant) with the density of the aerogels over a big range of frequencies. This was explained by the fact that there is a large amount of trapped gas and absorbed molecules within the pores of the material. Additional to this, the values of the dielectric constant have been observed to be low, this has been attributed to the low density of these materials which prevents electrical displacement (induction) [46]. Many of these materials also present low values of dielectric loss angles while maintaining low dielectric constants in great ranges of frequencies and temperatures; however, this property is greatly affected by the presence of water when measuring at low frequencies. These properties are useful since low dielectric constant materials are chemically stable and versatile and are used thoroughly on microelectronics because they help mitigate parasitic capacitances in circuits [47,48].

The research within the field of low and ultralow dielectric constant materials have yielded silica-based aerogels and aerogel composites that may be used over different ranges of frequencies with good results [46,49]. Recently, polymeric aerogels with ultralow dielectric constants have also been produced, for instance, polyimide and polystyrene aerogels have been produced with dielectric constants as low as $\kappa = 1.03$ [50–53].

One of the disadvantages of dielectric aerogels is their relatively low breakdown voltage at low frequencies in comparison with other materials used in microelectronics. In general, the magnitude of the breakdown electric field is around 100 kV/cm or lower, both for polymeric and silica-based aerogels [46,52,54]. However, some reports have found dielectric breakdown strengths of 500-1000 kV/cm for thin films of aerogels composed of silica and silica-based composites [10,53]. The higher the value of breakdown potential, the better the performance of the aerogel in high voltage applications, therefore, it is desirable to have higher breakdown voltages.

4.3 Aerogels with high conductivity

In stark contrast to the type of aerogels discussed previously, aerogels may be manufactured with conductive properties if the backbone materials are conductive. For instance, aerogels that contain carbon nanotubes or graphene-derived materials as well as aerogels manufactured with conductive metal oxides are conductive and show convenient characteristics for their use in microelectronics, batteries, capacitors, electrodes, sensors and piezoelectric materials.

Conductive metal-oxide aerogels have represented an important variety of materials since they are able to effectively intercalate lithium ions, an essential property in the production of electrodes for Li-ion batteries and to produce fuel cells [10]. Common metal oxide aerogels used for this purpose are transition metal oxides of vanadium(V), molybdenum(III), manganese(IV), and blends like ruthenium(IV) oxide titanium(IV) oxide [55,56]. Additional to this, aerogels composed of blends between metal oxides and conductive polymers such as polyaniline and polypyrrole have been synthesized as composites and interpenetrating networks to improve the electrical and mechanical properties of the materials [56–58].

Another important field is the carbon aerogels because they have been widely used in the manufacture of electrodes [47]. These materials are convenient for this purpose because of their big surface areas which allows for the creation of electrodes with extremely high capacitances in miniature sizes. This property makes these electrodes ideal for electrochemical probes and sensors that not only perform correctly, but also may be manufactured in miniature size or with ergonomic configurations [45,59–62].

5.   Optical properties

Aerogels have also been important because of their special optical properties. The most particular optical property of aerogels is their low refractive indexes which are just slightly higher than $n=1$ (typically from 1.05 to 1.1). In comparison with other materials, the refractive indexes for aerogels are between the values for gases and liquids, being also considerably lower than the values for any other solid material; therefore, these materials may be used in applications where low refractive indexes are required but working with pressurized gases or liquids is complicated. Apart from this, coupling the optical properties of these materials with the electrical properties of carbon aerogels has allowed producing efficient components for solar energy technologies. In the following sections, three main uses deriving of the optical properties of aerogels will be highlighted and some examples will be provided.

5.1   Radiators in Cherenkov counters

Cherenkov counters are devices that allow for the identification of the speed (and hence the energy) of charged particles that travel near the speed of light. For this purpose, these devices take advantage of the emission of Cherenkov radiation which is a well described effect in high energy particle physics that occurs when a charged particle is travelling through a physical medium faster than the speed of light would be through that medium (which is equal to c/n, where c is the speed of light and n is the refractive index of the medium), when this happens, radiation is emitted at an angle dependent on the speed of the particle (β) and the refractive index of the medium according to Eq.1 [63].

$$\cos\theta = \frac{1}{\beta n} \text{ , where } \frac{c}{n} < \beta < c \qquad (1)$$

According to electrodynamics, when a charged particle travels through a polarizable medium, it excites atoms and molecules of the medium which then relax to a basal energy state by emitting light. Although this effect is always present, when the particle is travelling below the speed of light, the amount of radiation produced through the wake of the particles is too low to be detected; nevertheless, when the particle is travelling at superluminal speeds through the medium, constructive interaction between the wave fronts of the wake are produced and the radiation is thus measurable, this radiation is called Cherenkov radiation. Cherenkov radiation is analogous to the effect produced during a "sonic boom" when aircraft travel faster than the speed of sound through air. An illustration of Cherenkov radiation can be seen in Fig. 3 [64].

Materials Research Forum LLC
https://doi.org/10.21741/9781644900994-7

Cherenkov Radiation

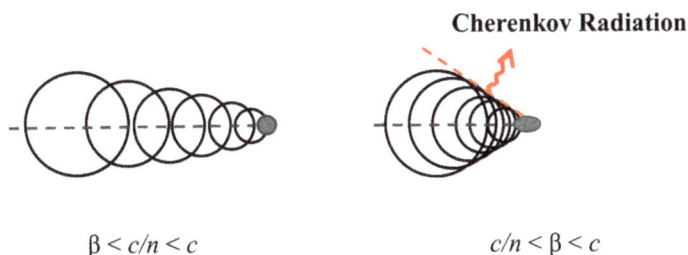

$$\beta < c/n < c \qquad\qquad\qquad c/n < \beta < c$$

Fig. 3 Cherenkov radiation produced when a charged particle travels at superluminal speeds through a polarizable medium [64].

According to Eq. 1, it is possible to detect the speed of high energy particles by measuring the angle of emission of Cherenkov radiation that is produced when a particle travels through a material. These materials, capable of emitting Cherenkov radiation effectively are called Cherenkov radiators. The choice of this material depends on the requirements of the experiment, namely, the velocity of the particles; therefore, many materials may be suitable for this purpose; however; the material must also be relatively transparent to the emitted radiation and present relatively low chromatic aberration (variation of the refractive index with the change in wavelength of emitted radiation) to be able to measure Cherenkov radiation. When this radiator is coupled with photomultipliers and detectors to measure the emission angle, the whole system is called a Cherenkov counter [47,63].

Aerogels have been important in this area since they have a very acceptable optical transmittances at visible wavelengths at which a lot of measurable Cherenkov radiation is produced. For instance, it has been found that silica aerogels have transmittances which are inversely proportional to the wavelength to the fourth power ($\propto 1/\lambda^4$) making them almost completely transparent at wavelengths above 500 nm. Another important property of these materials is that their values for refractive indexes are just between those of liquids and gases which allows for the measurement of particles with energies around 1-10 GeV which were difficult to quantify before the introduction of aerogels. Additionally, since these materials are solid, it is more practical to work with them instead of with pressurized gases or liquids. Finally, it is important to mention that the refractive index of most silica aerogels is linearly dependent on its density according to Eq. 2; therefore, by controlling the density of the final product at a synthetic level, it is possible to obtain a specific refractive index [47,63,65,66] .

$$n=1+k\rho, \text{ where } k\cong0.21 \text{ and is slightly dependent on } \rho \qquad (2)$$

5.2   Fiber optics

The low refractive index of aerogels has been also useful for the construction of optical fibers useful for information transmission. A usable optical fiber optics is composed of two main parts as shown in Fig. 4 [63], the core and the cladding.

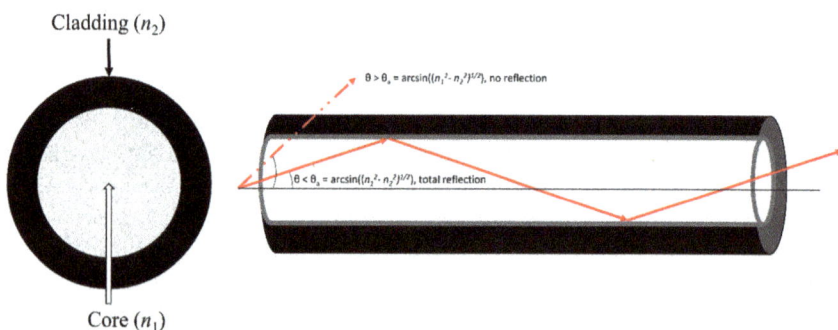

Fig. 4 Diagram depicting total reflection within an optical fiber with the acceptance angle given by the numerical aperture [63].

For a fiber to effectively transmit light through a large distance, light must be totally reflected within the core of the material thanks to the cladding. For this to happen, the incident angle of the light must be below a threshold value determined by the numerical aperture (Eq. 3, where NA is the numerical aperture, θ_a is the acceptance angle and n_1 and n_2 are the refractive indexes of the core and the cladding respectively); therefore, it is desirable that the refractive index of the cladding is lower than the cores'. Since aerogels have such low refractive indexes, they have been considered as cladding materials by some researchers [47,67–69].

$$NA = \sin\theta_a = \sqrt{n_1^2 - n_2^2} \; ; \qquad (3)$$

5.3   Non reflective materials

Another interesting application of carbonaceous aerogels is their use as non-reflective materials that may be used in optical equipment and solar energy collection. Carbon aerogels are very effective in absorbing light because of the intrinsic properties of graphene and carbon nanotubes to absorb light, and the highly porous and rough surface of the material which stops light from being reflected out of the material. This property has allowed for the production of materials with very small reflectance at a big range of wavelengths and incidence angles of light (directional hemispherical reflectance below even 1.0%) [47,70,71].

6.   Mechanical properties

Aerogels present a lot of properties that are used for several applications in thermal and acoustic insulation, catalysis, electronics, controlled drug delivery, etc.; nevertheless, these applications are often limited by the fragility of the systems. Generally, porous materials exhibit a brittle and elastic behavior, but also exhibit poor mechanical properties due to the minimal fraction of solid network [11].

Mechanical properties of aerogels are evaluated through common mechanical tests, (measurement of hardness, fracture, fatigue, tension, compression, and breakability). Using these techniques, it is possible to determine important aspects of the mechanical properties of aerogels, which almost always depend on the density of the material [72].

To characterize aerogels, there are some useful techniques that may be employed apart from common mechanical tests. For example, ultrasonic pulse and nanoindentation are used to determine Young's modulus and Poisson's ratio in silica-, carbon-, polymeric- and other aerogels.

The Young's modulus is a very important parameter in the characterization of any material because it quantifies the ability of a material to suffer deformations after applying tension stress (the formal definition of this parameter is the slope of the elastic deformation portion of a material's stress-strain behavior, a graphical representation of this parameter is presented in Fig. 5 [73]). In aerogels this parameter is often very small, meaning that the materials may deform greatly after applying mechanical stress to them. This parameter is very important to quantify because it demonstrates the elastic behavior of aerogels which allows them to deform easily even when suffering great mechanical tensions; however, this does not grant good mechanical properties since the low density of the materials still make them brittle [74].

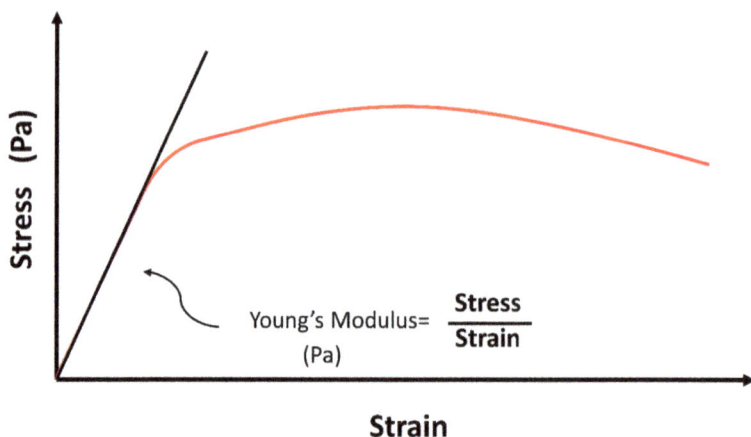

Fig. 5 Stress-Strain graph depicting the Young's modulus [73].

Additional techniques used to characterize aerogels are based on thermal analysis. This is especially important in the case of polymeric aerogels, where it is necessary to determinate glass transition temperatures and melting points. In this case the use of differential scanning calorimetry (DSC) and dynamic mechanical analysis (DMA) becomes very useful [75].

Current research in the mechanical properties of aerogels is mostly conducted to find materials with better mechanical properties, mainly to avoid fragility of aerogel samples. For example, silica aerogels tend to be brittle due not only to their low densities, but also to imperfections, the sizes of pores, and a disordered solid network morphology [76]. However, with the combination of a secondary material, such as a metal or a polymer, it is possible to obtain reinforced aerogel composites. This is convenient since the properties of silica aerogel are maintained when the ratio of both components (silica/other) is low, but the solid network density is increased [72,77].

In many cases, controlling some variables in the process of synthesis leads to the formation of stronger aerogels [78–80]. One of these approaches is the deposition of thin layers of polymers onto silica aerogel surfaces or nano-encapsulation, both of which have worked to improve mechanical properties of the final materials [81,82]. The same procedure is applied to enhance clay aerogels, where a secondary component such as

Materials Research Forum LLC
https://doi.org/10.21741/9781644900994-7

polymers (e. g. cellulose, poly(N-isopropylacrylamide) or polyimide) or carbon nanomaterials (e.g.carbon nanotubes) are added to the system [83–85]. As an example, in the case of cellulose/clay aerogels, the addition of cellulose favors the arrangement of the solid network, reaching effective stress transfer, and thus generating an increase in the compressive strength of the mixture [86].

The other way around, it is also possible to reinforce polymeric aerogels by using clay as a crosslinking agent. For example, polyimide/clay composites have been prepared as a proposal for improving overall mechanical properties, since polyimide aerogels already exhibit good thermal stability as well as mechanical properties compared with other polymeric aerogels [32]. As another example, in 2018 Fan and co-workers [90] designed a polyimide/carbon nanotube composite aerogel with enhanced mechanical properties. In these cases,carbon nanomaterials have been used to reinforce polymeric aerogels because of their resultingnano-porous structure [87,88].

The same strategy is applied to generate stronger metal oxide aerogels [89,90]. For instance, alumina aerogels (Al_2O_3) have been doped with titanium oxide (titania, TiO_2) to enhance the mechanical and thermal properties of the alumina [91]. Metals have also been used as dopants for other aerogels [92].

7. Acoustic properties

Aerogels also present interesting acoustic behavior since the sound velocity in these materials can be as low as 100 m s^{-1} over hearing frequencies and even lower speeds (20 m s^{-1}) at the ultrasonic ranges. These speeds are dramatically lower than the average speed of sound in air (343 m s^{-1}) and practically all other solid media [47]. Even when aerogels contain a great amount of trapped gases, since the speed of sound is a lot lower than that of air and other gases, it is possible to think that the sound is propagated through the solid network and not by the gases inside the pores in aerogels. The low speed of sound in these materials is related to their low Young's modulus, meaning that this material may be easily deformed when subjected to stress, even when this stress is coming from sound waves [74]. Because of this, aerogels also present good sound attenuation and low values of acoustic impedance. These materials have been readily used as acoustic insulators and have had application in acoustic wave generation [93–97].

8. Biocompatibility

Biocompatibility is the ability of a material to be accepted by a living system without generating an adverse effect such as irritation, inflammation, allergic reactions, or toxicity to the living system. This is the most important characteristic of biomaterials and must be determined as a prerequisite if a material is intended for use on or inside living organisms. ISO 10993 for the biological evaluation of medical devices is recognized by the U.S. Food and Drug Administration (FDA) and by the European Union as the standard for selecting the biological tests necessary for measuring the safety of a medical device [98]. This includes *in vivo* and *in vitro* methods that can be applied to a wide variety of materials. *In vitro* methods constitute the first step to test the biocompatibility of a material. These tests are conducted outside of an organism and are intended to measure either the number of dead cells or the cell growth on a controlled media to determine the interaction between the cells and material. The most used *in vitro* tests are cytotoxicity, hemocompatibility, genotoxicity, carcinogenicity, absorption/penetration and toxicity in three-dimensional models of skin, cornea and oral mucosa [99,100]. On the other hand, *in vivo* tests allow observing the performance of a material inside a living organism, and its inflammatory and immunological potential. Some examples of *in vivo* methods are the following: evaluation of dermal toxicity and implants acceptance (subcutaneous, bone, and intramuscular implants) [101,102].

For aerogels, biocompatibility depends on the identity of the backbone material. Usually, natural polymers and other biopolymers are very biocompatible and non-toxic, because of this these types of aerogels may be used for drug delivery, tissue engineering and scaffolds, biosensing, and storage of drugs [103].

Aerogels based in polysaccharides are the most studied for biomedical applications since these natural polymers show intrinsic biocompatibility [104,105]. For example, the biocompatibility of alginate gels and aerogels has been studied *in vitro* and *in vivo* showing positive results as novel drug delivery systems [106,107]. Chitosan aerogels have also shown low cytotoxicity and high biocompatibility in implants while also being responsive to pH [108,109]. Ubeyitogullari and co-workers [110] studied the digestibility of wheat starch aerogels and their temperature resistance for their use in foods. They found that the presence of starch aerogels in food products increases their digestibility due to its ultra-low density and high porosity. This effect could be used to decrease the caloric value of some foods [110].

Another biomedical application which has been extensively studied is the use of aerogels as drug delivery systems. In these applications, drugs can be attached and loaded in the pores of the material. For this purpose, many substrates have been tested, especially polysaccharide and silica derivatives. For example, Lovskaya, Lebedev, and Menshutina [111] developed aerogel microspheres useful for drug loading. In this work, they tested silica, sodium alginate, and starch aerogels. The adsorption of ibuprofen, loratadine, and other drugs was demonstrated to be successful due to the interaction between hydroxy and carboxylic groups present in the polymeric network of the aerogels and the drugs.

Materials Research Forum LLC
https://doi.org/10.21741/9781644900994-7

The delivery process of the drugs was effective in aqueous solution at different pH values. The release mechanism was assisted when water replaced trapped gases in the structure of the aerogel as represented in Fig. 6 [111].

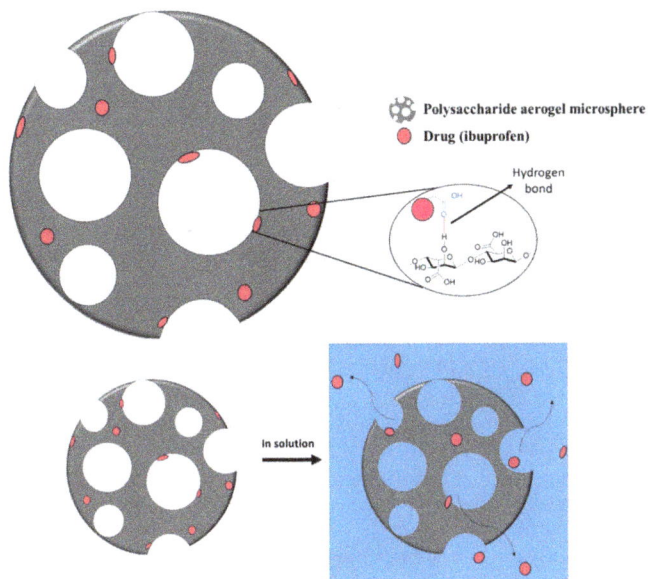

Fig. 6 Drug attachment into aerogel pores by hydrogen bonding and drug release in aqueous solution [111].

Conclusion

Aerogels are highly interesting materials which are used extensively due to their properties. Although the properties of aerogels depend greatly on the material that forms them, some of the characteristics of aerogels such as good thermal insulation, low Young's modulus, good sound isolation are mainstream for these materials and define the versatility of their uses. It is important to mention that current research focuses on improving the behavior of the materials in specific applications while also retaining good mechanical properties. With the development of new materials, it is probable that the properties of aerogels will continue to expand to cover more uses in many areas.

Aerogels I: Preparation, Properties and Applications   Materials Research Forum LLC
Materials Research Foundations **84** (2020) 172-200   https://doi.org/10.21741/9781644900994-7

Acknowledgements

Thanks to CONACyT for the doctoral scholarship provided for Marlene Alejandra Velazco Medel (696062/ 583700), Lorena Duarte Peña (887494) and the master's degree scholarship for Luis Alberto Camacho Cruz (916557).

References

[1]   S. Zhao, W.J. Malfait, N. Guerrero-Alburquerque, M.M. Koebel, G. Nyström, Biopolymer aerogels and foams: Chemistry, properties, and applications, Angew. Chem. Int. Ed. 57 (2018) 7580–7608. https://doi.org/10.1002/anie.201709014.

[2]   G. Reichenauer, Structural characterization of aerogels, in: M.A. Aegerter, N. Leventis, M.M. Koebel (Eds.), Aerogels handbook, Springer, New York, 2011: pp. 449–498. https://doi.org/10.1007/978-1-4419-7589-8-21.

[3]   D.W. Hua, J. Anderson, S. Hæreid, D.M. Smith, G. Beaucage, Pore morphology study of silica aerogels, MRS Proceedings. 346 (1994) 985. https://doi.org/10.1557/PROC-346-985.

[4]   J. jun Liao, P. zhao Gao, L. Xu, J. Feng, A study of morphological properties of SiO2 aerogels obtained at different temperatures, J. Adv. Ceram. 7 (2018) 307–316. https://doi.org/10.1007/s40145-018-0280-6.

[5]   L. Canham, Pore volume (porosity) in porous silicon, in: L. Canham (Ed.), Handbookof porous silicon. 1–2 (2018) 291–298. https://doi.org/10.1007/978-3-319-71381-6_13.

[6]   H. Maleki, L. Durães, An overview on silica aerogels synthesis and different mechanical reinforcing strategies, J. Non.-Cryst. Solids.385(2014)55–74. https://doi.org/10.1016/j.jnoncrysol.2013.10.017.

[7]   F. Zhu, Starch based aerogels: Production, properties and applications, Trends Food Sci. Technol. 89 (2019) 1–10. https://doi.org/10.1016/j.tifs.2019.05.001.

[8]   T. Budtova, Cellulose II aerogels: a review, Cellulose. 26 (2019) 81–121. https://doi.org/10.1007/s10570-018-2189-1.

[9]   C. Jiménez-Saelices, B. Seantier, B. Cathala, Y. Grohens, Spray freeze-dried nanofibrillated cellulose aerogels with thermal superinsulating properties, Carbohydr. Polym. 157 (2017) 105–113. https://doi.org/10.1016/j.carbpol.2016.09.068.

[10] A.C. Pierre, G.M. Pajonk, Chemistry of aerogels and their applications, Chem.

Rev. 102 (2002) 4243–4265. https://doi.org/10.1021/cr0101306.

[11]   T. Woignier, J. Reynes, J. Phalippou, Sintering of Silica Aerogels for Glass
       Synthesis: Application to Nuclear Waste Containment, in: M.A. Aegerter, N.
       Leventis, M.M. Koebel (Eds.), Aerogels handbook, Springer, New York,2011: pp.
       665-680. https://doi.org/10.1007/978-1-4419-7589-8_29.

[12]   C. Bi, G.H. Tang, Effective thermal conductivity of the solid backbone of aerogel,
       Int. J. Heat Mass Transf. 64 (2013)452–456.
       https://doi.org/10.1016/j.ijheatmasstransfer.2013.04.053.

[13]   M.A. Hasan, R. Sangashetty, A.C.M. Esther, S.B. Patil, B.N. Sherikar, A. Dey,
       Prospect of thermal insulation by silica aerogel: A brief review, J. Inst. Eng. India
       Ser. D. 98 (2017) 297–304. https://doi.org/10.1007/s40033-017-0136-1.

[14]   T.-Y. Wei, T.-F. Chang, S.-Y. Lu, Y.-C. Chang, Preparation of monolithic silica
       aerogel of low thermal conductivity by ambient pressure drying, J. Am. Ceram.
       Soc. 90 (2007) 2003–2007. https://doi.org/10.1111/j.1551-2916.2007.01671.x.

[15]   S.S. Kistler, A.G. Caldwell, Thermal conductivity of silica aërogel, Ind. Eng.
       Chem. 26 (1934) 658–662. https://doi.org/10.1021/ie50294a016.

[16]   A. Soleimani Dorcheh, M.H. Abbasi, Silica aerogel; synthesis, properties and
       characterization, J. Mater. Process. Technol. 199 (2008) 10–26.
       https://doi.org/10.1016/j.jmatprotec.2007.10.060.

[17]   G. Wei, Y. Liu, X. Zhang, X. Du, Radiative heat transfer study on silica aerogel
       and its composite insulation materials, J. Non.-Cryst. Solids. 362 (2013) 231–236.
       https://doi.org/10.1016/j.jnoncrysol.2012.11.041.

[18]   J.P. Zhao, D.T. Ge, S.L. Zhang, X.L. Wei, Studies on thermal property of silica
       aerogel/epoxy composite, Mater. Sci. Forum. 546–549 (2007) 1581–1584.
       https://doi.org/10.4028/www.scientific.net/MSF.546-549.1581.

[19]   J. Guo, B.N. Nguyen, L. Li, M.A.B. Meador, D.A. Scheiman, M. Cakmak, Clay
       reinforced polyimide/silica hybrid aerogel, J. Mater. Chem. A. 1 (2013) 7211.
       https://doi.org/10.1039/c3ta00439b.

[20]   J. Feng, D. Le, S.T. Nguyen, V. Tan Chin Nien, D. Jewell, H.M. Duong, Silica-
       cellulose hybrid aerogels for thermal and acoustic insulation applications, Colloids
       Surfaces A: Physicochem. Eng. Aspects 506 (2016) 298–305.
       https://doi.org/10.1016/j.colsurfa.2016.06.052.

[21]   J. Shi, L. Lu, W. Guo, J. Zhang, Y. Cao, Heat insulation performance, mechanics

and hydrophobic modification of cellulose–SiO2 composite aerogels, Carbohydr. Polym. 98 (2013) 282–289. https://doi.org/10.1016/j.carbpol.2013.05.082.

[22] J. Laskowski, B. Milow, L. Ratke, The effect of embedding highly insulating granular aerogel in cellulosic aerogel, J. Supercrit. Fluids. 106 (2015) 93–99. https://doi.org/10.1016/j.supflu.2015.05.011.

[23] A. Demilecamps, C. Beauger, C. Hildenbrand, A. Rigacci, T. Budtova, Cellulose-silica aerogels, Carbohydr. Polym. 122 (2015) 293–300. https://doi.org/10.1016/j.carbpol.2015.01.022.

[24] S. Motahari, G.H. Motlagh, A. Moharramzadeh, Thermal and flammability properties of polypropylene/silica aerogel composites, J. Macromol. Sci. B. 54 (2015) 1081–1091. https://doi.org/10.1080/00222348.2015.1078619.

[25] G. Horvat, T. Fajfar, A. Perva Uzunalić, Ž. Knez, Z. Novak, Thermal properties of polysaccharide aerogels, J. Therm. Anal. Calorim. 127 (2017) 363–370. https://doi.org/10.1007/s10973-016-5814-y.

[26] S. Takeshita, S. Yoda, Chitosan aerogels: Transparent, flexible thermal insulators, Chem. Mater. 27 (2015) 7569–7572. https://doi.org/10.1021/acs.chemmater.5b03610.

[27] X. Lu, R. Caps, J. Fricke, C.T. Alviso, R.W. Pekala, Correlation between structure and thermal conductivity of organic aerogels, J. Non.-Cryst. Solids. 188 (1995) 226–234. https://doi.org/10.1016/0022-3093(95)00191-3.

[28] H. Maleki, L. Durães, A. Portugal, Synthesis of lightweight polymer-reinforced silica aerogels with improved mechanical and thermal insulation properties for space applications, Microporous Mesoporous Mater. 197 (2014) 116–129. https://doi.org/10.1016/j.micromeso.2014.06.003.

[29] D.B. Mahadik, H.-N.-R. Jung, W. Han, H.H. Cho, H.-H. Park, Flexible, elastic, and superhydrophobic silica-polymer composite aerogels by high internal phase emulsion process, Compos. Sci. Technol. 147 (2017) 45–51. https://doi.org/10.1016/j.compscitech.2017.04.036.

[30] P. Paraskevopoulou, D. Chriti, G. Raptopoulos, G.C. Anyfantis, Synthetic polymer aerogels in particulate form, Materials. 12 (2019) 1543. https://doi.org/10.3390/ma12091543.

[31] X. Lu, M.C. Arduini-Schuster, J. Kuhn, O. Nilsson, J. Fricke, R.W. Pekala, Thermal conductivity of monolithic organic aerogels, Science. 255 (1992) 971–

972. https://doi.org/10.1126/science.255.5047.971.

[32] H. Guo, M.A.B. Meador, L. McCorkle, D.J. Quade, J. Guo, B. Hamilton, M. Cakmak, G. Sprowl, Polyimide aerogels cross-linked through amine functionalized polyoligomeric silsesquioxane, ACS Appl. Mater. Interfaces. 3 (2011) 546–552. https://doi.org/10.1021/am101123h.

[33] M.A.B. Meador, C.R. Alemán, K. Hanson, N. Ramirez, S.L. Vivod, N. Wilmoth, L. McCorkle, Polyimide aerogels with amide cross-links: A low cost alternative for mechanically strong polymer aerogels, ACS Appl. Mater. Interfaces. 7 (2015) 1240–1249. https://doi.org/10.1021/am507268c.

[34] J. Feng, X. Wang, Y. Jiang, D. Du, J. Feng, Study on Thermal conductivities of aromatic polyimide aerogels, ACS Appl. Mater. Interfaces. 8 (2016) 12992–12996. https://doi.org/10.1021/acsami.6b02183.

[35] N. Diascorn, S. Calas, H. Sallée, P. Achard, A. Rigacci, Polyurethane aerogels synthesis for thermal insulation - textural, thermal and mechanical properties, J. Supercrit. Fluids. 106 (2015) 76–84. https://doi.org/10.1016/j.supflu.2015.05.012.

[36] J. Yamashita, T. Ojima, M. Shioya, H. Hatori, Y. Yamada, Organic and carbon aerogels derived from poly(vinyl chloride), Carbon. 41 (2003) 285–294. https://doi.org/10.1016/S0008-6223(02)00289-0.

[37] J. Biener, M. Stadermann, M. Suss, M.A. Worsley, M.M. Biener, K.A. Rose, T.F. Baumann, Advanced carbon aerogels for energy applications, Energy Environ. Sci. 4 (2011) 656–667. https://doi.org/10.1039/c0ee00627k.

[38] S. Araby, A. Qiu, R. Wang, Z. Zhao, C.H. Wang, J. Ma, Aerogels based on carbon nanomaterials, J. Mater. Sci. 51 (2016) 9157–9189. https://doi.org/10.1007/s10853-016-0141-z.

[39] M. Wiener, G. Reichenauer, S. Braxmeier, F. Hemberger, H.-P. Ebert, Carbon aerogel-based high-temperature thermal insulation, Int. J. Thermophys. 30 (2009) 1372–1385. https://doi.org/10.1007/s10765-009-0595-1.

[40] M. Wiener, G. Reichenauer, F. Hemberger, H.-P. Ebert, Thermal conductivity of carbon aerogels as a function of pyrolysis temperature, Int. J. Thermophys. 27 (2006) 1826–1843. https://doi.org/10.1007/s10765-006-0086-6.

[41] V. Bock, O. Nilsson, J. Blumm, J. Fricke, Thermal properties of carbon aerogels, J. Non.-Cryst. Solids. 185 (1995) 233–239. https://doi.org/10.1016/0022-3093(95)00020-8.

[42] X. Lu, O. Nilsson, J. Fricke, R.W. Pekala, Thermal and electrical conductivity of monolithic carbon aerogels, J. Appl. Phys. 73 (1993) 581–584. https://doi.org/10.1063/1.353367.

[43] M.B. Bryning, D.E. Milkie, M.F. Islam, L.A. Hough, J.M. Kikkawa, A.G. Yodh, Carbon nanotube aerogels, Adv. Mater. 19 (2007) 661–664. https://doi.org/10.1002/adma.200601748.

[44] Q. Zhang, F. Zhang, S.P. Medarametla, H. Li, C. Zhou, D. Lin, 3D printing of graphene aerogels, Small. 12 (2016) 1702–1708. https://doi.org/10.1002/smll.201503524.

[45] G. Gorgolis, C. Galiotis, Graphene aerogels: A review, 2D Mater. 4 (2017). https://doi.org/10.1088/2053-1583/aa7883.

[46] L.W. Hrubesh, R.W. Pekala, Dielectric properties and electronic applications of aerogels, in: Y.A. atta (Ed.), Sol-gel processing andapplications, Springer, Boston, MA, 1994: pp. 363–367. https://doi.org/10.1007/978-1-4615-2570-7_31.

[47] Y.K. Akimov, Fields of application of aerogels (review), Instruments and Experimental Techniques 46 (2003) 287–299. https://doi.org/10.1023/A:1024401803057.

[48] H. Seung, A. S., K. Baek, S. Sang, Low dielectric materials for microelectronics, in: M. A. Silaghi (Ed.), Dielectricmaterial, InTech, 2012. https://doi.org/10.5772/51499.

[49] H. Qi, E. Mäder, J. Liu, Electrically conductive aerogels composed of cellulose and carbon nanotubes, J. Mater. Chem. A. 1 (2013) 9714–9720. https://doi.org/10.1039/c3ta11734k.

[50] J. Kim, J. Kwon, M. Kim, J. Do, D. Lee, H. Han, Low-dielectric-constant polyimide aerogel composite films with low water uptake, Polym. J. 48 (2016) 829–834. https://doi.org/10.1038/pj.2016.37.

[51] A.M. Joseph, B. Nagendra, P. Shaiju, K.P. Surendran, E.B. Gowd, Aerogels of hierarchically porous syndiotactic polystyrene with a dielectric constant near to air, J. Mater. Chem. C. (2018) 360–368. https://doi.org/10.1039/c7tc05102f.

[52] X.M. Zhang, J.G. Liu, S.Y. Yang, Synthesis and characterization of flexible and high-temperature resistant polyimide aerogel with ultra-low dielectric constant, Express Polym. Lett. 10 (2016) 789–798. https://doi.org/10.3144/expresspolymlett.2016.74.

[53] P. Breüsch, F. Stucki, T. Baumann, P. Kluge-Weiss, B. Brühl, L. Niemeyer, R. Strümpler, B. Ziegler, M. Mielke, Electrical and infrared dielectric properties of silica aerogels and of silica-aerogel-based composites, Appl. Phys. A Solids Surfaces. 57 (1993) 329–337. https://doi.org/10.1007/BF00332286.

[54] J.K. Yoo, R. Wagle, C.W. Lee, E.Y. Lee, Synthesis of silica aerogel thin sheets and evaluation of its thermal, electrical, and mechanical properties, Int. J. Appl. Ceram. Technol. 16 (2019) 832–842. https://doi.org/10.1111/ijac.13125.

[55] K.E. Swider, C.I. Merzbacher, P.L. Hagans, D.R. Rolison, Synthesis of ruthenium dioxide-titanium dioxide aerogels: Redistribution of electrical properties on the nanoscale, Chem. Mater. 9 (1997) 1248–1255. https://doi.org/10.1021/cm960622c.

[56] D.R. Rolison, B. Dunn, Electrically conductive oxide aerogels: New materials in electrochemistry, J. Mater. Chem. 11 (2001) 963–980. https://doi.org/10.1039/b007591o.

[57] H.P. Wong, B.C. Dave, F. Leroux, J. Harreld, B. Dunn, L.F. Nazar, Synthesis and characterization of polypyrrole/vanadium pentoxide nanocomposite aerogels, J. Mater. Chem. 8 (1998) 1019–1027. https://doi.org/10.1039/a706614g.

[58] J.H. Harreld, B. Dunn, L.F. Nazar, Design and synthesis of inorganic-organic hybrid microstructures, Int. J. Inorg. Mater. 1 (1999) 135–146. https://doi.org/10.1016/S1466-6049(99)00022-7.

[59] M.A. Worsley, P.J. Pauzauskie, T.Y. Olson, J. Biener, J.H. Satcher, T.F. Baumann, Synthesis of graphene aerogel with high electrical conductivity, J. Am. Chem. Soc. 132 (2010) 14067–14069. https://doi.org/10.1021/ja1072299.

[60] H. Zhuo, Y. Hu, Z. Chen, X. Peng, L. Liu, Q. Luo, J. Yi, C. Liu, L. Zhong, A carbon aerogel with super mechanical and sensing performances for wearable piezoresistive sensors, J. Mater. Chem. A. 7 (2019) 8092–8100. https://doi.org/10.1039/c9ta00596j.

[61] F. Guo, Y. Jiang, Z. Xu, Y. Xiao, B. Fang, Y. Liu, W. Gao, P. Zhao, H. Wang, C. Gao, Highly stretchable carbon aerogels, Nat. Commun. 9 (2018) 1–9. https://doi.org/10.1038/s41467-018-03268-y.

[62] B. Lee, S. Lee, M. Lee, D.H. Jeong, Y. Baek, J. Yoon, Y.H. Kim, Carbon nanotube-bonded graphene hybrid aerogels and their application to water purification, Nanoscale. 7 (2015) 6782–6789. https://doi.org/10.1039/c5nr01018g.

[63] Y.N. Kharzheev, Use of silica aerogels in Cherenkov counters, Phys. Part. Nuclei

39 (2008) 107–135. https://doi.org/10.1134/s1063779608010085.

[64] M.F. L'Annunziata, Radiation physics and radionuclide decay, in:M.F. L'Annunziata (Ed.), Handbook ofradioactivityanalysis, Elsevier, Oxford, 2012: pp. 1–162. https://doi.org/https://doi.org/10.1016/C2009-0-64509-8.

[65] M. Tabata, I. Adachi, H. Kawai, M. Kubo, T. Sato, Recent progress in silica aerogel Cherenkov radiator, Phys. Procedia. 37 (2012)42–649. https://doi.org/10.1016/j.phpro.2012.02.410.

[66] L. Cremaldi, D.A. Sanders, P. Sonnek, D.J. Summers, J. Reidy, A cherenkov radiation detector with high density aerogels, IEEE Trans. Nucl. Sci. 56 (2009) 1475–1478. https://doi.org/10.1109/TNS.2009.2021266.

[67] T.A. Birks, M.D.W. Grogan, L.M. Xiao, M.D. Rollings, R. England, W.J. Wadsworth, Silica aerogel in optical fibre devices, 2010 12th Int. Conf. Transparent Opt. Networks, Ict. 2010. (2010) 1–4. https://doi.org/10.1109/ICTON.2010.5549052.

[68] K. Tsubaka, T. Kamae, H. Yokoyawa, M. Yokoyawa, K. Sonoda, Optical fiber with silica aerogel cladding,Panasonic Electric Works Co Ltd. EP0779523A3, 1996.

[69] R. Hui, M. O'Sullivan, Fundamentals of optical devices, in: Fiber Optic Measurement Techniques, 2009: pp. 1–128. https://doi.org/10.1016/b978-0-12-373865-3.00001-x.

[70] C.I. Merzbacher, S.R. Meier, J.R. Pierce, M.L. Korwin, Carbon aerogels as broadband non-reflective materials, J. Non.-Cryst. Solids. 285 (2001) 210–215. https://doi.org/10.1016/S0022-3093(01)00455-0.

[71] S.R. Meier, M.L. Korwin, C.I. Merzbacher, Carbon aerogel: a new nonreflective material for the infrared, Appl. Opt. 39 (2000) 3940-3944. https://doi.org/10.1364/ao.39.003940.

[72] J.C.H. Wong, H. Kaymak, P. Tingaut, S. Brunner, M.M. Koebel, Mechanical and thermal properties of nanofibrillated cellulose reinforced silica aerogel composites, Microporous Mesoporous Mater. 217 (2015) 150–158. https://doi.org/10.1016/j.micromeso.2015.06.025.

[73] T. Ficker, Young's modulus of elasticity in student laboratories, Phys. Educ. 34 (1999) 376–383. https://doi.org/10.1088/0031-9120/34/6/407.

[74] J. Fricke, ed., Aerogels, Springer, Berlin, Heidelberg, 1986.

https://doi.org/10.1007/978-3-642-93313-4.

[75] J. Lei, J. Hu, Z. Liu, 116. Mechanical properties of silica aerogel - A molecular dynamics study, 2013 World Congr. Adv. Struct. Eng. Mech. (2013) 778–785.

[76] K.E. Parmenter, F. Milstein, Mechanical properties of silica aerogels, J. Non.-Cryst. Solids. 223 (1998) 179–189. https://doi.org/10.1016/S0022-3093(97)00430-4.

[77] K.A.D. Obrey, K. V. Wilson, D.A. Loy, Enhancing mechanical properties of silica aerogels, J. Non.-Cryst. Solids. 357 (2011) 3435–3441. https://doi.org/10.1016/j.jnoncrysol.2011.06.014.

[78] T. Woignier, J. Primera, A. Alaoui, P. Etienne, F. Despestis, S. Calas-Etienne, Mechanical properties and brittle behavior of silica aerogels, Gels. 1 (2015) 256–275. https://doi.org/10.3390/gels1020256.

[79] C. Li, X. Cheng, Z. Li, Y. Pan, Y. Huang, L. Gong, Mechanical, thermal and flammability properties of glass fiber film/silica aerogel composites, J. Non.-Cryst. Solids. 457 (2017) 52–59. https://doi.org/10.1016/j.jnoncrysol.2016.11.017.

[80] S. He, X. Chen, Flexible silica aerogel based on methyltrimethoxysilane with improved mechanical property, J. Non.-Cryst. Solids. 463 (2017) 6–11. https://doi.org/10.1016/j.jnoncrysol.2017.02.014.

[81] N. Leventis, S. Mulik, X. Wang, A. Dass, V.U. Patil, C. Sotiriou-Leventis, H. Lu, G. Churu, A. Capecelatro, Polymer nano-encapsulation of templated mesoporous silica monoliths with improved mechanical properties, J. Non.-Cryst. Solids. 354 (2008) 632–644. https://doi.org/10.1016/j.jnoncrysol.2007.06.094.

[82] N. Leventis, A. Palczer, L. McCorkle, G. Zhang, C. Sotiriou-Leventis, Nanoengineered silica-polymer composite aerogels with no need for supercritical fluid drying, J. Sol-Gel Sci. Technol. 35 (2005) 99–105. https://doi.org/10.1007/s10971-005-1372-7.

[83] O.A. Madyan, M. Fan, L. Feo, D. Hui, Physical properties of clay aerogel composites: An overview, Compos. B Eng. 102 (2016) 29–37. https://doi.org/10.1016/j.compositesb.2016.06.057.

[84] O.A. Madyan, M. Fan, L. Feo, D. Hui, Enhancing mechanical properties of clay aerogel composites: An overview, Compos. Part B Eng. 98 (2016) 314–329. https://doi.org/10.1016/j.compositesb.2016.04.059.

[85] K. Haraguchi, H.-J. Li, Mechanical properties and structure of polymer–clay

nanocomposite gels with high clay content, Macromolecules. 39 (2006) 1898–1905. https://doi.org/10.1021/ma052468y.

[86] W. Chen, H. Yu, Q. Li, Y. Liu, J. Li, Ultralight and highly flexible aerogels with long cellulose I nanofibers, Soft Matter. 7 (2011) 10360. https://doi.org/10.1039/c1sm06179h.

[87] P. Liu, T.Q. Tran, Z. Fan, H.M. Duong, Formation mechanisms and morphological effects on multi-properties of carbon nanotube fibers and their polyimide aerogel-coated composites, Compos. Sci. Technol. 117 (2015) 114–120. https://doi.org/10.1016/j.compscitech.2015.06.009.

[88] C. Jiao, J. Xiong, J. Tao, S. Xu, D. Zhang, H. Lin, Y. Chen, Sodium alginate/graphene oxide aerogel with enhanced strength–toughness and its heavy metal adsorption study, Int. J. Biol. Macromol. 83 (2016) 133–141. https://doi.org/10.1016/j.ijbiomac.2015.11.061.

[89] A. Feinle, N. Hüsing, Mixed metal oxide aerogels from tailor-made precursors, J. Supercrit. Fluids. 106 (2015) 2–8. https://doi.org/10.1016/j.supflu.2015.07.015.

[90] A. Benad, F. Jürries, B. Vetter, B. Klemmed, R. Hübner, C. Leyens, A. Eychmüller, Mechanical Properties of Metal Oxide Aerogels, Chem. Mater. 30 (2018) 145–152. https://doi.org/10.1021/acs.chemmater.7b03911.

[91] M. Gao, B. Liu, P. Zhao, X. Yi, X. Shen, Y. Xu, Mechanical strengths and thermal properties of titania-doped alumina aerogels and the application as high-temperature thermal insulator, J. Sol-Gel Sci. Technol. 91 (2019) 514–522. https://doi.org/10.1007/s10971-019-05057-5.

[92] M. Zhong, Z. Fu, L. Yuan, H. Zhao, J. Zhu, Y. He, C. Wang, Y. Tang, A solution-phase synthesis method to prepare Pd-doped carbon aerogels for hydrogen storage, RSC Adv. 5 (2015) 20966–20971. https://doi.org/10.1039/C4RA16505E.

[93] E. Moretti, F. Merli, E. Cuce, C. Buratti, Thermal and acoustic properties of aerogels: preliminary investigation of the influence of granule size, Energy Procedia. 111 (2017) 472–480. https://doi.org/10.1016/j.egypro.2017.03.209.

[94] T. Yang, X. Xiong, M. Venkataraman, R. Mishra, J. Novák, J. Militký, Investigation on sound absorption properties of aerogel/polymer nonwovens, J. Text. Inst. 110 (2019) 196–201. https://doi.org/10.1080/00405000.2018.1472540.

[95] K. Matsumoto, K. Ohmori, S. Abe, K. Kanamori, K. Nakanishi, Ultrasound propagation in dense aerogels filled with liquid ^{4}He, J. Phys. Conf. Ser. 400 (2012).

https://doi.org/10.1088/1742-6596/400/1/012045.

[96] Y. Xie, J. Beamish, Ultrasonic properties of silica aerogels at low temperatures, Phys. Rev. B. 57 (1998) 3406–3410. https://doi.org/10.1103/PhysRevB.57.3406.

[97] J. Gross, J. Fricke, Ultrasonic velocity measurements in silica, carbon and organic aerogels, J. Non.-Cryst. Solids. 145 (1992) 217–222. https://doi.org/10.1016/S0022-3093(05)80459-4.

[98] S. Ramakrishna, L. Tian, C. Wang, S. Liao, W.E. Teo, 6 - Safety testing of a new medical device, in: S. Ramakrishna, L. Tian, C. Wang, S. Liao, W.E. Teo (Eds.), Med. Devices, Woodhead Publishing, 2015: pp. 137–153. https://doi.org/https://doi.org/10.1016/B978-0-08-100289-6.00006-5.

[99] V. Narayan, 19 - Alternate antioxidants for orthopedic devices, in: S.M. Kurtz (Ed.), UHMWPE Biomaterials Handbook, William Andrew Publishing, Oxford, 2016: pp. 326–351. https://doi.org/https://doi.org/10.1016/B978-0-323-35401-1.00019-3.

[100] V.R. Sastri, 4 - Material requirements for plastics used in medical devices, in: V.R. Sastri (Ed.), Plast. Med. Devices, William Andrew Publishing, Oxford, 2014: pp. 33–54. https://doi.org/https://doi.org/10.1016/B978-1-4557-3201-2.00004-5.

[101] V. Vitcheva, Biological sata in the light of toxicological risk assessment, in: D. Neagu, A.N. Richarz (Eds.), Big data in predictive toxicology, The Royal Society of Chemistry,2020, pp: 38–68. https://doi.org/10.1039/9781782623656-00038.

[102] J. Hadley, J. Hirschman, B.I. Morshed, F. Sabri, RF Coupling of Interdigitated Electrode array on aerogels for in vivo nerve guidance applications, MRS Adv. 4 (2019) 1237–1244. https://doi.org/10.1557/adv.2019.141.

[103] F.P. Soorbaghi, M. Isanejad, S. Salatin, M. Ghorbani, S. Jafari, H. Derakhshankhah, Bioaerogels: Synthesis approaches, cellular uptake, and the biomedical applications, Biomed. Pharmacother. 111 (2019) 964–975. https://doi.org/10.1016/j.biopha.2019.01.014.

[104] R. Mavelil-Sam, L.A. Pothan, S. Thomas, Polysaccharide and protein based aerogels: An introductory outlook, in: S. Thomas, L.A. Pothan, R. Mavelil-Sam (Eds.) Biobased aerogels: Polysaccharideand protein-based Materials, The Royal Society of Chemistry, 2018: pp. 1–8. https://doi.org/10.1039/9781782629979-00001.

[105] A. Veronovski, Ž. Knez, Z. Novak, Preparation of multi-membrane alginate

aerogels used for drug delivery, J. Supercrit. Fluids. 79 (2013) 209–215. https://doi.org/10.1016/j.supflu.2013.01.025.

[106] G. Orive, A.M. Carcaboso, R.M. Hernández, A.R. Gascón, J.L. Pedraz, Biocompatibility evaluation of different alginates and alginate-based microcapsules, Biomacromolecules. 6 (2005) 927–931. https://doi.org/10.1021/bm049380x.

[107] S. Fernández, A. León, F. Gude, M. Castaño, Biocompatibility of agarose gel as a dermal filler: Histologic evaluation of subcutaneous implants, Plast. Reconstr. Surg. 120 (2007) 1161–1169. https://doi.org/10.1097/01.prs.0000279475.99934.71.

[108] S. Rodrigues, M. Dionísio, C.R. López, A. Grenha, Biocompatibility of Chitosan Carriers with Application in Drug Delivery, J. Funct. Biomater. 3 (2012) 615–641. https://doi.org/10.3390/jfb3030615.

[109] S.B. Rao, C.P. Sharma, Use of chitosan as a biomaterial: Studies on its safety and hemostatic potential, J. Biomed. Mater. Res. 34 (1997) 21–28. https://doi.org/10.1002/(SICI)1097-4636(199701)34:1<21::AID-JBM4>3.0.CO;2-P.

[110] A. Ubeyitogullari, S. Brahma, D.J. Rose, O.N. Ciftci, In vitro digestibility of nanoporous wheat starch aerogels, J. Agric. Food Chem. 66 (2018) 9490–9497. https://doi.org/10.1021/acs.jafc.8b03231.

[111] D.D. Lovskaya, A.E. Lebedev, N. V. Menshutina, Aerogels as drug delivery systems: In vitro and in vivo evaluations, J. Supercrit. Fluids. 106 (2015) 115–121. https://doi.org/10.1016/j.supflu.2015.07.011.

Aerogels I: Preparation, Properties and Applications
Materials Research Foundations 84 (2020) 201-213

Materials Research Forum LLC
https://doi.org/10.21741/9781644900994-8

Chapter 8

Tailor-Made Aerogels

Pallavi Jain[1], Sapna Raghav[2], Praveen Kumar Yadav[3], Dinesh Kumar*,[4]

[1]Department of Chemistry, SRM Institute of Science & Technology, Delhi-NCR Campus, Modinagar-210204, India

[2]Department of Chemistry, Banasthali Vidyapith, Banasthali, Tonk 304022, Rajasthan, India

[3]CSIR-National Physical Laboratory, Dr K.S. Krishnan Marg, New Delhi-110012, India

[4]*School of Chemical Sciences, Central University of Gujarat, Gandhinagar, India

*dinesh.kumar@cug.ac.in

Abstract

This chapter emphasizes modern progress in aerogel researches and its technological implementation. The methods that allow introducing required characteristics in aerogels to meet the necessary requirements in their implementations are given prime attention. The determining factors of aerogels regarding the already present and capable implementation areas are provided in brief. Many customizing techniques like modulating the pore structure, changing the surface, coating of the surface, and post-treatment is described by the outcomes of the last ten years. Regarding the commercial uses of aerogels and its products, an unbroken view of industrial aerogel suppliers is provided and a discussion of plausible substitute sources for raw materials and precursors. Last, the chapter summarizes opinions and potential points regarding the aerogels.

Keywords

Tailor-made Aerogels, Pore Structure, Surface Coating, Post-Treatment, Technical Implementations

Contents

1.   Introduction

Aerogels are defined as materials having high porosities along with low densities and large inner surface areas [1]. One main factor of aerogels is that all elements present within the nanoscale regime given a distinct permutation of attributes that make them potential elements for many implementations like catalysis, drug delivery, adsorbents, thermo-receptors, sensors or piezoelectrics. Kistler prepared the first aerogel in 1931, but the modification of its implementations started in the early 1970s in consideration with the quick sol-gel process development. After that, various chemical constituents like pure inorganic gels, carbides, metal-doped or mixed oxides like inorganic-organic hybrids, organic polymeric gels like polyurethane (PU), resorcinol/methanol, conducting polymers, polyamide, and many others, carbon aerogels (CAgs), graphene aerogels (GAgs), are developed.

Aerogels are considered being solid materials having very little weight and are prepared at a large scale in industries. They have interested researchers since its advent [2]. They have low-density and have a highly porous structure. They are prepared by drying wet gels under specified conditions, especially those sustaining their volume.The gels can be prepared in the form of monoliths or particles or varied structures. Based on the category and mechanical strength of the gels, varying processes of drying are considered and implemented. Usually, a supercritical fluid is substituted for the solvent, which is being filled in the pores, and later, it is vented out. The forces acting on the capillaries are removed, which destroys the aerogel nanostructure during evaporative drying. This outcome can also be brought about with the help of freeze-drying, the specified condition is that the gel structure can sustain the growth of ice-crystal or capillary forces. The collapsing of the pores can also be prevented in gels that have developed surfaces that show spring-back during evaporative drying. The gel produced initially takes from any inorganic or organic precursor, or their mixtures can form a stable 3D network within a potential solvent. So, there are two degrees of freedom during the production of aerogels, first being the initial aerogel prepared, second requires its processing. Because of these variations, there is still no generality in the definition of aerogel.

According to IUPAC, an aerogel is, "a gel comprising a microporous solid in which the dispersed phase is a gas" [3]. Only zeolites and microporous glass are considered suitable for this definition. So, to define almost all categories of aerogels, two new definitions are devised based on their processes of production. One, aerogel is a nonfluid colloidal network, has an open structure and a polymer network which is spread across the whole volume with the help of a gas, which is prepared by eliminating the agents of swelling from a gel with no significant reduction in volume or compaction of a network. The second definition stresses the final properties that are, aerogels are solids that have very little density and large specific area and comprise a reasonable open porous 3D network having loosely bound fibers [4]. Both statements, when combined, give a full-fledged meaning to the aerogels based on its production and properties. Aerogel is an open nonfluid polymeric or colloidal 3D network that consisting of loosely bound fibers that is spread across its volume by gas and thus shows large specific areas and low density. Aerogels are prepared by eliminating all the swelling agents from the gel prepared initially with no significant decrease in volume or compaction in a network.

2.   Existing and potential applications of aerogels

Being the lightest and the possibility of varying attributes and methods of preparation of distinct aerogels, aerogels have a great capability in a lot of implementations in which huge pore volume and large surface area play important parts. At present, the production of aerogels at an industrial scale is restricted to systems based on silica, carbon, and a few amounts of organic aerogels. The main implementation of aerogels is in thermal insulation. This is because of aerogels being the best insulating material for thermal insulation apart from the vacuum insulation board [5]. In recent times, BASF has taken measures towards the usage of organic aerogels on a commercial scale. The huge attraction towards biopolymer-based aerogels allows implementations in life science like tissue engineering, drug delivery, food, and cosmetics. Aerogels depending on carbon [6], conductive, give way to a lot of new potential in their implementations like batteries and supercapacitors [7]. A noble metal derived aerogels and aerogels based on a chalcogenide, and similar hybrids are studied deeply for further uses [8]. The recent modifications in a surface that enable customization of aerogels hydrophobicity have given rise to applications for the environment [9]. The capability to alter hydrophobicity allows significant implementations in separation technologies. The usage of graphene oxide in its reduced form as membranes was proposed so as to eliminate oils from water surfaces. Contrary to the old experiments, it was proven that the floating oil could be consequently absorbed and released out of the aerogel with a simple peristaltic pump. For many implementations, some properties remained fixed that is, pore structure, functional surface, and size of the groups for modification of aerogels.

2.1   Pore engineering

Three major amounts that attribute to the pore structure of the aerogels have been reported. They include specific surface area, density, and specific pore volume. The determination of the initial two is done by nitrogen adsorption/desorption isotherms. Mostly, the nitrogen porosimetry permits the rebuild of the size of the pores and their distribution. Pore size distribution is unimodal. It can entail a radius of range between 5 and 20 nm. These attributes depict a common biopolymer or metal oxide aerogel. Many parameters of the process are affected by these quantities. Many implementations depend on the large surface area of aerogels besides a specific structure of pore, especially with a pore size distribution of choice. In the discussion below, we review pore architecture and the modulations that can be done on the structure of the pore.

2.2   Customizable surface and coating

The most important prospect to change the functional attributes of the aerogel with no significant change in the structural characteristics of the materials is by coating. The coating is done using a layer comprising a material having potential attributes like a polymer. The main problem is to complete the coating in a way that only a thin homogeneous layer is prepared by casting the whole surface of aerogel with no pore blockage or any additional growth in bulk density. This method leads to varying last states of the aerogels that are modified.

One is that either only the thin layer of polymer is coated on the exterior surface. And other properties like chemical characteristics and structure of the pristine aerogel are conserved. Second, the interior structure, which includes the pore surfaces and the intersection among them, and it is concealed with a coating material's homogenous layer having a thickness of few nanometers. Based on the implementation, both final outcomes can be attained.

From a molecular perspective, someone can differentiate many categories of interactions among the aerogel surface and coating. The coating material can either be fastened to the surface with the help of interactions that are noncovalent. They can chemically react with the chosen functional groups of the aerogel, giving covalent bonds between both of the materials. In chemical reactions, the homogeneity and thickness of the layer based on the amount of functional groups present on the surface of aerogel and the coating material's capability to self-polymerization whereas non-specific associations lead to a coverage of total surface based at most on the process parameter, for example, the coating material's quantity, deposition cycle's number and time of coating. We suggest a differentiation among two cases and, with the help of the terms coated aerogel and coating for non-specific noncovalent associations and the expressions, aerogel, which is modified for layers grafted chemically. Both the processes (surface modification and coating) can happen in a liquid or gas or supercritical phase when in touch with a dried aerogel and wet aerogel consequently.

2.3   Hybrid aerogels (HAgs): Influence of the sol-gel process on final properties

The common word **hybrid aerogels (HAgs)** shows that at least two predecessors are engaged in the step of gelation. In most cases, this entails scenarios ranging from those where predecessors hold their uniqueness in the aerogel obtained as the product to excel the chemical homogeneity with the help of a variety of intermediate cases. In addition, aerogel backbones can be crystalline or amorphous, comprising amorphous or crystalline domains. Without taking into regards, the chemical homogeneity level, HAgs show a pleasing and a new class of materials which take from the single components and exhibit a synergy permitting implementations that would not be by synergy only. Many features of this topic were discussed partially [10, 11]. The process of transformation from sol to gel of mixed systems is more difficult and complex in contrast to that of single aerogels. Keeping the synergy as the outcome in the aerogel properties, the reactivities of the predecessors used, are so different that without careful governance over hydrolysis, and hetero- and homo-condensation rates only ambiguous structures can be produced [12].

Therefore, chemical reactivity firmly decides the final aerogel properties, specifically mechanical and catalytic ones. So here, we discuss the approaches or methods of aerogels. One method to improvise the homogeneity of chemicals is by spatial separation of the precursors at the hydrolysis step, and the condensation is the next step [13]. This method enhances the catalytic activity of TiO_2/SiO_2 aerogels by three factors in contrast to non-prehydrolyzed samples in the isomerization of 1-butene. Depending on higher cis/trans isomerization ratios of products for the non-prehydrolyzed aerogels, it was deduced that there was segregation of silica and titania domains. But the segregation in the mixed aerogels should not be viewed in a negative view. It was concluded that a series of TiO_2/SiO_2 aerogels having photocatalytic activities toward salicylic acid degradation in water. Pre-hydrolyzed that is, aerogels having more chemical homogeneity showed lesser catalytic activity. The separate and easily available anatase particle's presence led to the high activity [14].

Second approach to govern the homogeneity of chemicals like mixed oxide aerogels by inhibiting the rate of reaction like ligand complexation with chelating ligands or the occurrence of sol-gel processes in a highly viscous medium. The homogeneity can also be maintained by engaging the precursors with many metal centers. The customized precursors, even though they are restricted in their stoichiometry, provide a lot of governance over the chemical homogeneity and also hold functional groups. Apart from the oxide aerogels, biopolymer aerogels can be cross-linked using covalent or noncovalent bonding. Similarly, like oxide aerogels, a double polymer can make domains of different sizes and can also be distributed statistically.

Polymers like starch, pectin, cellulose, gelatin, lignin, and chitosan can be mixed with one another and withinorganic substrates to make hybrid aerogels [15-17]. Because even more HAgs are known in the food science, we expect this area to grow significantly over the next few years. Owing to the less severe conditions of supercritical drying and the inertness of supercritical-CO_2, a huge range of insoluble materials can be mixed at the step of gelation.

3.   Applications of Tailor-made aerogels

These days more focus has been given to customizable aerogels that are formed using additional components that are engaged in particular reactions involving catalysts. A new heterogeneous catalyst, iron-doped on alumina was developed for degrading Rhodamine B. The aerogel catalyst was prepared by a sol-gel approach, which is easier and prevents leaching of the active components. Also, the catalyst controls the highly active reaction for the Rhodamine B degradation within a decent range of pH values. The Fe-O-Al bond can be increased using heat treatment. The effects of various factors like dosages of hydrogen peroxide, pH values, reaction temperature, and concentration on Rhodamine B degradation were researched. Fourier transform infrared spectroscopy (FTIR) and inductively coupled plasma mass spectroscopy (ICP) analysis indicate that the catalyst entails lesser leaching of iron, higher catalytic efficiency, and good structural stability after its usage. The biodegradability and biocompatibility of starch, a natural polysaccharide derived substance, permits the usage of starch as a microsphere in form of a carrier or as a chemical for science implementations. Although, the latest techniques of making starch microspheres use different methods like chemical cross-linkers and different drying methods that lead to a decrease in the degradability problems of the matrix and lower specific surface areas along with chemical loading capacities. Here, corn starch aerogel microspheres are a specialized class of materials that are nanoporous and were made by the addition of supercritical drying and emulsion-gelation method. The starch aerogels obtained had characteristics determined with the help of nitrogen adsorption-desorption magnitudes, CHN elemental analyses, and helium pycnometer along with scanning electron microscopy and thermogravimetry. Aerogel textural attributes were affected basically by the temperature of gelation used. The morphology of particles was dependent on the three parameters of processing research.

Graphene acts as a host, and it has potential in a lot of implementations like holding sulfur in Lithium-Sulfur (Li-S) batteries. Still, there are many challenges to overcome the modulation of graphene's nanostructure so as to increase the performance. The self-closure Gags with builtin baffle plates (SGA) were made by an added method that engages an electrostatic assembly, polydopamine, hydrothermal fixing, annealing, and coating. The promotion of electron transport, enhancing the sulfur loading, and dissolution confinement are done by using as-made SGA as the host material for the Li-S batteries.

For energy-saving windows, silica aerogels with low thermal conductivity and high optical transparency are fit for the purpose. These applications need exclusive governance of the microstructure. A number of silica aerogels having constant density and composition, but varied attributes and microstructures were made by changing the amount of the tetramethoxysilane (TMOS). Initially, the decrease in the mean pore size and thermal conductivity of the silica aerogels was noticed with the decrease of the amount of the TMOS added before bending out and then ultimately enhanced. On the other hand, properties like specific surface area, pore-volume, and light transmittance show reverse a change in a trend. In comparison, the age-old initial placing of TMOS, the properties like a ratio of transparency, and specific surface area increase from 71% (550 nm) to 88% (550 nm) and from 845to 1060 m^2/g, respectively. Also, silica aerogel's thermal conductivity decreases from 24.6 to 20.2 mW $m^{-1}K^{-1}$) with the mean pore size reduces from 20.7 to 16.6 nm. This technique shows a novel approach to make the structure and tailor properties of silica aerogels [18].

Methyl modified silica aerogels are made with the help of methyltriethoxysilane (MTES) and tetraethylorthosilicate (TEOS) as co-precursor using a double step sol-gel reaction preceded by supercritical drying and exchange of solvent. They have customized dual-modal pore structure comprising mesopores from the stacking of TEOS-based clusters and small pores through stacking and filling of minute MTES based clusters. The $Si-CH_3$ groups act as a substitute to Si-OH groups on the surface of aerogel, and hence they have high hydrophobicity with a contact angle of 148°. The hydrophobicity and double modal pore structure give the $Me-SiO_2$ aerogels an excellent adsorbing capability of quality factor Q up to 23 g for C_2H_5OH and selectively adsorbing oil from water.

A lot of changes in the pH levels, protein type, and concentration levels were determined in the observation of protein derived aerogels from heat-set whey and egg white. The pH was customized, and this led to the brittle hydrogels with already known syneresis in the pH region about the isoelectric point. Aerogels were fragile and didn't have the typical large specific interior surface area of aerogels. In an acid, hydrogels were rigid but also had fragility. The specific area was large. In an alkaline medium, the hydrogels were elastic and soft. Hence the aerogels obtained as products were stable and had a large surface area. The addition of calcium chloride or sodium chloride made the same changes on the protein solution, like pH lessening. The supercritical drying is implemented to sodium caseinate with different concentrations of protein. Last, in the method of emulsion to prepare aerogels derived from proteins useful for microencapsulation. All major protein types make individual spherical aerogel particles with easy to differentiate aerogel attributes in contrast to the aerogel monolith. The protein aerogel particle attributes can be customized according to the usage and desires in food, pharmaceutical, and cosmetic implementations [19].

Co-Pt/MgO-Al$_2$O$_3$ is an aerogel catalyst bimetallic and is prepared using a sol-gel process followed by supercritical drying. The catalysts were determined by x-ray diffraction, high-angle annular dark-field scanning transmission electron microscopy (STEM-HAADF), temperature-programmed reduction of hydrogen (H$_2$-TPR), thermal analysis technique (TG/DSC), field emission scanning electron microscopy (FESEM) and their performances as catalysts in methane oxidative carbon dioxide (CO$_2$) reformation were checked. The hydrogen spillover effect among platinum and cobalt increased the catalyst reducibility, whereas the strong interaction of metal support within the bimetallic catalyst restricted the metal particle agglomeration. Pt/Co ratio has been playing a major part in the surface metal species in existing, leading to varied performance of the catalyst. The Pt/Co ratio was at an optimum, and a 50 % higher activity was observed in terms of methane conversion in contrast to monometallic Pt or Co aerogel catalysts prepared. These catalysts also exhibited a high resistance towards the formation of inactive carbon. The temperature of oxidation of the carbon species that were added on the Pt/Co catalyst used was around 275 °C and had no graphite carbon or filamentous carbon was investigated. It revealed that the inactive carbon formation was reduced owing to the synergy present between Co and Pt and the effect of SMSI [20].

Tetraethoxysilane (TES) and methyltrimethoxysilane (MTMS) are utilized as precursors to prepare a couple of two-precursor aerogel using ambient pressure drying. As the concentration was increased of NH$_3$.H$_2$O, the density reduced at first and then increased significantly. On the other hand, porosity exhibited a reverse trend while the size of the cluster was reduced and the pores quantity increased drastically. These distinctions

Materials Research Forum LLC
https://doi.org/10.21741/9781644900994-8

caused variations in the couple aerogel's thermal properties. With the $NH_3.H_2O$ amounts enhancing, the thermal conductivity approximated linear decrease to 0.027 $Wm^{-1}k^{-1}$, the specific heat demonstrated algorithmic reduction, and the diffusivity using heat had the same change in a trend like that of porosity. These things proved that the coupled aerogel's thermal properties could be adjusted [21].

Nitrogen-doped graphene aerogels and ultralight are prepared using a hydrothermal method in the environment of dopamine, graphene, L-arginine. Dopamine is said to be useful for graphene functionalization. It also attached nitrogen atoms on the graphene sheets with the help of pyrolysis. The pH values affect the ultralight and graphene aerogels and on their preparation. By customizing the amount of dopamine and L-arginine in the precursor mixture, the aerogel of density 2.54 mg/cm^3 can be formed. This aerogel shows a high absorption capacity for various oils owing to less density,large surface area, and N-doping [22].

Iron alginate aerogel, which is a crosslinked polymer, were synthesized and filled with ibuprofen with the help of adsorptive deposition from supercritical-CO_2. Formulations were synthesized and co-infused using ascorbic acid and ibuprofen. The ibuprofen excretion from the iron alginate of pH 7.4 is more rapid than that of pH 2.0. By lowering the bead size and using a greater G content alginate, the rate of release can be enhanced slightly. 3D GAg, which is a self-assembled anode material with some defects in the surface, has been made successfully with the help of a superficial approach using precursor-like graphene oxide. The textural properties and morphology of prepared GAg were evaluated by Scanning electron microscope (SEM) and transmission electron microscope (TEM). 3D porous network with enhanced defect density, along with the considerable electrical conductivity, gave the output in the excellent electrochemical performance of the as-made GAg anodes in Li-ion batteries [23].

Compressed and supercritical fluid technology gives a powerful device to design particles and in engineering. Aerogels are nanoporous materials sustaining many global classes attribute. New kinds of starch-based aerogels were produced and were investigated. They had their specific target in life sciences like tissue engineering, cosmetics, agriculture, pharmaceutics, and biotechnology. The effect of micronization on the kinetics of the supercritical drying of the wet gel was also investigated. Thus, the aerogels produced were checked for their usage as active agent carries regarding the capacity of filling with the help of ketoprofen as a model compound [24].

Silicon oxide carbon-ceramic (SiOC) aerogels have been prepared from a linear polysiloxane crosslinked with divinylbenzene using the reaction of hydrosilylation in the presence of platinum catalyst and a solvent in the form of acetone. The wet gels obtained are subjected to aging using the drying method under supercritical-CO_2. The aerogels that are pre-ceramic are then put through pyrolysis at 1000 °C in the presence of a maintained argon environment. The SiOC prepared had 43 % by weight of free carbon, which is separated within the amorphous SiOC matrix [25-27].

Conclusions

The aerogel field has seen faster developments and modifications, and it still has a potential to grow further. The important factor is that the development should be that it doesn't significantly increase the cost amount for that might be a drawback. The cost prices should be economical for usage in the academic and the industrial fields. Owing to the huge number of single aerogel particles that are obtainable, consequent development should be solely based on implementation. Major attributes of product and limitations are to be investigated while studying a particular aerogel or tailor-made aerogel. This can be done by partnership and cooperation among academic researchers and industrialists. The following topics should be investigated further to advance this field.

> ➢ Carbon Aerogel
> ➢ Hybrid Aerogel
> ➢ Aerogel for segregation
> ➢ Process development
> ➢ Substitute raw materials for the synthesis of aerogels
> ➢ Surface modulation

Acknowledgments

Dr. Pallavi Jain is grateful to SRM Institute of Science & Technology, Modinagar and Dinesh Kumar is also thankful DST, New Delhi, for financial support to this work (sanctioned vide project Sanction Order F. No. DST/TM/WTI/WIC/2K17/124(C).

References

[1]   M.A. Aegerter, N. Leventis, M.M. Koebel, Aerogels handbook, Springer, New York, 2011.

[2]   S. Mulik,C.Sotiriou-Leventis, G. Churu,H. Lu, N. Leventis, Crosslinking 3D assemblies of nanoparticles into mechanically strong aerogels by surface-initiated free-radical polymerization, Chem. Mater. 20 (2008) 5035–5046. https://doi.org/10.1021/cm800963h

[3] D.H. Everett, Manual of symbols and terminology for physicochemical quantities and units, appendix ii: definitions, terminology and symbols in colloid and surface chemistry, Pure Appl. Chem. 31 (1972) 577–638.https://doi.org/10.1351/pac197231040577

[4] L. Falk,A. Nikita, S.Christian,P. Antje, R. Thomas, Bacterial cellulose aerogels: from lightweight dietary food to functional materials, Funct. Mater. Renew. Sources 1107 (2012) 57–74. https://doi.org/10.1021/bk-2012-1107.ch004

[5] M. Venkataraman,R. Mishra, T.M. Kotresh, J. Militky, H.Jamshaid, Aerogels for thermal insulation in high-performance textiles, Text. Prog. 48 (2016) 55–118. https://doi.org/10.1080/00405167.2016.1179477

[6] M.M. Titirici, R.J. White, N.Brun, V.L.Budarin, D.S. Su, Sustainable carbon materials, Chem. Soc. Rev. 44 (2014) 250–290. https://doi.org/10.1039/C4CS00232F.

[7] S. Araby, A.Qiu, R. Wang, Z. Zhao, C.H. Wang, J. Ma, Aerogels based on carbon nanomaterials, J. Mater. Sci. 51 (2016) 9157–9189. https://doi.org/10.1007/s10853-016-0141-z

[8] C. Zhu, D.Du, A. Eychmuller, Y. Lin, Engineering ordered and non ordered porous noble metal nanostructures: synthesis, assembly, and their applications in electrochemistry, Chem. Rev. 115 (16) 8896– 8943. https://doi.org/10.1021/acs.chemrev.5b00255

[9] H. Maleki, Recent advances in aerogels for environmental remediation applications: a review, Chem. Eng. J. 300 (2016) 98– 118. https://doi.org/10.1016/j.cej.2016.04.098

[10] O.A. Madyan, M. Fan, L. Feo, D. Hui, Physical properties of clay aerogel composites: an overview, Compos. B Eng. 102 (2016) 29– 37. https://doi.org/10.1016%2Fj.composites b.2016.06.057

[11] L. Zuo,Y. Zhang, L. Zhang, Y.E. Miao, W. Fan, T. Liu, Polymer/carbon-based hybrid aerogels: preparation, properties and applications, Materials 8 (2015) 6806–6848. https://doi.org/10.3390/ma8105343

[12] A. Feinle, N. Husing, Mixed metal oxide aerogels from tailor-made precursors. J. Supercrit. Fluids 106 (2015) 2–8. https://doi.org/10.1016/j.supflu.2015.07.015.

[13] J.B. Miller, S.T. Johnston, E.I. Ko, Effect of prehydrolysis on the textural and catalytic properties of titania-silica aerogels, J. Catal. 150 (1994) 311–320. https://doi.org/ 10.1006/jcat.1994.1349

[14] K. Brodzik, J. Walendziewski, M. Stolarski, L.V. Ginneken, K. Elst, V. Meynen, The influence of preparation method on the physicochemical properties of titania silica aerogels : part two, J. Porous Mater. 15 (2007) 541–549. https://doi.org/10.1007/s10934- 007-9130-6

[15] S.P. Raman, P. Gurikov, I. Smirnova, Hybrid alginate based aerogels by carbon dioxide induced gelation: novel technique for multiple applications, J. Supercrit. Fluids 106 (2015) 23–33. https://doi.org/10.1016/j.supflu.2015.05.003

[16] N. Pircher, D. Fischhuber, L. Carbajal, C.. Strauß,J.M. Nedelec, C. Kasper, T. Rosenau, F. Liebner,. Preparation and reinforcement of dual-porous biocompatible cellulose scaffolds for tissue engineering, Macromol. Mater. Eng. 300 (2015) 911–924. https://doi.org/10.1002/mame.201500048

[17] A. El Kadib, M. Bousmina, Chitosan bio-based organic-inorganic hybrid aerogel microspheres, Chem. Eur. J. 18 (2012) 8264–8277. https://doi.org/10.1002/chem.201104006.

[18] T. Xia, H. Yang, J. Li, C. Sun, C. Lei, Z. Hua, Y.Zhanga, Tailoring structure and properties of silica aerogels by varying the content of the tetramethoxysilane added in batches, Micropor. Mesopor. Mat. 280 (2019) 20–25.https://doi.org/10.1016/ j.micromeso.2019.01.038

[19] C.Kleemann, I. Selmer, I. Smirnova, U.Kulozik, Tailor made protein based aerogel particles from egg white protein, whey protein isolate and sodium caseinate: Influence of the preceding hydrogel characteristics, Food Hydrocoll., 83 (2018) 365-374. https://doi.org/10.1016/j.foodhyd.2018.05.021

[20] L. Chen, Q. Huang, Y. Wang, H. Xiao, W. Liu, D. Zhang, T. Yang, Tailoring performance of Co-Pt/MgO-Al2O3 bimetallic aerogel catalyst for methane oxidative carbon dioxide reforming: Effect of Pt/Co ratio, Inter. J. Hyd. Energy 44 (2019) 19878-19889. https://doi.org/10.1016/j.ijhydene.2019.05.201

[21] Z. Li, X. Cheng, S. He, X. Shi, H. Yang, H. Zhang, Tailoring thermal properties of ambient pressure dried MTMS/TEOS co-precursor aerogels, Mater. Letters 171 (2016) 91–94. https://doi.org/10.1016/j.matlet.2016.02.025

[22] Y. Xue, W. Pei-wen, L. You-chang, W. Liang, F. Li-Juan, L. Chun-Hu, Dopamine and L-arginine tailored fabrication of ultralight nitrogen-doped graphene aerogels for oil spill treatment, J. Fuel Chem. Technol. 45 (2017) 1230-1235. https://doi.org/10.1016/S1872-5813(17)30055-5

[23] H. Shan, D. Xiong, X. Li, Y. Sun, B. Yan, D. Li, S. Lawes, Y. Cui, X. Sun, Tailored lithium storage performance of graphene aerogel anodes with controlled

surface defects for lithium-ion batteries, Appl. Surface Sci. 364 (2015) 651-659. https://doi.org/10.1016/j.apsusc.2015.12.143

[24] C.A. García-González, I. Smirnova, Use of supercritical fluid technology for the production of tailor-made aerogel particles for delivery systems, J. Supercrit. Fluids 79 (2013) 152–158. https://doi.org/10.1016/j.supflu.2013.03.001

[25] V.S. Pradeep, D.G. Ayana, M. Graczyk-Zajac, G.D. Soraru, R. Riedel, High rate capability of SiOC ceramic aerogels with tailored porosity as anode materials for Li-ion batteries, Electrochim. Acta 157 (2015) 41–45. https://doi.org/10.1016/j.electacta. 2015.01.088

[26] C. Zhao, C. Yu, M. Zhang, J. Yang, S. Liu, M. Li, X. Han, Y.Dong, J. Qiu, Tailormade graphene aerogels with inbuilt baffle plates by charge-induced template directed assembly for high-performance Li-S batteries, J. Mater. Chem. A, 3 (2015) 21842-21848. https://doi.org/10.1039/C5TA05146K

[27] C. Zhao, C. Yu, M. Zhang, J. Yang, S. Liu, M. Li, X. Han, Y.Dong, J. Qiu,Tailor-made graphene aerogels with inbuilt baffle plates by charge-induced template-directed assembly for high-performance Li-S batteries, J. Mater. Chem. A, 3 (2015) 21842-21848.https://doi.org/10.1039/C5TA05146K

Aerogels I: Preparation, Properties and Applications
Materials Research Foundations **84** (2020) 214-229

Materials Research Forum LLC
https://doi.org/10.21741/9781644900994-9

Chapter 9

Aerogels Envisioning Future Applications

Rafael Henrique Holanda Pinto*, Jhonatas Rodrigues Barbosa, Lucas Cantão Freitas, Ivonete Quaresma da Silva de Aguiar, Flávia Cristina Seabra Pires, Ana Paula de Souza e Silva and Raul Nunes de Carvalho Junior

Extraction Laboratory, Faculty of Food Engineering, Program of Post-Graduation in Food Science and Technology, Federal University of Pará, Rua Augusto Corrêa S/N, Guamá, 66075-900, Belém, Pará, Brazil

* rafael_holanda90@hotmail.com

Abstract

Aerogels are nanoporous structures with low thermal conductivity, specific mass and refractive index. These materials can be divided into organic, inorganic and hybrid groups. The advantages of aerogels include their synthesis with raw materials from different natures, such as macronutrients, metal oxides, polymers, carbon allotropes and transition metals. Prospects for the future applications of aerogels focus on the synthesis of functional foods, medicines, impact absorbing materials, catalytic supports, and aerospace components.

Keywords

Aerogels, Nanoporous, Structures, Perspectives, Applications

Contents

1.   Introduction

Aerogels are solid, nanoporous, hollow-space materials that have their own physicochemical characteristics, such as surface areas and pore volumes specific for coating/encapsulation of bioactive compounds or drugs that have certain active principles. In addition, aerogels have low thermal conductivity, specific mass and refractive index. Because of these characteristics, aerogels are structures with potential for use in various segments, including the synthesis of functional foods, thickeners, food stabilizers, medicines, energy storage, environmental cleaning, water treatment filters, fuel cell electrodes and thermal insulation. These materials can be divided into organic, inorganic and hybrid groups [1].

The advantages of aerogels include the synthesis of these materials using different types of raw materials such as carbohydrates, specifically polysaccharides, as well as other macronutrients such as proteins and polymers extracted from biomass, which gives to these materials the nomenclature of bioaerogels, mainly due to the biocompatibility associated with these structures due to low levels of toxicity. Other matrices used in aerogel production include metal oxides, polymers, carbon allotropes and transition metals [1,2].

Different methods are used in the production of aerogels. The best mechanism should be defined considering the three-dimensional formation of these structures. The most applied conventional method in the preparation of aerogels is called sol-gel reaction, which consists of a wet chemical synthesis, with precursor mixing, hydrolysis, polycondensation, gelation, aging and drying steps. Each step is adjusted by parameters such as temperature, medium pH, time, solvent type and precursor concentration [3-6]. From all the steps, drying is the main one, as the solvent removal is expected to preserve the 3D structure of the aerogel. Supercritical fluid technology represents an alternative to solvent removal, once it can maintain the characteristics of nanoporous structures [7].

In this context, this chapter aims to show the future perspectives of aerogels applications from different natures, in order to show the potential of these structures in different research areas, through synthesis techniques and previous studies.

2.   Future applications of bioaerogels

The aerogels synthesis has drawn the interest of different segments due to the characteristics that these materials present which are high surface area, porous structure and low weight [7]. Bioaerogels can be formed through organic aerogels based on nanofiber polysaccharides of cellulosic materials, pectin, β-glucan, chitosan, chitin and starch [7,1] covering bioactive compounds. The interest in the use of renewable, biodegradable and biocompatible raw materials has been increasing, especially in the production areas of functional foods and medicines, in the direction of the synthesis of phytochemical transporters, aiming to improve controlled drug release mechanisms [7].

2.1   Bioaerogels applied as functional foods

Ubeyitogullari and Ciftci [8], in their research, developed the synthesis of nanoporous starch bioaerogels impregnated with phytosterols via supercritical CO_2 under the following conditions: temperature 70-120 °C, pressure 450 bar and volumetric flow rate of 1 L min^{-1}. The impregnated phytosterols presented sizes ranging from 59-87 nm. The impregnation capacity reached 99 mg phytosterols/g bioaerogel. The produced nanoparticles represent a potential functional food, consisting of a consumption alternative from its dissolution in water, facilitating its practicality by changing the nature of the product (coating of phytosterols from hydrophilic substance). The presence of phytosterols in foods is interesting because of the anti hypercholesterolemic activity associated with these substances [9].

2.2   Bioaerogels applied as thickeners and stabilizers

The study by Ubeyitogullari and Ciftci [10] developed the production of bioaerogels from camelin seed mucilage via supercritical CO_2 under the following operating conditions: temperature 40 °C and pressure 100 bar. The bioaerogels presented a surface area of 240

m^2 g^{-1}, with a size of 6 nm, a density of 0.05 $g.cm^{-3}$ and a porosity of 0.94. These materials presented an open and porous 3D fibrillary network structure with good thermal stability, high viscosity due to high polysaccharide concentration, with satisfactory rheological properties, indicating that these materials are potential thickeners and stabilizers for food products.

2.3　Bioaerogels applied as medicines and scaffolding in tissue repair

Bioaerogels have physicochemical and biological properties that enhance their use as medicine, mainly because they have large superficial areas, high surface / volume ratios, and excellent biocompatibility, giving these materials the carrying capacity of bioactive compounds or drugs [1]. Polysaccharide-based bioaerogels have desirable intrinsic characteristics that allow their use in drug production, as well as perspective in the synthesis of tools that simulate extra cellular matrices, consisting of bioaerogels that can be applied in the pharmaceutical industry, as well as in scaffolding in tissue engineering [1,11]. Alginate and chitosan represent polysaccharides that have mucoadhesive properties of interest, which makes these carbohydrates potential materials to be used in medicated bioaerogels in order to prolong the permanence of drug control systems on mucosal surfaces, contributing to the improved drug absorption. Cellulose fiber-based nanoparticulate materials are applicable in tissue repair [1].

Tresoldi et al. [12] synthesized hybrid bioaerogels of chitosan, carboxymethyl cellulose and graphene oxide. These materials can be applied to controlled delivery of 5-fluorouracil under specific temperature and pH conditions. Frindy et al. [13] produced collagen-based bioaerogels that can be applied as a biological dressing and promote cell proliferation.

3.　Future applications of polymeric aerogel

Polymeric aerogel has been widely exploited in the last decade, especially from natural polymeric matrices. Polysaccharides extracted from plants, fungi, algae, animals and lignocellulosic biomass have increasingly taken up space in the development of new technologies [14]. Natural polysaccharides are an inexpensive and abundant option for the production of polymeric aerogel. The polysaccharides such as cellulose, starch, chitosan, alginate, carrageenan and pectin have been successfully applied in the release of drugs and bioactive compounds [15].

Polymeric aerogels have interesting structural and molecular qualities and can be used for purposes other than those already applied. The polymeric aerogels qualities such as nanopores, mechanical strength, structural network composed of molecular interlacings and the possibility of being malleable makes these biomaterials attractive for possible future applications such as impact absorbing materials, catalytic supports, dielectrics for electronic materials and aerospace components.

3.1　Polymeric aerogel as impact absorbing materials

Polymeric cellulose aerogels are quite resistant, thus depending on the type of cellulose various applications can be explored. Nanofibrillated cellulose polymeric aerogels have fibers ranging from 5 to 70 nm and three-dimensional polymer networks can be introduced to increase the surface area. These aerogels are resistant to external forces and can be applied in the production of impact absorbing materials [16,17].

Cellulose nanocrystalline polymeric aerogels are even more interesting for impact absorbing material applications. High mechanical properties are achieved by crystalline network composed of cellulose nanocrystals. In addition, the structure of cellulose nanocrystalline polymeric aerogel is known to be reinforced by high modulus hydrogen bonding of nanowhiskers, improving barrier properties and mechanical properties [18]. These aerogels could be used as impact absorbing materials, such as absorption pad coverings. Several possibilities could be explored, for example, special tools or methods for implanting or extracting artificial joints, accessories and bone grafts.

3.2　Polymeric aerogels used as catalyst supports

Heterogeneous catalysis has been applied in the chemical industries, in energy production, and in the process of environmental remediation. This process occurs on the surface of a catalyst particle, and two types can be identified depending on the type of catalysis, thus they can occur on solid-gas or solid-liquid interfaces for thermal catalysis, and solid-liquid or solid-liquid-gas for electrocatalysis [19].

Most catalysts are nanomaterials, except when enzymes are used; in these cases they are biomolecules with high molecular weight. In cases where the catalysts are nanomaterials, polymeric aerogel can be used as a support material for loading these catalysts [19]. Understanding aerogel structure can help in architectural engineering for adding catalysts. The addition of small molecules as metal catalysts in the polymeric aerogel structure can be considered, since the challenges for the application of this technology may be related to catalytic capacity. Thus, polymeric aerogels could be used as platforms for controlled release of catalysts, managing the catalytic capacity of the process. In addition to small catalyst molecules, others such as enzymes could also be explored and released from catalytic bases. Polymeric aerogels could serve as molecular coupling bases for enzymes. Enzymes could be attached to aerogel by grafting reactions as part of the basic structure of aerogel. These enzyme-grafted polymeric aerogels could be used to remediate reactions in a controlled manner, i.e. these could be applied to all industrial processes such as drug and food production.

3.3 Polymeric aerogels can be used as aerospace components

Technological development and space exploration have both been side by side. Since the beginning of space exploration the production of innovative technologies has been the essence of success. The production of space exploration components, from electronics to food, has undergone several changes. The changes are influenced by the new geopolitical, socioeconomic and environmental reality of the world [20].

With the development of a new mindset in space exploration research, new, more sustainable, inexpensive and versatile features have been applied. Among these features, polymeric aerogel can be used in a variety of ways. Several technologies could be developed based on natural and modified polymeric aerogel. Just remember that every scientific development needs human minds interested in exploring unimaginable challenges. Among the possibilities of polymer aerogel applications, food formulation and sustainable packaging would be interesting [21].

Food to be used by astronauts has been a focus of research since the beginning of space exploration. Most of the food is dehydrated and packed in little reusable containers. As a proposal for polymeric aerogel applications in food, the production of edible packaging would be interesting. The production of packaging with polymeric aerogel has been evaluated, however, it is yet to be accessed under aerospace operating conditions. The production of smart packaging and adaptive space exploration conditions could maintain the quality of food and make it more viable [21].

4. Future applications of carbon aerogel

The use of carbon aerogel (CA) has expanded over the past decade due to the worldwide trend towards green and sustainable appeal. Thus, the development of carbon polymer-based aerogel has attracted broad interest from researchers seeking new materials and applications [22]. The development of new technologies and the production of new hybrid materials will include the development of carbon aerogel.

Carbon aerogels (CAs) are another class of porous, nanostructured carbon materials with high surface areas and large volumes of mesopores obtained by the carbonization of organic aerogels prepared by sol-gel polycondensation of certain organic monomers or by methods without the use of sol-gel chemistry models [23]. Carbon aerogels are materials prepared by pyrolysis of bioerogels and polymeric aerogels. These materials become attractive due to their characteristics such as non-toxicity, low cost and availability. In addition, organic polysaccharide-based aerogels have several beneficial characteristics compared to inorganic aerogels, such as high surface areas and very porous structures, excellent electrical conductivity, high performance over a wide temperature range, good chemical stability in strongly acidic and alkaline solutions, besides being a biodegradable material [24].

Carbon aerogels have been used in several areas such as environmental, civil, electrochemical and food engineering. In addition, studies suggest that aerogel characterization and its life cycle should aim to increase the production from bench scale to pilot scale for commercial applications [25–27]. For the use of carbon aerogel as encapsulation method, comparative studies with other types of encapsulation are required [27].

According to Smirnova and Gurikov [24], the primary scope of carbon aerogel research should be to establish quantitative relationships between aerogel properties and the chemical nature of raw materials, as well as their combinations. In addition, the search for appropriate raw materials should aim for sustainability and lower prices, especially for commodities applications.

4.1 Future applications of carbon aerogels as photocatalytic components, electrodes and supercapacitor

Carbon aerogel has already been used in applications such as photocatalytic components, electrodes and supercapacitors. However, these studies are very recent and the development of these new technologies will still undergo fundamental changes. According to Justh et al. [28], resorcinol-formaldehyde polymeric aerogel grafted with monolithic structured TiO_2 airgel/aerorgel composites were synthesized by the sol-gel technique and dried by supercritical CO_2. Polymeric aerogel was transformed into carbon aerogel by pyrolysis at 900 °C in dry N_2. The composite carbon aerogels from TiO_2-polymer and TiO_2-carbon, prepared by atomic layer deposition, have photocatalytic properties.

The study with electrodes and supercapacitors has been developed together, in a scientific effort to advance in these new biomaterials. In recent years, several studies have explored the potential of metal grafted carbon aerogel. Future applications with these materials suggest that industrial development may reach the technological level of highly efficient, inexpensive and environmentally friendly components. As reported in the study of Abdelwahab et al. [29], nickel-doped carbon aerogel can be used as supercapacitor electrodes. In the developed work, a series of Ni-grafted carbon aerogels at various oxidative stages were prepared and characterized by Raman and x-ray excited photoelectron spectroscopy (XPS). Although no micropore aerogels were produced, a large increase in mesopore volume was observed. In addition, all Ni-doped carbon aerogels have high gravimetric capacitance due to the formation of graphite clusters.

Although the development of supercapacitor carbon aerogel is quite advanced, some issues such as compression resilience, mechanical strength characteristics, electrochemical performance and conductivity have yet to be solved [30]. In the study developed by Zhang et al. [31], elastic and hierarchical carbon nanofiber aerogels and hybrids with carbon nanotubes and cobalt oxide nanoparticles were designed. The aerogels and their hybrids were produced from lyophilized, carbonized and CO_2-activated polyvinylpyrrolidone electro-spun polyacrylonitrile nanofibers. The new electrodes showed high specific capacitance, reversible compressibility and excellent electrical conductivity. The authors conclude that the use of carbon and hybrid aerogel can be applied in the production of high performance, high energy density asymmetric supercapacitor. Yin et al. [32], reported the first use of organic material such as carbonized *Aloe spp* juice to be used as a carbon matrix in the production of a hierarchical 3D porous structure for devices such as asymmetric capacitors. Three-dimensional composites of Co_3O_4/C in network form and *Aloe spp* juice as carbon substrate for carbon aerogel production were hydrothermally manufactured and later calcined. The results show that the material has excellent pseudo-capacitive performances and high electrical conductivity from the organic base and 3D hierarchical pores.

4.2 Materials against electromagnetic interference, lipid adsorbents and scaffolds for polymers

Projected three-dimensional architectures based on carbon aerogel and constituents such as graphene and carbon nanotubes are preferable. The properties of these materials have been attracting the attention of researchers due to their excellent electrical conductivity, superelasticity, high porosity, flexibility and low density. These materials could be applied in many ways such as electronic components, protection against electromagnetic interference, electrode current collector, fabrication of adsorption materials, protection and adsorption of contaminants such as oils. [33, 34].

Although lightweight, three-dimensional carbon aerogels with superelasticity and mechanical strength are highly desirable, and designing these structures remains a major challenge. The main challenges are those related to the multifunctionality of the structure, which are a composition that offers fast negatively charged particles transport [35]. In general, the performance of carbon aerogels and doped cellular foams of other materials such as graphene and carbon nanotubes is determined by their shape, characteristics of chemical composition and grafts, as well as interphase properties between pores and leaves. [36].

Liu et al. [37], synthesized graphene-based superelastic and multifunctional aerogel by interfacial reinforcement with graphite carbon. An interfacial reinforcement approach was designed. Heat treatment was applied to grapheme oxide sheets to convert this material to graphene, integrating it with turbostratic carbon graphite. Mechanical and functional properties of these materials are satisfactory, in addition to excellent shielding against electromagnetic interference from electrical conductivity, low density, reversible compressibility and resistant to stress. These materials hold great promise for various applications as previously described. Future development of this technology may lead to the production of new intelligent and safer technologies.

5.   Future applications of inorganic aerogels

Inorganic aerogels are solid structures that have pores in the nanometer scale, resulting in a percentage of 90 to 99.8% of internal voids. As a consequence of these peculiar characteristics, they present specific surface area and pore volume which may vary from 400 to 900 m^2 g^{-1} and 0.25 to 1.25 cm^3 g^{-1}, respectively. In addition to these characteristics, these materials have low thermal conductivity of approximately 0.012W $m^{-1}K^{-1}$ and reduced specific mass from 0.06 to 0.103g.cm^{-3} [38]. Due to their physical characteristics, inorganic aerogel structures enable applications in several areas, such as cell electrodes, water filters for effluent treatment and thermal insulation as the main applications used until today [39].

There are three types of aerogels: inorganic, organic and hybrid. In the case of inorganic aerogels, several substances may form these products, for example silicon dioxide (SiO_2), aluminum oxide (Al_2O_3), transition metal and lanthanide oxides, calcogenide metals, organic and inorganic polymers, and carbon [38]. The most common types of inorganic aerogel are made using silica, carbon and metal oxides, but silica is the material most often used in experimental applications. Silica should not be confused with silicon, which is the semiconductor used in microprocessors, while silica is a glassy material commonly applied in thermal insulation [40,41].

5.1   Inorganic aerogels used as fuel cells

Fuel cell applications for the production of more economical and high performance cars are a possibility of using carbon aerogel. The use of condensers to transform them into supercondensers may also be considered for the use of this type of aerogel. Another application that presents a possible solution to social drinking water problems is the use of carbon aerogel for water desalination, making it suitable for human ingestion [42].

The use of inorganic aerogel could help to boost green technology because of its high energy storage capacity, but the impasse that currently exists for the use of these materials is their high production cost, which prevents large-scale production of the material. Therefore, it is necessary to carry out studies to reduce production costs to make it more accessible and consequently enable the emergence of new technologies [38].

5.2   Inorganic aerogels used as catalysts

Inorganic aerogels can be produced from metallic structures and are applied in reactions of substance transformations. Other applications include the production of carbon nanotubes for use as sensors for medical diagnostics and treatments. Inorganic aerogels can also be found in the area of explosives production. Metal aerogel can have magnetic characteristics, which also offers the possibility of several other practical applications [43-45].

The brilliant color range differentiates metal oxide aerogels (iron oxide and chromium oxide) from the most well-known silica aerogel. When transformed into aerogel, iron oxide has a metal characteristic rust color. On the other hand, chromium oxide aerogel has a blue or dark green color. Each type of structure results in different color [46-49].

The silica aerogel is bluish in color due to the characteristic of its molecules (larger size in relation to the wavelengths of light). This material spreads shorter wavelengths of light more easily compared to larger wavelengths. For this reason, the blue and violet colors reflect more compared to other shades in the visible spectrum. Human vision perceives wavelengths in the form of colors, and blue wavelengths are more noticeable than violets [38, 40].

Conclusion

Prospects for future applications of aerogel focus on the synthesis of functional foods, medicines, impact absorbing materials, catalytic supports and aerospace components. These applications highlight the importance of these structures as an alternative for the production of materials through natural resources, providing indispensable tools to different research and industry segments. Although studies indicate such applications, it is necessary to emphasize that the elaborated products must be subjected to more rigorous tests in order to guarantee their quality. In addition, it is necessary to know the largest number of specific characteristics according to the assigned purpose, especially with regard to the food shelf life, the medicines validity period and the useful life of accessories and equipment.

Aerogels I: Preparation, Properties and Applications
Materials Research Foundations **84** (2020) 214-229

Materials Research Forum LLC
https://doi.org/10.21741/9781644900994-9

Acknowledgements

The authors thank the Coordination for the Improvement of Higher Education Personnel (CAPES) and National Council of Scientific and Technological Development (CNPq) for funding this research.

References

[1]   F. P. Soorbaghi, M. Isanejad, S. Salatin, M. Ghorbani, S. Jafari, H. Derakhshankhah, Bioaerogels: Synthesis approaches, cellular uptake, and the biomedical applications, Biomedicine & Pharmacotherapy 111 (2019) 964–975. https://doi.org/10.1016/j.biopha.2019.01.014

[2]   S.W. Ruban, Biobasedpackaging-application in meatindustry, Vet. World 2 (2)(2009) 79–82.https://doi.org/10.5455/vetworld.2009.79-82

[3]   L.L. Hench, J.K. West, The sol-gel process, Chem. Rev. 90 (1) (1990) 33–72.https://doi.org/10.1021/cr00099a003

[4]   A.S. Dorcheh, M. Abbasi, Silicaaerogel; synthesis, propertiesandcharacterization, J. Mater. Process. Technol. 199 (1–3) (2008) 10–26.https://doi.org/10.1016/j.jmatprotec.2007.10.060

[5]   S. Jafari, H. Derakhshankhah, L. Alaei, A. Fattahi, B.S. Varnamkhasti, A.A. Saboury, Mesoporoussilicananoparticles for therapeutic/diagnosticapplications, Biomed.Pharmacother. 109 (2019) 1100–1111.https://doi.org/10.1016/j.biopha.2018.10.167

[6]   M. Alnaief, ProcessDevelopment for ProductionofAerogelswithControlledMorphology as PotentialDrug Carrier Systems, TechnischeUniversität Hamburg, 2011.

[7]   A. Ubeyitogullari, O. N. Ciftci, Formation of nanoporous aerogels from wheat starch, Carbohydrate Polymers 147 (2016) 125–132. https://dx.doi.org/10.1016/j.carbpol.2016.03.086

[8]   A. Ubeyitogullari, O. N. Ciftci, Generating phytosterol nanoparticles in nanoporousbioaerogels via supercritical carbon dioxide impregnation: Effect of impregnation conditions, Journal of Food Engineering xxx (2017) 1-9. https://dx.doi.org/10.1016/j.jfoodeng.2017.03.022

[9]   C. Gupta, D. Prakash, Phytonutrients as therapeutic agents, J Complement Integr Med. 2014; 11(3): 151–169. https://dx.doi.org/10.1515/jcim-2013-0021

[10]   A. Ubeyitogullari, O. N. Ciftci, Fabrication of bioaerogels from camelina seed mucilage for food applications, Food Hydrocolloids 102 (2020) 105597. https://doi.org/10.1016/j.foodhyd.2019.105597

[11]   R. Mavelil-Sam, L.A. Pothan, S. Thomas, Polysaccharide and protein based aerogels: an introductory outlook, Biobased Aerogels (2018) 1–8 978-1-78262-765-4.https://doi.org/10.1039/9781782629979-00001

[12]   C. Tresoldi, D.P. Peneda Pacheco, E. Formenti, R. Gentilini, S. Mantero, P. Petrini, Alginate/gelatin hydrogels to coat porous tubular scaffolds for vascular tissue engineering, Eur. Cells Mater. 33 (2017).

[13]   S. Frindy, A. Primo, H. Ennajih, A. el KacemQaiss, R. Bouhfid, M. Lahcini, E.M. Essassi, H. Garcia, A. El Kadib, Chitosan–graphene oxide films and CO_2-dried porous aerogel microspheres: interfacial interplay and stability, Carbohydr. Polym. 167 (2017) 297–305. https://dx.doi.org/10.1016/j.carbpol.2017.03.034

[14]   J.R. Barbosa, M.M.S. Freitas, L.H.S. Martins, R.N.C. Junior, Polysaccharides of mushroom Pleurotusspp: New extraction techniques, biological activities and development of new technologies, Carbohydrate Polymers. 229 (2019) 115550. https://doi.org/10.1016/j.carbpol.2019.115550

[15]   J.P. Vareda, A. Lamy-Mendes, L. Durães, A reconsideration on the definition of the term aerogel based on current drying trends, Micropor. Mesopor. Mat. 258 (2018) 211-216. https://doi.org/10.1016/j.micromeso.2017.09.016

[16]   C. Wan, Y. Jiao, S. Wei, L. Zhang, Y. Wu, J. Li, Functional nanocomposites from sustainable regenerated cellulose aerogels: A review, Chem. Eng. J. 359 (2019) 459–475. https://doi.org/10.1016/j.cej.2018.11.115

[17]   J.P. Oliveira, G.P. Bruni, S. L. M. el Halal, F. C. Bertoldi, A.R.G. Dias, E.R. Zavareze, Cellulose nanocrystals from rice and oat husks and their application in aerogels for food packaging, Int. J. Biol. Macromol. 124 (2019) 175-184. https://doi.org/10.1016/j.ijbiomac.2018.11.205

[18]   L. Heath, W. Thielemans, Cellulose nanowhisker aerogels, Green Chem. 12 (2010) 1448–1453. https://doi.org/10.1039/c0gc00035c

[19]   L. Nguyen, F.F. Tao, Y. Tang, J. Dou, X.J. Bao, Understanding Catalyst Surfaces during Catalysis through Near Ambient Pressure X ray Photoelectron Spectroscopy, Chem. Rev. 119 (2019) 16822-6905. https://doi.org/10.1021/acs.chemrev.8b00114

[20] M.R. Mansor, A.H. Nurfaizey, N. Tamaldin, M.N.A. Nordin, Natural fiber polymer composites: utilization in aerospace engineering, in: D. Verma, E. Fortunati, S. Jain, X. Zhang (Eds.), Biomass, Biopolymer-Based Materials, and Bioenergy: Construction, Biomedical, and other Industrial Applications, Elsevier Ltd., Amsterdã, 2019, pp. 203-224. https://doi.org/10.1016/B978-0-08-102426-3.00011-4

[21] E. Fortunati, J.M. Kenny, L. Torre, Lignocellulosic materials as reinforcements in sustainable packaging systems: processing, properties, and applications, in: D. Verma, E. Fortunati, S. Jain, X. Zhang (Eds.), Biomass, Biopolymer-Based Materials, and Bioenergy: Construction, Biomedical, and other Industrial Applications, Elsevier Ltd., Amsterdã, 2019, pp. 87-102. https://doi.org/10.1016/B978-0-08-102426-3.00005-9

[22] C. Rudaz, R. Courson, L. Bonnet, S. Calas-Etienne, H. Sallée, T. Budtova, Aeropectin: Fully biomass-based mechanically strong and thermal superinsulating aerogel, Biomacromolecules. 15 (2014) 2188–2195. https://doi.org/10.1021/bm500345u

[23] C. Wang, S.Yang, Q. Ma, X. Jia, P.C. Ma, Preparation of carbon nanotubes/graphene hybrid aerogel and its application for the adsorption of organic compounds. Carbon, 118 (2017), 765-771. https://doi.org/10.1016/j.carbon.2017.04.001

[24] I. Smirnova, P. Gurikov, Aerogel production: Current status, research directions, and future opportunities, J. Supercrit. Fluids. 134 (2018) 228–233. https://doi.org/10.1016/j.supflu.2017.12.037

[25] Z. Ulker, C. Erkey, An emerging platform for drug delivery: Aerogel based systems, J. Control. Release. 177 (2014) 51–63. https://doi.org/10.1016/j.jconrel.2013.12.033

[26] M.E. El-Naggar, S.I. Othman, A.A. Allam, O.M. Morsy, Synthesis, drying process and medical application of polysaccharide-based aerogels, Int. J. Biol. Macromol. In press (2019) 1–14. https://doi.org/10.1016/j.ijbiomac.2019.10.037

[27] Y. Wang, Y. Su, W. Wang, Y. Fang, S.B. Riffat, F. Jiang, The advances of polysaccharide-based aerogels: Preparation and potential application, Carbohydr. Polym. 226 (2019) 1–13. https://doi.org/10.1016/j.carbpol.2019.115242

[28] N. Justh, G. J. Mikula, L. P. Bakos, B. Nagy, K. László, B. Parditka, J.Mikula, Z. Erdélyi, V.Takáts, J. Mizse, I. M. Szilágyi. Photocatalytic properties of TiO_2

polymer and TiO_2 carbon aerogel composites prepared by atomic layer deposition. Carbon, 147 (2019), 476-482.

[29] A. Abdelwahab, J. Castelo-Quibén, M. Pérez-Cadenas, F. J.Maldonado-Hódar, F. Carrasco-Marín, A. F. Pérez-Cadenas,Insight of the effect of graphitic cluster in the performance of carbon aerogels doped with nickel as electrodes for supercapacitors, Carbon, 139 (2018), 888-895. https://doi.org/10.1016/j.carbon.2018.07.034

[30] M. Serrapede, A. Rafique, M. Fontana, A. Zine, P. Rivolo, S. Bianco, L. Cetibi, E.Tresso& A. Lamberti. Fiber-shaped asymmetric supercapacitor exploiting rGO/Fe2O3 aerogel and electrodeposited MnOx nanosheets on carbon fibers. Carbon, 144 (2019), 91-100. https://doi.org/10.1016/j.carbon.2018.12.002

[31] M. Zhang, D. Yang, S. Zhang, T. Xu, Y. Shi, Y. Liu, W. Chang &Z. Z.Yu., Elastic and hierarchical carbon nanofiber aerogels and their hybrids with carbon nanotubes and cobalt oxide nanoparticles for high-performance asymmetric supercapacitors. Carbon (2019). https://doi.org/10.1016/j.carbon.2019.11.071

[32] Q.Yin, L.He, J.Lian, J.Sun, S.Xiao, J.Luo, D. Sun, A.Xie & Lin, B. The synthesis of Co_3O_4/C composite with aloe juice as the carbon aerogel substrate for asymmetric supercapacitors. Carbon, 155 (2019), 147-154. https://doi.org/10.1016/j.carbon.2019.08.060

[33] Y. Qin, Q. Peng, Y. Ding, Z. Lin, C. Wang, Y. Li, F. Xu, J. Li, Y. Yuan, X. He&Y. Li. Lightweight, superelastic, and mechanically flexible graphene/polyimide nanocomposite foam for strain sensor application. ACS nano, 9 (2015), 8933-8941. https://doi.org/10.1021/acsnano.5b02781

[34] H. L. Gao, Y. B. Zhu, L. B. Mao, F. C. Wang, X. S. Luo, Y. Y. Liu, Y. Lu, Z. Pan, J. Ge, W. Shen, Y.R. Zheng, L. Xu, L.J.Wang, W.H.Xu, H.A. Wu &Y. R. Zheng. Super-elastic and fatigue resistant carbon material with lamellar multi-arch microstructure. Nature communications, 7 (2016), 12920. https://doi:10.1038/ncomms12920

[35] Qiu, L., Liu, J. Z., Chang, S. L., Wu, Y., & Li, D. Biomimetic superelastic graphene-based cellular monoliths. Nature communications, 3 (2012), 1241. https://doi:10.1038/ncomms2251

[36] Y. Wu, N. Yi, L. Huang, T. Zhang, S. Fang, H. Chang, N. Li, J. Oh, J. A. Lee, M. Kozlov, A. C. Chipara, H. Terrones, P. Xiao, G. Long, Y. Huang, F. Zhang, L. Zhang, X. Lepró, C. Haines, M. D. Lima, N. P. Lopez, L. P. Rajukumar, A. L.

Elias, S. Feng, S. J. Kim, N. T. Narayanan, P. M. Ajayan, M. Terrones, A. Aliev, P. Chu, Z. Zhang, R. H. Baughman & Y. Chen. Three-dimensionally bonded spongy graphene material with super compressive elasticity and near-zero Poisson's ratio. Nature communications, 6 (2015), 6141.
https://doi.org/10.1038/ncomms7141

[37] J. Liu, Y. Liu, H. B. Zhang, Y. Dai, Z. Liu, & Z. Z. Yu. Superelastic and multifunctional graphene-based aerogels by interfacial reinforcement with graphitized carbon at high temperatures. Carbon, 132 (2018), 95-103.
https://doi.org/10.1016/j.carbon.2018.02.026

[38] C. Ziegler, A. Wolf, W. Liu, A. K. Herrmann, N.Gaponik& A.Eychmüller. Modern Inorganic Aerogels. AngewandteChemieInternational Edition, 16, 56(43) (2017),13200-13221. https://doi.org/10.1002/anie.201611552

[39] G. Gan, X. Li, S. Fan, L. Wang, M. Qin, Z. Yin & G. Chen, Carbon Aerogels for Environmental Clean-Up, European Journal of Inorganic Chemistry, 27 (2019), 3126-3141. https://doi.org/10.1002/ejic.201801512

[40] S. Rezaei, A. Jalali, A. M. Zolali, M. Alshrah, S. Karamikamkar& C. B. Park, Robust, Ultra-Insulative and Transparent Polyethylene-based Hybrid Silica Aerogel with a Novel Non-particulate Structure, Journal of Colloid and Interface Science (2019). https://doi.org/10.1016/j.jcis.2019.04.028

[41] S. Rezaei, A. M. Zolali, A. Jalali& C. B. Park, Novel and simple design of nanostructured, super-insulative and flexible hybrid silica aerogel with a new macromolecular polyether-based precursor, J. Colloid Interface Sci. (2019).https://doi.org/10.1016/j.jcis.2019.11.072

[42] Y. Shen, D. Li, B. Deng, Q. Liu, H. Liu & T. Wu, Robust polyimide nano/microfibre aerogels welded by solvent-vapour for environmental applications, Royal Society Open Science, 6, 8 (2019), https://doi.org/10.1098/rsos.190596

[43] J. Yang, Y. Li, Y. Zheng, Y. Xu, Z. Zheng, X. Chen & Wei Liu, Versatile aerogels for sensors, Small, 15 (2019). https://doi.org/10.1002/smll.201902826

[44] N. Hebalkar, K. S. Kollipara, Y. Ananthan& M. K.Sudha, Nanoporous Aerogels for Defense and Aerospace Applications, Handbook of Advanced Ceramics and Composites, 5-1, (1-43), (2019). https://doi.org/10.1007/978-3-319-73255-8

[45] García-González, Budtova, Durães, Erkey, D.Gaudio, Gurikov, Koebel, Liebner, Neagu and Smirnova, An opinion paper on aerogels for biomedical and environmental applications, Molecules, 24, 9, (1815),

(2019).https://doi.org/10.3390/molecules 24091815

[46] A. Benad, F. Jürries, B. Vetter, B. Klemmed, R. Hübner, C.Leyens&
A.Eychmüller, Mechanical properties of metal oxide aerogels. Chem. Mater.1
(2018), 145-152. https://doi.org/10.1021/acs.chemmater.7b03911

[47] B. Trepka, J. Stiegeler, I. Wimmer, M. Fonin and S. Polarz, Eurogels: A
ferromagnetic semiconductor with a porous structure prepared via the assembly of
hybrid nanorods, Nanoscale (2018).https://doi.org/10.1039/C8NR06536E

[48] Z. Qian, M. Yang, R. Li, D. Li, J. Zhang, Y. Xiao, C. Li, R. Yang, N. Zhao & J.Xu,
Fire-resistant, ultralight, superelastic and thermally insulated polybenzazole
aerogels, J. Mater. Chem. A, (2018). https://doi.org/10.1039/C8TA07204C

[49] T.Berestok, P. Guardia, R. Du, J. B. Portals, M. Colombo, S. Estradé, F. Peiró, S.
L. Brock & A. Cabot, Metal oxide aerogels with controlled crystallinity and
faceting from the epoxide-Driven cross-linking of colloidal nanocrystals, ACS
Appl. Mater. Interfaces, 10, 18 (2018), 16041-
16048.https://doi.org/10.1021/acsami.8b03754

Aerogels I: Preparation, Properties and Applications
Materials Research Foundations **84** (2020) 230-249

Materials Research Forum LLC
https://doi.org/10.21741/9781644900994-10

Chapter 10

Recent Patents on Aerogels

Sapna Raghav[1], Pallavi Jain[2], Praveen Kumar Yadav[3], Dinesh Kumar*,[4]

[1]Department of Chemistry, Banasthali University, Banasthali, Tonk 304022, Rajasthan, India

[2]Department of Chemistry, SRM Institute of Science & Technology, Delhi-NCR Campus, Modinagar-201204, India

[3]Academy of Scientific and Innovative Research, CSIR-National Physical Laboratory, Dr. K.S. Krishnan Marg, New Delhi-110012, India

[4]School of Chemical Sciences, Central University of Gujarat, Gandhinagar, India

*dinesh.kumar@cug.ac.in

Abstract

Aerogels possess low density, large surface area, and high porosity, as a result these are considered as the fascinating materials of the 21st century. Aerogel materials have a broad range of extraordinary properties and, therefore, have a remarkable number of applications. Several industrial applications of aerogels are under advance development, like particle detectors, catalysts, and thermal insulators. As the researchers realized their peculiar and extraordinary properties, newer applications of aerogels started emerging. There are technical and scientific applications, besides increasing industrial applications of aerogels. This chapter addresses the recent patents on aerogels.

Keywords

Granted Patents, Cellulose Aerogel, Patents, Industrial Applications

Contents

1.   Introduction

Aerogels are the most important and interesting materials of the 21st century. The exclusive synthesis approach develops materials possessing low density and thermal conductivities, high specific surface area, porosity, and dielectric strengths. Because of such properties (Fig. 1) [1], aerogels are used in sensors, coatings, aerospace, energy generation, energy storage, implantations, and biomedical instruments. Aerogels are the lightest solid materials known and produced nowadays on an industrial scale. They are low density (typically <0.2 g/cm^3), high-porosity (>90% v/v) nanostructured solids obtained by drying almost any kind of wet gel under conditions that preserve their volume. The gels can be formed as monoliths or particles of different shapes and sizes. Depending on the type and mechanical stability of the gels, different drying processes can be used. Usually, the solvent filling the pores is converted or exchanged by supercritical fluid and subsequently vented out, eliminating the capillary forces that would collapse the nanostructure during a simple evaporative drying. However, sometimes, this effect can also be achieved by freeze-drying or even evaporative drying if the gel structure is strong enough to withstand ice crystal growth or capillary forces. Additionally, gels with modified surfaces may exhibit a spring-back that allows for evaporative drying without significant pore collapse [2-4].

The first aerogels were developed by S. Kistler in the 1930s [3]. He utilized various materials like alumina, silica, derivatives of cellulose, and rubber to synthesize aerogels. The manufacturing process developed by him is still active today. Initially, a hydrogel was developed, which undergoes several solvent exchange steps to replace the water with an organic solvent to produce an organogel. Aerogel is formed after processing and drying of organogel under a supercritical atmosphere to retain extremely high porosity. It follows that aerogel is a highly porous material composed of many pores that may be empty or filled, or partially filled, with gas (usually air) [5].

In the literature, a broad range of aerogels have been recorded, developed from carbon, silica, chromia, alumina, and tin oxide. Additionally, metal oxides, resins, biopolymers, and chalcogenides are some materials that are used to produce aerogels. Aerogels are employed in the nanotechnology field, integrating a range of nano-substances in the matrix of aerogel to develop composite aerogels. Aerogels are a fascinating synthesized solid material and their appearance attracts attention, the way light is dispersed through the material, giving them a translucent and smoke appearance similar to that of glass with silica aerogels. Some such as aerogels prepared from biopolymers look like soft, thin foams, while others such as carbon-based aerogels look like crunchy black bodies. They feel fragile and light to the touch because they are formed by up to 90-99% of gas or air [5-10].

Since the early 2000s, because of the rapid development of processing techniques, new compositions, structural reinforcement strategies, drying methods, and the sustainability of their synthesis, aerogels have become an emerging class of material with great potential of advanced technological applications [11-13]. The field of aerogel is exploding; most recent potential applications of aerogels are tabulated in Table 1 [14-30]several patents have been granted in diverse applications of aerogels throughout the world, which are listed in Table 2 [31-59].

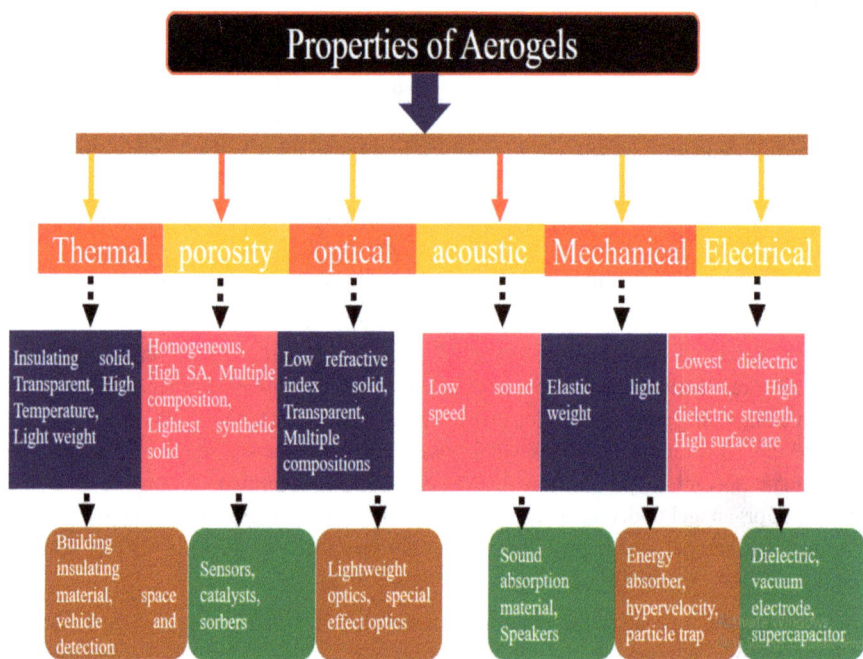

Fig. 1 *Properties of aerogels with applications [1, (A.C. Pierre, A. Rigacci, SiO2 aerogels, in: M.A. Aegerter, N. Leventis, M.M. Koebel (Eds.), Aerogels Handbook, 2011, Springer, New York, NY, pp. 21–45)].*

Table 1 Potential applications of aerogels [14-30].

Application	Aerogel type	Decisive application	Reference
Thermal insulation in form of monoliths; Insulation panels for textiles, pipelines, winter apparel;	Silica aerogels, carbon aerogels, biopolymers aerogels such as pectin, alginate, and cellulose	Thermal conductivity and Higher mechanical stability,	[14-16]
Drug delivery	Biopolymer aerogels such as chitosan, cellulose, pectin, starch	Higher surface area,	[17, 18]
Tissue engineering	Biopolymer aerogels	Biocompatibility; meso and macro porosity in same material	[19, 20]
Medical implant	Natural biopolymer aerogels, polymer (polyurea) crosslinked aerogels, alginate-starch aerogels	Porous structure; liquid stability; biocompatibility;	[21-23]
Cosmetics	Silica particles	Higher absorption capacity; amorphous backbone aerogels	[24]
Biosensors	Gold nanoparticles/carbon aerogels, silica aerogels, graphene and its derivative aerogels	Surface functionalization to ensure the selectivity to target molecules, open porosity	[25-28]
Bio catalysis	Silica aerogels	Enzyme compatibility; pore size adjustments	[29, 30]

Table 2 List of patents in the different areas of aerogel applications [31-59].

Application No	Patent No	Title	Claims	Priority date	Ref.
09/355,074	US 6,598,358	Use of Aerogels for deadening structure-borne and/or impact sound	7		[31]
10/745,331	US 7,364,553	Breath aerosol management and collection system	32		[32]
10/861,554	US 7,540,286	Multiple-dose condensation aerosols device and methods of forming condensation	55		[33]

		aerosols			
PCT/US2003/0277 29	-	Aerosol generating device and method of use thereof	43	September 6, 2002	[34]
PCTUS/2003/0277 30	-	Aerosol generating devices and methods for generating aerosols having controlled particle sizes	40	September 6, 2002	[35]
PCT/US2007/0686 92	-	Compressed gas propellants in plastic aerosols	18	May 31, 2006	[36]
PCT/US99/26799	-	Aerosols comprising nanoparticles drug	28	November 12,1998	[37]
PCT/US2005/0003 49	-	Ormosil aerogels containing silicon polymethacrylate	19	January 6, 2004	[38]
PCT/US1998/0006 02	2006/34	A method for on-line analysis of polycyclic aromatic hydrocarbons in aerosols	24	January 29, 1997	[39]
EP 1 158 292	2005/45	Aerosol hazard characterization and early warning network	44	May 23, 2000	[40]
EP1503204	-	Detection and analysis of chemical and biological aerosols	14	July 29, 2003	[41]
EP174527	2015/46	Aerosol generators and methods for producing aerosols	16	April 23, 2004	[42]
11/592,604	US 7,708,803	Method and apparatus for the enhanced removal of aerosols from a gas stream	20	November 3, 2006	[43]
11/315.951	US 7,802,569	Aerosol processing and inhalation method and system for high dose rate aerosol drug delivery	35	December 22, 2005	[44]
12/776,088	US 8.258,251	Highly porous ceramic oxide aerogels having	51	May 7, 2010	[45]

		improved flexibility			
12/490,102	US 8,506,935	Respiratory drug condensation aerosols and methods of making and using them	14	June 23, 2009	[46]
12/455,236	US 8,834,849	Medicinal aerosols and methods of delivery thereof	6	May 29, 2009	[47]
10/842,977	-	Small particle liposome aerosols for delivery of anti-cancer drugs	5	May 11, 2004	[48]
11/372,925	-	Transparent assemblies with ormosl aerogels	12	March 11, 2006	[49]
11/180,038	-	High strength, nanoporous bodies reinforced with fibrous materials	33	July 12, 2005	[50]
11/743,944	-	Organic aerogels reinforced with inorganic aerogel fillers	20	May 3, 2007	[51]
11/761,924	-	Aerogel-foam composites	46	June 12, 2007	[52]
12/951,323	-	Aerogel composites and methods for making and using them	59	November 22, 2010	[53]
13/206,087	-	Silica aerogels and their preparation	35	August 9, 2011	[54]
PCT/US2005/000295	-	Ormosil aerogels containing silicon bonded linear polymer	21	January 6, 2004	[55]
PCT/GB20 10/05 1542	-	Cellulose nanoparticle aerogels, hydrogels and organogels	34	September 14, 2009	[56]
PCT/US20 14/0229 19	-	Aerogel blanket and method of production	34	March 15, 2013	[57]
11/592,604	US 7,708,803	Method and apparatus for the enhanced removal of aerosol from a gas stream	20	Nov. 3, 2006	[58]

2. Applications

The peculiar aforementioned physical properties derived from their microstructure and composition, and the possibility of controlling these properties by regulating the synthesis parameters during preparation, has introduced aerogels as highly efficient versatile materials that can be outstanding alternatives to materials that are commonly used today. Some of the applications are, however, in the development phase, and their use on a large scale is still limited. However, there are a few commercial technologies based on aerogels that make use of their super-insulating properties and thus reduce energy losses (Table 3) [59]. The capture and/or catalysis of gases or liquids can also be carried out by aerogel systems with an aim at capturing or neutralizing pollutants by controlling the composition of surface and structure. Systems based on aerogels have also been developed for the cleaning of water as chemical/ physical adsorbents and clarification of sea and brackish water by adsorption and capacitive deionization. Some compositions of aerogels have also been studied as advanced materials for life science and biomedical applications. In this context, aerogels can be designed from the molecular level to their macroscopic structure to be used, for example, as scaffolds in bone tissue engineering, absorbable implants for controlled release of drugs and substances, and as biological sensors. The most important applications of relevant aerogels are in energy savings, environmental remediation, and tissue engineering [59].

2.1 Patents on aerogel generators(WO 2004/022242 Al)

There are various fields in which aerogels are having many applications. These fields cover medicinal and nonmedicinal fields. While talking about uses of aerogels in medicinal fields, the use of medicated liquid in the aerogel form may serve as an example. An aerogel generator is used for producing aerogels, which then get inhaled into the patient's lungs. Besides above, dispensing insecticides, air fresheners, lubricants, and paint delivery may be cited as the use of aerogels in nonmedicinal fields.

In addition, the devices of different sizes were also designed for the development of variable size aerogels. Since the application of aerogels varies according to their size. For instance, suppose we have to monitor the drug delivery in human lung, in this case, the desired mass median aerodynamic diameter (MMAD) of an aerogel should accord to that portion of the lung where the drug has to be delivered. MMAD reveals that at a particular diameter, the 50% of the particles by mass are larger and remaining are smaller. Usually, the aerogels of smaller MMAD are more effective compared to that of larger MMAD, as the aerogels are of smaller MMAD can cause extreme lung penetration as compared to the other ones [34].

Table 3 Insulation branches where aerogel products have economic relevance [59].

S. No.	Application	Installation cost	Operating costs	Economic potential
1.	Aeronautics	Overall design is simplified; Reduction in size; Light production; Lesser material cost	Smaller gross weight and ensuing additional capacity or fuel saving	High
2.	Appliances and Apparel	Significantly more complex than standard technology.	Energy saving increased thermal comfort for lightweight extreme performance clothes	Middle
3.	Building insulation	More elaborated due to lack of experience; Comparable to conventional insulation;	Decrease of cooling or heating energy and larger usable space	High
4.	Cryogenic	Easier installation; Smaller overall pipe diameter or exterior dimensions	Increase energy, lifetime, and building spaces; Decreased sensitivity to cryoembrittlement;	Middle
5.	High-temperature insulation	Smaller overall pipe diameter; Exterior dimension; Easier installation;	Lower radiation losses Decrease area per unit length; Enhanced resistance and lifetime;	High
6.	Offshore gas and oil	Low weight; Fewer trips; More pipes per installation round trip; Smaller pipe diameter;	Superior lifetime; Enhanced degradation resistance	High

Some aerogel generators have a heated tube that can vaporize the liquid e.g. U.S. Patent No. 5,743, 251 discloses an aerogel generator having a tube with an open end and a capillary sized fluid passage. This tube can be heated up to a temperature so as to get liquid vaporized. This vaporized liquid spreads out from the open end of the tube and combines with the ambient air resulting in aerogel. The heater which is connected to a power supply, is arranged near the tube. The tube also has an inlet end from which the liquid material can be introduced [60].

Other exemplary aerogel generator, includes a heated tube for vaporizing liquids to produce a condensation aerogel, are disclosed in commonly assigned U.S. Patent: Application No. 6234167 and in normally assigned U. S. Patent Nos.10/003437 filed December 6, 2001, and 09/956,966 filed September 21, 2001. Another US patent application number 2003270321 also discloses a method of generating aerogels in controlled particle sizes with the help of aerosols generators [61, 62].

2.2   Aerogel blanket and its production (PCT/US2014/022919)

Insulative materials having low thermal conductivity and density are very useful. These materials are formed from aerogels. More particularly, silica aerogels are the best for making insulative materials. Those aerogels can either be mixed with some other materials like foams, fibers, and adhesives or sometimes can themselves be used as an insulating material. Using some techniques like solvent substitution combined with ambient pressure drying or via supercritical drying techniques, these aerogels are produced. While some silica aerogels are typically hydrophilic; however, the use of a specific treating agent can turn them as hydrophobic.

An insulative aerogel blanket fabrication process is as follows. This method is effective in producing an insulative aerogel blanket with improved properties like lower corrosivity, a uniform and consistent construction, reduced levels of dust, and improved thermal insulation. These insulative aerogel blanket is known as wet-laid aerogel blankets. During the wet-laid process the particles in the blanket or the fibers are suspended in water. In these processes, optional materials such as pacifiers and fibers, aerogel particles are as well flocculated (floe).By developing a two-phase system i.e., the supernatant of substantial water and flocculated aerogel particles, the floe can be separated from the solution very easily. The blanket can be formed from the flocculated materials having an even distribution of the different particles. This uniform composition results in a stable aerogel blanket in which the distribution of the particles is consistent everywhere. This aerogel blanket is very useful in those applications where the temperature range is very high (650°C or even higher). Thus, under various standard conditions suitable for industries, including the above temperature, these aerogel blankets generate reduced or sometimes even no dust. The aerogel blanket produced in this manner may be noncorrosive and also have very little shrinkage at the high-temperature range. These kinds of aerogel blankets may sometimes be rigid or adaptable to various conditions and come in various shapes.

Some other kinds of aerogel blanket can be made up of aerogel particles having a low density. These low-density aerogel particles have been gelled previously and dried. In this way, a single blanket having numerous types of aerogel can be formed. This method of development of aerogel blanket can eradicate the problem arises during the formation of aerogel on a fibrous matrix. For making these aerogels blanket, a hydrophobic aerogel is being mixed with other additives like opacifiers, fibers, etc. or rather it can be mixed with a combination of different aerogels. The aerogel blanket can be formed without getting aerogel gelled and dried if fibers are getting mixed to hydrophobic aerogel.

Usually, the aerogels can be produced using alumina. However, silica also has an important place in the formation of aerogels. Silica aerogels are always preferred over any other material in view of their thermal insulation properties. Further, silica aerogels are very light and also have poor thermal conductivity. These silica aerogels can be turned to hydrophobic by treating with certain agents as mentioned earlier and thus producehydrophobic silica aerogels that are very beneficial for the insulating applications as they are immune to moisture. In addition, there is no need for any additional treatment. These aerogels particles are found in the unbounded form. Sometimes these unbounded particles are bound by using some additional material or an adhesive or a substrate or by using a network of fibers. These aerogels are further having many applications in industrial processes as they can withstand high temperatures. Also, these aerogel blankets are best suitable for wherever there is a need for insulation in industrial application like pipe insulation, chemical processing equipment, machinery, a further example could be insulation for aircraft and its parts, aerospace insulation, insulated glass unit, building insulation, automotive insulation, and daylighting etc. [63].

Wet laid processes are very useful for inserting aerogel particles into an insulation blanket. U.S. Patent Nos, 7635411 [64] (both of which are incorporated by reference herein), describe wet laid processes. As per this described process, an aqueous slurry is used to disperse the various additives, aerogel particles, and fibers. A layer of various additives, aerogel particles, and fibers is formed by dehydrating the slurry. This layer can be dried and thus resulting in a nonwoven aerogel blanket.

2.3   Cellulose aerogels PCT/GB2010/051542

Till now, there are number of aerogels have been developed using different materials like carbon, silica, chromia, alumina, and tin oxide etc.Recently, due to biodegradable and renewable properties, cellulose has been utilized for the manufacturing of aerogels. Tan and co-workers [65] produced the first cellulose aerogels. These aerogels displayed a specific surface area of 300 m^2/g with bulk densities in the range 0.1-0.35 g/cm^3, whereas other aerogels displayed specific surface area between 600 and 000 m^2/g with bulk densities in the range 0.004-0.500 g/cm^{-3} and porosities up to 99%. Different manufacturing processes, along with different starting materials, result in different structural properties of aerogels such as a sound insulator, catalysts, support for catalysts, particle filters, thermal insulators, storage for gases, and many more. Especially, cellulose aerogels have discovered as a useful material in the pharmaceutical and medicinal field, due to biocompatible and biodegradable properties.

Cotton may be used as a natural source of cellulose. The amount of hemicellulose, lignin, and natural waxes/oils varies from species to species, with cotton being almost purely cellulose. Cellulose is fibrillar and is a semi-crystalline polymer where the crystalline sections have nanosized dimensions. There are two distinct regions observed within cellulose, the crystalline and amorphous. In the crystalline regions, the chains of β-D-glucopyranosyl units are held together in a highly ordered array by van der Waals and hydrogen bonds. The remaining amorphous regions display randomly arranged cellulose chains and act as structural defects.

The acid hydrolysis of native cellulose under controlled conditions yields cellulose nanoparticles by preferential hydrolysis of the amorphous regions. The hydronium ions penetrate the amorphous regions in the cellulose and hydrolytically cleave the (1 - 4) glycosidic bonds. Mass transfer control (diffusion limitations) dictate the hydrolysis of amorphous regions before the crystalline ones. Therefore, under controlled conditions, such as temperature, time, and acid strength, preferentially, the glycosidic bonds in the amorphous regions are hydrolytically cleaved, leaving the crystalline sections mostly intact. Acids such as sulphuric acid may be used as catalysts to catalyze the production of cellulose nanoparticles from cotton. The hydrolysis of cotton typically yields high-aspect-ratio nanoparticles that are highly crystalline i.e. > 70% crystalline by weight of cellulose nanoparticles. The dimensions of the crystals can vary significantly depending on, the time of hydrolysis, source of cellulose, and the location of cellulose bearing plant growth. The dimensions of the nanoparticles can be controlled by varying the source material of cellulose and the hydrolysis conditions and duration [5].

The surface of the nanoparticle bears many hydroxyl functions. When the cellulose is hydrolyzed in sulphuric acid, a fraction of these hydroxyl functions is converted to sulfate groups. The hydroxyl groups enable network formation by hydrogen bonding. The deprotonation of the sulfate groups can cause some repulsion of the nanoparticles, and stable dispersions of the nanoparticles is formed.

2.4   Some miscellaneous patents

The modified aerogel's production processes and their use (WO1996022942A1):

The discovery is mostly inclined towards the manufacturing process for reformed SiO_2 aerogels [66], which includes the following steps:

a) Production of silicatic lyogel

Materials Research Forum LLC
https://doi.org/10.21741/9781644900994-10

b) Solvent exchanging of produced Silicaticlyogel to give another organic solvent

c) The reaction of gel received from the first two steps with one chlorine-free silylation agent

d) Subcritical drying of the gel obtained in the third step and use thereof.

Method for producing organically modified, permanently hydrophobic aerogels (EP0946277A2): The discovery is mostly inclined towards the manufacturing process for originally reformed aerogels having permanent hydrophobic surface groups [67], which includes the following steps:

a) Providing of a lyogel

b) Washing of provided lyogel with an organic solvent

c) Silylation of the gel surface obtained in the second step

d) Drying of silylated surface gel

A disiloxane of the formula (I) $R_3Si-O-SiR_3$ is used as a silylating agent in the third step where the radicals' R mean individually, being the same or different, either an H-atom or a non-reactive organic linear, cyclic, branched aromatic/heteroatomic, saturated/unsaturated radical.

Organically modified aerogels, a method for their production by surface modification of the aqueous gel without previous solvent exchange and subsequent drying and the use thereof (WO1998023366A1): The present creation is related to the manufacturing process for originally reformed aerogels [68], which includes the following steps:

a) Providing a hydrogel

b) Modification of provided hydrogel surface

c) Drying of the modified surface of hydrogel obtained in the second step

d) Drying of silylated surface gel and use thereof

The present invention also relates to novel organically modified wet gels, a method for the production and the use thereof.

Organically reformed aerogels preparation process with alcohols, wherein the resultant salts are precipitated (WO1997018161A1): The present discovery is related to the manufacturing process for originally reformed aerogels [69], which includes the following steps:

Silicic acid sol production from an aqueous solution of K_2O_3Si utilizing at least one organic and/or inorganic acid

Aerogels I: Preparation, Properties and Applications   Materials Research Forum LLC
Materials Research Foundations **84** (2020) 230-249   https://doi.org/10.21741/9781644900994-10

a)Polycondensation of produced silicic acid to develop SiO_2 gel, by adding a base

b) Washing of produced gel by an organic solvent (water content in the gel must be $\leq$ 5 wt%)

c)Surface modification of gel obtained in the third step by one C1-C6 alcohol at least

d) Drying of surface-modified gel

The process is distinguished by the fact that at least one acid is formed by the potassium silicate cation salts which is insoluble in the silicic acid sol. The resulting salts are difficult to dissolve and are precipitated out before the second step and isolated from the sol of silica acid.

Method for preparing lignocellulose aerogel by using ionic liquid (CN102702566B): Utilizing ionic liquids prepared lignocellulosic aerogel method, relates to a process for preparing lignocellulosic aerogel. This method is developedto overcome the problems arises in the conventional method where the aerogels are prepared by directly using the timber e.g. problems arises due to direct use of timber and problem related to the disposal of raw materials [70].

The steps involved in this method are as follow:

a)Pulverisation and sieving of timber to obtain the fine powder

b) Heating and stirring of mixture of ionic liquid and wood powder to obtain the mix solution

c)Freeze drying (-196 to -20°C) of mix solution followed by its melting at ~40°C for ~10h

d) Repetition of third step for ~9 times to obtain the lignocellulose gel

e)Critical point drying of obtained gel to produce the lignocellulose aerogel

Aerogel composite with fibrous batting (CN1306993C): Having a bulky batt reinforced aerogel composite material, preferably in combination with one or two types of short microfibers and each conductive layer of randomly oriented, which exhibit one or all of the following improved properties: flexibility, drape ability durability, seizure resistance, thermal conductivity XY, XY conductivity, RFI-EMI attenuation, and/or burnthrough resistance [71].

Acknowledgments

Dr. Sapna Raghav is thankful to the Department of Chemistry, Banasthali Vidyapith. One author Dr. Pallavi Jain, is grateful to Dr. (Prof) D. K. Sharma, Dean, SRM Institute of

Science & Technology, Modinagar for the encouragement and for providing the research facilities. Dr. Dinesh Kumar is thankful to DST, New Delhi, for the financial support extended. (vide Sanction Order F. No. DST/TM/WTI/WIC/2K17/124(C).

References

[1] S.K.Montesa, G.H. Maleki, Aerogels and their applications, in S. Thomas, A.T. Sunny, P. Velavudhan (Eds.), Colloidal Metal Oxide Nanoparticles, Elsevier 2002, pp.337–399. https://doi.org/10.1016/B978-0-12-813357-6.00015-2

[2] S. Mulik, C. Sotiriou-Leventis, G. Churu, H. Lu, N. Leventis, Cross-linking 3D assemblies of nanoparticles into mechanically strong aerogels by surface-initiated free-radical polymerization, Chem. Mater. 20 (2008) 5035–5046. https://doi.org/10.1021/cm800963h

[3] S.S. Kistler, Coherent expanded aerogels and jellies, Nature 127 (1931) 741–741. https://doi.org/ 10.1038/127741a0

[4] I. Smirnova, P. Gurikov, Aerogels in chemical engineering: strategies toward tailor-made aerogels, Annu. Rev. Chem. Biomol. Eng. 8 (2017) 14.1–14.28. https://doi.org/10.1146/annurev-chembioeng-060816-10145810.1146/annurev-chembioeng-060816-101458

[5] W.A.W.I. Thielemans, R. Davies, Cellulose nanoparticle aerogels, hydrogels and organogels, PCT/GB2010/051542, 2009.

[6] J.V. Aleman, A.V. Chadwick, J. He, M. Hess, K. Horie, R.G. Jones, P. Kratochvíl, I. Meisel, I. Mita, G. Moad, S. Penczek, Definitions of terms relating to the structure and processing of sols, gels, networks, and inorganic-organic hybrid materials (IUPAC recommendations 2007), Pure Appl. Chem. 79 (2007) 1801–1829. https://doi.org/10.135/pac200779101801

[7] S. Araby, A. Qiu, R. Wang, Z. Zhao, C.H. Wang, J. Ma, Aerogels based on carbon nanomaterials, J. Mater. Sci. 51 (2016) 9157–9189. https://doi.org/10.1007/s10853-016-0141-z

[8] A.C. Pierre, History of aerogels,in M. Aegerter, N. Leventis, M. Koebel, (Eds.),Aerogels handbook. advances in sol-gel derived materials and technologies, Springer: New York, NY, USA, 2011; pp. 3–18. https://doi.org/10.1007/978-1-4419-7589-8_1

[9] S. Teichner, G.A. Nicolaon, M.A. Vicarini, G.E.E. Gardes, Inorganic oxide aerogels, Adv. Colloid Interf. Sci. 5 (1976) 245–273. https://doi.org/10.1016/0001-8686(76)80004-8

[10] R. Saliger, U. Fischer, C. Herta, J. Fricke, High surface area carbon aerogels for supercapacitors, J. Non-Cryst. Solids 225 (1998) 81–85. https://doi.org/10.1016/S0022-3093(98)00104-5

[11] N. H€using, U. Schubert, Aerogels—airy materials: chemistry, structure, and properties, Angew. Chem. Int. Ed. 37 (1998) 22–45. https://doi.org/10.1002/(SICI)1521-3773(19980202)37:1/2<22::AID-ANIE22>3.0.CO;2-I

[12] B. Wicklein, A. Kocjan, G. Salazar-Alvarez, F. Carosio, G. Camino, M. Antonietti, L. Bergström, Thermally insulating and fire-retardant lightweight anisotropic foams based on nanocellulose and graphene oxide, Nat. Nanotechnol. 10 (2015) 277–283. https://doi.org/10.1038/NNANO.2014.248

[13] C. Li, X. Yang, G. Zhang, Mesopore-dominant activated carbon aerogels with high surface area for electric double-layer capacitor application, Mater. Lett. 161 (2015) 538–541. https://doi.org/10.1016/j.matlet.2015.09.003

[14] S. Araby, A. Qiu, R. Wang, Z. Zhao, C.H. Wang, J. Ma, Aerogels based on carbon nanomaterials, J. Mater. Sci. 51 (2016) 9157–9189. https://doi.org/10.1007/s10853-016-0141-z

[15] M. Venkataraman, R. Mishra, T.M. Kotresh, J. Militky, H. Jamshaid, Aerogels for thermal insulation in high-performance textiles, Text. Prog. 48 (2016) 55–118. https://doi.org/10.1080/00405167.2016.1179477

[16] M. Koebel, A. Rigacci, P. Achard, Aerogel-based thermal superinsulation: an overview, J. Sol-Gel Sci. Technol. 63 (2012) 315–339. https://doi.org/10.1007/s10971-012-2792-9

[17] Z. Ulker, C. Erkey, An emerging platform for drug delivery: aerogel based systems, J. Control. Release 177 (2014) 51–63. https://doi.org/10.1016/j.jconrel.2013.12.033

[18] H. Maleki, L. Duraes, C.A. Garcıa-Gonzalez, P. del Gaudio, A. Portugal, M. Mahmoudi, Synthesis and biomedical applications of aerogels: possibilities and challenges, Adv. Colloid Interface Sci. 236 (2016) 1–27. https://doi.org/10.1016/j.cis.2016.05.011

[19]   M. Martins, A.A. Barros, S. Quraishi, P. Gurikov, S.P. Raman, I. Smirnova, A.R.C. Duarte, R.L. Reis, Preparation of microporous alginate-based aerogels for biomedical applications, J. Supercrit. Fluids 106 (2015) 152–159. https://doi.org/10.1016/j.supflu.2015.05.010

[20]   S. Quraishi, M. Martins, A.A. Barros, P. Gurikov, S.P. Raman, I. Smirnova, A.R.C. Duarte, R.L. Reis, Novel non-cytotoxic alginate-lignin hybrid aerogels as scaffolds for tissue engineering, J. Supercrit. Fluids 105 (2015) 1–8. https://doi.org/10.1016/j.supflu.2014.12.026

[21]   A.A. Barros, A. Rita, C. Duarte, R.A. Pires, B. Sampaio-Marques, P. Ludovico, E. Lima, J.F. Mano, R.L. Reis, Bioresorbable ureteral stents from natural origin polymers, J. Biomed. Mater. Res. B Appl. Biomater. 103 (2015) 608–617. https://doi.org/10.1002/jbm.b.33237

[22]   F. Sabri, J.A. Cole, M.C. Scarbrough, N. Leventis, Investigation of polyurea-crosslinked silica aerogels as a neuronal scaffold: a pilot study,Plos One 7 (2012) e33242. https://doi.org/10.1371/journal.pone.0033242

[23]   F. Sabri, D. Gerth, G.R.M. Tamula, T.C.N. Phung, K.J. Lynch, Boughter JD Jr. Novel technique for repair of severed peripheral nerves in rats using polyurea crosslinked silica aerogel scaffold, J. Investig. Surg. 27 (2014) 294–303. https://doi.org/10.3109/08941939.2014.906688

[24]   R. Lorant, Cosmetic composition comprising silica aerogel particles, a gemini surfactant and a solid fatty substance, US Patent No. 20140302105 A1, 2014.

[25]   D. Wen, W. Liu, A.K. Herrmann, A. Eychmuller, A membrane less glucose/O2 biofuel cell based on Pd aerogels, Chemistry 20 (2014) 4380–4385. https://doi.org/10.1002/chem.201304635

[26]   B. Wang, S. Yan, Z. Lin, Y. Shi, X. Xu, L. Fu, J. Jiang, Fabrication of graphene aerogel/platinum nanoparticle hybrids for the direct electrochemical analysis of glucose, J. Nanosci. Nanotechnol. 16 (2016) 6895–6902. https://doi.org/10.1166/jnn.2016.11359

[27]   Z. Yu, Y. Kou, Y. Dai, X. Wang, H. Wei, D. Xia, Direct electrochemistry of glucose oxidase on a three-dimensional porous zirconium phosphate-carbon aerogel composite, Electrocatalysis 6 (2015) 341–347. https://doi.org/10.1007/s12678-015-0249-y

[28]   A. Harley-Trochimczyk, T. Pham, J. Chang, E. Chen, M.A. Worsley, M.A., A. Zettl, W. Mickelson, R. Maboudian, Platinum nanoparticle loading of boron nitride aerogel and its use as a novel material for low-power catalytic gas sensing, Adv. Funct. Mater. 26 (2016) 433. https://doi.org/10.1002/adfm.201503605

[29]   C.E. Barao, L.D. de Paris, J.H. Dantas, M.M. Pereira, L.C. Filho, de H.F. Castro, G.M. Zanin, de F.F. Moraes, C.M. Soares, Characterization of biocatalysts prepared with Thermomyces lanuginosus lipase and different silica precursors, dried using aerogel and xerogel techniques, Appl. Biochem. Biotechnol. 172 (2013) 263–274. https://doi.org/10.1007/s12010-013-0533-3

[30]   A. Karout, P. Buisson, A. Perrard, A.C. Pierre, Shaping and mechanical reinforcement of silica aerogel biocatalysts with ceramic fiber felts, J. Sol-Gel Sci. Technol. 36 (2005) 163–171. https://doi.org/10.1007/s10971-005-5288-z

[31]   F. Schwertfeger, M. Schmidt, D. Frank, Use of aerogels for deadening structure-borne and/or impact sounds, US 6598358 B1, 2003.

[32]   F.M. Paz, D. Howson, M.V. Wiernicki, Breath aerosol management and collection system, US 7364553 B2, 2008.

[33]   S.D. Cross, M. Herbette, A.J.G. Kelly, D.J. Myers, W.W. Shen, R.D. Timmons, C. Tom, J.M. Virgili, M.J. Wensley, Multiple dose condensation aerosol devices and methods of forming condensation aerosols, US 7540.286 B2, 2009.

[34]   G.E. Grollimund, D.D. Mcrea, W.A. Nichols, T.T. Nguyen, K.A. Cox, U. Smith, D.L. Brookman, Aerosol generating device and method of use thereof, W O 2004/022242 Al, 2004.

[35]   D.D. McRae, K.A. Cox, W.A. Nichols, R. Gupta, Aerosol generating devices and methods for generating aerosols having controlled particle sizes, WO J240 04/022243 Al, 2004.

[36]   D. Shiekh, D.A. Huitt, J.T. Kennedy, R. Ruiz De Gopegui, Compressed gas propellants in plastic aerosols, WO 2007/143330, 2007.

[37]   W.H. Bosch, K.D. Ostrander, E.R. Cooper, Aerogels comprising nanoparticles drugs, WO 2000/27363 A1, 2000.

[38]   D.L. Duan, G.L. Gould, C.J. Stepania, Ormosil aerogels containing silicon bonded polymethacrylate, WO2005/098553 A2, 2005.

[39]   Schechter, A method for on-line analysis of polycyclic aromatic hydrocarbons in aerosols, PCT/US1998/000602, 1998.

[40]   P.J. Wyatt, Aerosol hazard characterization and early warning network, EP 1 158 292, 2005.

[41]   M.S. Chou, Detection and analysis of chemical and biological aerosols, EP1503204, 2005.

[42]   W.A. Nicholas, R. Gupta, G.G. Faison, K.A. Cox, Aerosol generators and methods for producing aerosols, EP174527, 2004.

[43]   M.S. Berry, R. Chang, Method and apparatus for the enhanced removal of aerosols from a gas stream, US 7,708,803 B2, 2006.

[44]   D.B. Yeates, J. Yi, G. Li, Aerosol processing and inhalation method and system for high dose rate aerosol drug delivery, US 7,802,569 B2, 2005.

[45]   M.A.B. Meador, B.N. Nguyen, H. Guo, Highly porous ceramic oxide aerogels having improved flexibility, US 8.258,251, 2010.

[46]   R.L. Hale, P.M. Lloyd, A.T. Lu, J.D. Rabinowitz, M.J. Wensley, Respiratory drug condensation aerosols and methods of making and using them, US 8,506,935 B2, 2009.

[47]   F.C. Millar, Medicinal aerosols and methods of delivery thereof, US 8,834,849, 2009.

[48]   B.C. Giovanella, J.V. Knight, J.C. Waldrep, N. Koshkina, B. Gilbert, C.W. Wellen, Small particle liposome aerosols for delivery of anti-cancer drugs, US 2004/0208935 A1, 2004.

[49]   W.E. Rhine, W.L. Gourd, R. Begag, J.H. Sonn, D.L. Ou, Transparent assemblies with ormosl aerogels, US 2006/0246806, 2006.

[50]   G.L. Gourd, J.K. Lee, C.J. Stepania, K.P. Lee, High strength, nanoporous bodies reinforced with fibrous materials, US 2007/0222116 A1, 2005.

[51]   J.K. Lee, Organic aerogels reinforced with inorganic aerogel fillers, US 2007/0259979 A1, 2007.

[52]   Y. Tang, A. Polli, C.A. Bilgerian, D.R. Young, W.E. Rhine, G.L. Gould, Aerogel-foam composites, US 2009/0029147 A1, 2009.

[53]   D.A. Doshi, T.M. Miller, J.A. Chase, C.A. Norwood, Aerogel composites and methods for making and using them, US 2011/0206471 A1, 2010.

[54]   T. Zhang, Y. Zhao, G. Chen, Silica aerogels and their preparation, US 2012/0128958 A1, 2011.

[55]   D.L. OU, G.L. Gourd, Ormosil aerogels containing silicon bonded linear polymer, WO2005/068361.

[56]   W.A.W. Thielemans, R. Davies, Irene Cellulose nanoparticle aerogels, hydrogels and organogels, PCT/GB20 10/05 1542, 2010.

[57]   S. Samata, F.P. Pescator, B.P. Thomas, Aerogel blanket and method of production, WO2014/150310 A1, 2013.

[58]   M.S. Berry, R. Chang, Method and apparatus for the enhanced removal of aerosol from a gas stream, US 7,708,803, 2006

[59]   M. Koebel, A. Rigacci, P. Achard, Aerogel-based thermal superinsulation: anoverview, J. Sol-Gel. Sci. Technol. 63 (2012) 315–339. HTTPS://DOI.ORG/10.1007/s10971-012-2792-9.

[60]   K.A. Cox, R. Gupta, D.D. Mcrae, W.A. Nicholas, Aerosol generating devices and methods for generating aerosols having controlled particle sizes, U.S. Patent No. 5,743, 251, 2002.

[61]   K.A. Cox, T.P. Beane, W.R. Sweeny, Aerosol generator and methods of making and using an aerosol generator, U.S6234167, 1998.

[62]   K.A. Cox, R. Gupta, D.D. Mcrae, W.A. Nicholas, Aerosol generating devices and methods for generating aerosols having controlled particle sizes, AU2003270321B2

[63]   S. Samanta, F.P. Pescatore, B.P. Thomas, Aerogel blanket and its production PCT/US2014/022919, 2013.

[64]   S.F. Rouanet, R.K. Massey, J. Menashi, Aerogel containing blanket, US 7635411, 2004.

[65]   C. Tan, B.M. Fung, J.K. Newman, C.Vu, Organic aerogels with very high impact strength, Adv.Mater., 13(2002) 644. HTTPS://DOI.ORG/10.1002/1521-4095(200105)13:9<644::AID-ADMA644>3.0.CO;2-%23

[66]   F. Schwertfeger, A. Zimmerann, The modified aerogel's production processes and their use, WO1996022942A1, 1995.

[67]   F. Schwertfeger, Method for producing organically modified, permanently hydrophobic aerogels, EP0946277A2, 1997.

[68]   F. Schwertfeger, D. Frank, Organically modified aerogels, a method for their production by surface modification of the aqueous gel without previous solvent exchange and subsequent drying and the use thereof, WO1998023366A1, 1991.

[69]   F. Schwertfeger, A. Zimmermann, Organically reformed aerogels preparation process with alcohols, wherein the resultant salts are precipitated, WO1997018161A1, 1995

[70]   L. Yun, L. Jiang, S. Qingfeng, L. Yixing, Method for preparing lignocellulose aerogel by using ionic liquid, CN102702566B, 2012.

[71]   R. Begag, Aerogel composite with fibrous batting CN1306993C, 2002.

Aerogels I: Preparation, Properties and Applications
Materials Research Foundations **84** (2020) 250-271

Materials Research Forum LLC
https://doi.org/10.21741/9781644900994-11

Chapter 11

State-of-the-Art and Prospective of Aerogels

Kaushalya Bhakar[a], Parul Khurana[b], Sheenam Thatai[c], Dinesh Kumar[a]*

[a]School of Chemical Sciences, Central University of Gujarat, Gandhinagar, India

[b]Department of Chemistry, G. N. Khalsa College of Arts, Science and Commerce, University of Mumbai, Mumbai, India

[c]Amity Institute of Applied Sciences, Amity University, Noida, Sector-125, India

*dinesh.kumar@cug.ac.in

Abstract

The term aerogel is used for the cluster of materials, which exhibits definite geometry and certain properties with no set chemical formula. The astonishing physical properties such as non-toxic, high porosity with a large surface area, comparatively lightweight with other solids, inflammable insulation material. The sol-gel process implies the synthesis of aerogel. The composition of aerogel is controlled by some factors such as starting material, catalyst, and conditions of the preparation method. The physical and chemical properties of aerogel make it an interesting material for textiles, aerospace engineering, construction material and especially for energy-efficient retrofitting opportunities of residential buildings. In this chapter preparation methods of aerogel, state-of-the-art properties, and preparation of aerogel and future perspective of aerogels are discussed.

Keywords

Aerogel, Nanoporous, Sol-Gel Process, Thermal Insulation, Surface Modification.

Contents

Aerogels I: Preparation, Properties and Applications   Materials Research Forum LLC
Materials Research Foundations **84** (2020) 250-271   https://doi.org/10.21741/9781644900994-11

1.   Introduction

Aerogel is a broad term that is used for a group of materials used in space travel since 1960 but now its availability increases in industries because of new findings related to these materials [1]. The superior characteristics of aerogel such as high porosity, the three-dimensional network structure of connected particles simulating a pearl necklace, with pores containing 99% air volume, is considered as the 'marvel material for the 21st century [2]. Their remarkable physical properties make them more attractive among scientists and research centres. These are non-toxic, inflammable thermal insulation material with lightweight nanoporous solids containing fine, open-pore structure resulting in low density in the range between 0.003 and 0.15 kg/m^3 with high porosity and the large surface area of approximately 500-1000 m^2/g. Mostly aerogels are known for their extraordinarily lightweight. Aerogels are considered to be one of the lightest materials synthesized in the laboratory, including the silica aerogel that is only 2-3 times heavier than air and could get even lighter by removing the air from its pore structures as shown in Fig. 1 [3]. This unique arrangement of a solid particle in network structure gives aerogels the status of the lowest density solid materials with an unlimited range of applications and finding their way in different branches such as thermal insulation, kinetic energy, filtration, paints and cosmetics, electronics, optics, nanotechnology, sensing, radiators in Cherenkov detectors, carriers, and biomedicine among others [4-7]. A variety of different terms used for aerogel including frozen smoke, solid smoke, solid air, solid cloud, and blue smoke owing to its translucent nature and the light scattering properties of the material. It feels like delicate expanded polystyrene to touch and is extremely light. The illustrative structure of aerogel is shown below.

Materials Research Forum LLC
https://doi.org/10.21741/9781644900994-11

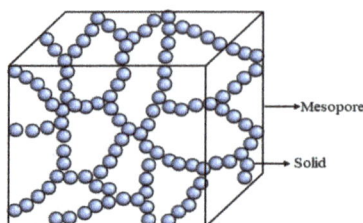

Fig.1 Explanatory structure of the aerogel [3].

To describe an aerogel a lot of different definitions have been discussed in the last few years. According to the terminology of the IUPAC, aerogels are gels comprising nanoporous solids in which the diffused phase is a gas [8]. The term aerogel is not used for a single material, but is used for a group of a material that exhibits definite geometry and certain properties with no set chemical formula and the physical properties of aerogels do not resemble with their name since it may be related to materials which compose aerogel. Recently, aerogel has become a material of attraction to scientists because of its exclusive physical composition that gives it the capability to improve technologies in a variety of fields [9]. The synthesis of these materials came in identity at the beginning of the 1930s, and since that time, many products have been produced, mainly using silica as a starting material. The aerogel first synthesized by Kistler in 1931and who first used the term "aerogel" because of the liquid component of the wet gel exchange by the air without denting the solid microstructure [10]. Silica aerogel is the first synthesized aerogel by the sol-gel method using Na_2SiO_3 as a precursor of silica but saw little development for several decades. After the development of inorganic aerogel, including SiO_2, Al_2O_3, WO_3, Fe_2O_3, SnO_2, and other inorganic oxide aerogels have become one of the most studied compositions. Further improvement comes up with a new method to replacing alcohol by liquid CO_2 before the sample undergoes the supercritical drying process to avoid the dangerous path by heating pure alcohol at high temperature and pressure above its critical point and allow for the separation of liquid from the gel using supercritical-extraction process [11]. In 1980, Pekala [12] expanded the class of

Materials Research Forum LLC
https://doi.org/10.21741/9781644900994-11

aerogel by developing organic and carbon aerogel using the sol-gel process from organic polymer resorcinol formaldehyde. Since the early 2000 easy and cheap methods, new composition, structural reinforcement strategies are developed for their synthesis and aerogel becomes an emerging class of material with great potential of advanced technological application [13]. Organic aerogel exhibits higher mechanical properties as compared to inorganic aerogel [14]. The superior characteristics of the physical and chemical properties of aerogel make it an interesting material for clothing, aerospace application, construction material and especially an insulating material in residential buildings for retrofitting purpose, still, there is a lot of scope to application of aerogel in other technological application [15]. In the current development of new insulating materials, aerogel can be a promising alternative. The aerogels are deemed as one of the most promising families of materials for the insulating application because of their high thermal insulation properties [9]. Specifically, silica aerogel is the most studied aerogel because of its unique physical and mechanical properties, such as the lowest density among all aerogel and high thermal insulation. Traditionally, the synthesis process of all aerogel takes place through three steps: gel formation, aging, or solvent exchange and drying. Additional procedures can be combined to influence the arrangement and properties of the final product. A range of methods is involved in the synthesis of silica aerogel.

2.1 Synthesis of aerogels

The most common procedure used for the synthesis of silica aerogel is the sol-gel method. There are various parameters such as the activity of metal alkoxide, pH of solution, temperature, and nature of the solvents which can influence the sol-gel process. By varying these parameters, the material with a different functionalized surface can be obtained. Various salts, oxides, hydroxide, alkoxides, and amines are used as precursors provided these are soluble in a solvent. The main precursor for silica aerogel is silicon alkoxide. The sol-gel structure starts from the chemical breakdown of precursor in the presence of water and subsequent condensation reaction as shown in Fig. 2 [16]. The gelation is carried out by the addition of a catalyst and may take several days. To decrease the gelation time from several days to an hour; acid or base catalyst is added during this reaction [9]. The gelation is achieved when a continuous network structure is formed, and the solution no longer flows under the influence of gravity [16].The sol-gel reaction can be summarized according to the scheme where the breakdown of the compound takes place because of the reaction with water and condensation occurs parallel.

Materials Research Forum LLC
https://doi.org/10.21741/9781644900994-11

Hydrolysis

$$R_3M-OR + H_2O \leftrightarrow R_3-M-OH + R-OH$$

Water condensation

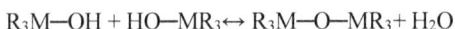

$$R_3M-OH + HO-MR_3 \leftrightarrow R_3M-O-MR_3 + H_2O$$

Alcohol condensation

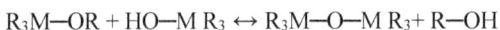

$$R_3M-OR + HO-M R_3 \leftrightarrow R_3M-O-M R_3 + R-OH$$

After the sol reaches to the gel point, the silica spine contains several unreacted alkoxide groups, which can hydrolyse and condense for further strengthening of the network. The continuous process of chemical bond formation and hardening because of additional condensation reactions so as to change the physical state of matter is called aging. Drying is an essential step in the aerogel production, and relies on the extraction of the solvent from the matrix without disturbing the gel network producing a highly porous solid material with unaffected size and shape [17]. During drying, the break-down of the gel network occurs due to the capillary forces which arise because of intramolecular forces between the liquid-vapor interfaces set up in the narrow pores. This small diameter of the pores generates a large solid-liquid interface, which makes interfacial capillary forces capable to collapse the microstructure. The evaporation of solvent from wet gel to get the final product is the most crucial step. There are several drying techniques used, such as supercritical drying, ambient pressure drying freeze-drying. Supercritical extraction is the first and most effective process used for the preparation of aerogel. In this process, the liquid components of the wet gel are transformed into a gaseous state in the absence of surface tension and with minimal structural changes [13]. The supercritical CO_2, the liquid state of carbon dioxide is used as a drying agent to replace the organic solvent in the aerogel with various advantages such as the recovery and reuse of pure solvent and supercritical CO_2 can be expanded to the gaseous state avoiding unwanted phase change from liquid to gas. In contrast, when the solvent is removed from the gel using other traditional methods like evaporation under normal non-supercritical conditions, the product is known as xerogel. In this drying process, it may keep its original structure but often cracks because of the extreme shrinkage caused by intramolecular forces between the vapor-liquid surface. In the supercritical drying process, controlling the heat and pressure at the critical point of the drying medium is very difficult and expensive. So, in recent years, ambient pressure drying has become popular in the scientific community. Many hydroxyl clusters present on the surface of the gel, which is responsible for the hydrophilic behaviour of the wet gel. These hydrophilic groups absorb moisture and because of the absorption of moisture, a large amount of attraction generates between the groups present on the surface area which damages the brittle 3-D composition of the wet gel. To maintain the composition of the wet gel, surface modification should be performed to change the hydrophilic behaviour of the wet gel.

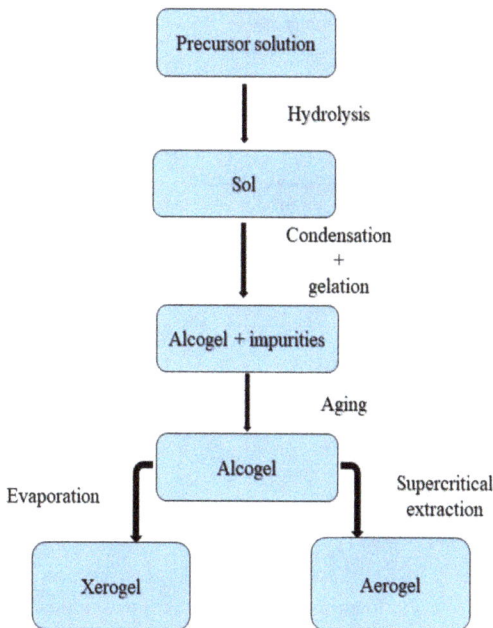

Fig. 2 Preparation of aerogel [16].

The classification of aerogel is based upon their appearance, preparation method and chemical structure and classification based on the chemical structure can be further divided into single element aerogel and composite aerogel as given in Fig. 3 [18]. The

Materials Research Forum LLC
https://doi.org/10.21741/9781644900994-11

single element aerogel includes inorganic aerogel such as silica-based aerogel and organic aerogel. The composite aerogel contains organic-inorganic hybrid aerogel and described as the material in which organic fragments are incorporated within the structural component of inorganic materials. The primary purpose behind this combination (organic-inorganic aerogel) is to improve the properties of aerogel.

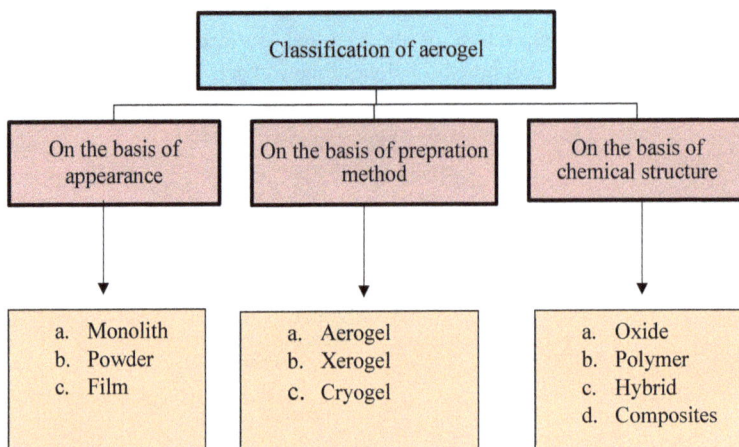

Fig.3 Classification of aerogel [18].

3.   State-of-the-art of aerogel

Since their first synthesis in 1930, aerogel has undergone tremendous advancement and changes throughout the years. Continuous development to synthesized numerous type of composite aerogel with extraordinary features increase their potential application in the various field especially in the building to make sound insulation rooms, air purificator to remove indoor contaminants, aerogel blankets and also open the door to the new ranges of opportunities in other fields like textile, electronics, aerospace, coating, and so on [19].

3.1   State-of-the-art properties of aerogel

Since the development of aerogel, various methods, precursors, and techniques have been used to improve their thermal, mechanical, and physical properties. Traditionally synthesized SiO_2 aerogel has poor mechanical properties and is difficult to use in thermal insulation and energy storage. Wang and co-workers [20] synthesized polyvinylpolymethylsiloxane aerogel with improved thermal conductivity and high surface area through free radical polymerization and hydrolytic polycondensation process. This silica aerogel holds up with an organic component that exhibits better mechanical strength with a high specific surface area. This doubly crosslinked polyvinyl polymethylsiloxane (PVPMS) aerogel resolves the brittleness of silica aerogel and inhibits cracking during the drying process [21-23]. Usually, silica aerogel has poor mechanical properties that limit its application commercially. The various processes developed to improve its mechanical properties [24-26] include the combination of fibers with aerogel [27-29], nanomaterial [30], changing precursor content [31,32], crosslinking by isocyanates [33-38] and polystyrene [39]. However, the main drawback of using fibers with silica aerogel is the dust drop due to the diameter difference between aerogel and fibers so fiber surface cannot be covered with silica aerogel and creates a void which increases the thermal conductivity of aerogel [40]. Safi and co-workers [41] developed a new method by impregnating silica gel on the composite of silica aerogel and glass fiber to improve its mechanical strength and heat insulation property. The group used silica gel particles for binding and improved the strength of the network of siloxane among the fibers by formulating a chemical bond with silica aerogel through the hydroxyl group of silica gel surface. The cavities formed between the silica aerogel and glass fiber composites were filled with silica gel, which not only improves its mechanical properties but also gives low thermal conductivity. The characterization of synthesized silica aerogel is done by using scanning electron microscopy (SEM), fourier transmission infrared (FTIR) spectroscopy, universal tensile testing machine, and thermal conductivity is measured by thermal constants analyzer. These synthesized aerogels possess excellent mechanical properties with super ultra-low heat insulation. Resorcinol formaldehyde (RF) aerogel is the most popular organic aerogel synthesized by sol-gel polycondensation of resorcinol and formaldehyde [42,43] and possess higher mechanical strength than silica aerogel. These properties are also influenced by the molar ratio of resorcinol, formaldehyde, and catalyst. The cross-linking of aerogel using cross-linking agent isocyanate is one way to improve its mechanical strength. The three-dimensional porous network of aerogel is mainly formed by primary particle and their size lies between 7-10 nanometres. By the aggregation of these particles spherical shaped secondary particles formed. The compression forces broke the interconnected network of secondary particles while the primary particles do not affect by these forces. To enhance the strength of these particles, the aerogel is cross-linked with isocyanate, which leads to stronger aerogel. Aghabararpour and co-workers [35] prepared isocyanate cross-linked RF aerogel and used two different methods in situ and ex situ for cross-linking. During the in situ

Materials Research Forum LLC
https://doi.org/10.21741/9781644900994-11

process, the isocyanate which is responsible for crosslinking was added during the preparation of sol while in ex situ method, the wet gel was placed into the solution of the crosslinking agent. In the course of in situ process water is present in formaldehyde, which is responsible for the formation of urea linkage besides urethane groups. The urea linkage is stronger than urethane and forms a more robust aerogel in comparison to the ex situ process in which only the urethane linkage is present. Aerogels are excellent thermal insulator but their application in many fields restricted due to their poor mechanical properties. Researchers are continuously working in the direction to improve the mechanical strength of these materials. Aghabararpour and co-workers (2019) [44] investigated the effect of various crosslinker molecule structures on the mechanical and thermal properties of RF aerogel. In an earlier study; Aghabararpour [35], the researchers used only one crosslinking agent isocyanate to study the mechanical properties of RF aerogel but in the present study [44] three crosslinking agent were used: 1) crosslinking reaction by the addition of isocyanate, 2) crosslinking via the reaction of mono ethylene glycol(EG) and isocyanate groups(NCO), and 3) crosslinking by the reaction of glycerol(GL) and NCO groups. The sequencing of the addition of the crosslinking agent is the major difference between the two studies. In the previous study crosslinking agent is added in the wet gel after the step of aging, while in the present report, the process of crosslinking is carried out on final dried aerogel. In this process, the crosslinking agent disperses into the sample quickly and determination of mass diffusion is possible quantitatively but in the other studies, the diffusion of the crosslinking agent in the wet gel takes several days and only a qualitative analysis of a diffused mass is carried out. The morphology of synthesized aerogel is characterized by using SEM analysis and Brunauer-Emmett-Teller (BET) theory. Several organic components are used for the surface modification of the inorganic aerogel which is hazardous for the environment [45]. Organic aerogels derived from biocompatible and renewable materials have attained large interest in recent years. The nanocellulose based aerogels have drawn great attention in the research area. The high thermal and chemical stability and biodegradability make cellulose nanofiber-based aerogel good heat insulator. Zhou et al. [46] synthesized polyvinyl alcohol/cellulose nanofiber (CNF) aerogel/gelatin crosslinked hybrid organic aerogel by freeze-drying which is an eco-friendly method and studied the mechanical performance and thermal stability. The crosslinking is one of the appropriate ways for increasing the mechanical strength of aerogel. In this study gelatine, a biobased nontoxic and inexpensive material was used as a crosslinking agent and combined with the polyvinyl alcohol and cellulose nanofiber through hydrogen bonding. The freeze-drying technique is used for the drying process to avoid the collapsing of the network due to capillary pressure and it is an inexpensive and environment-friendly process for preparing CNF aerogel as compare to the supercritical drying method. The mechanical properties, thermal conductivity, density morphology examined. The composites provided outstanding properties and will have considerable applications as a thermally insulating material.

3.2 State-of-the-art of preparation of aerogel

Aerogel has attracted the attention of the researcher community after 1980; since then different material has been used for the synthesis of numerous aerogel composites. Nowadays, to develop a highly porous material of aerogel, several types of chemical components are used which are dangerous for living species and removal of these components from the environment is very difficult [47]. The development of a green method for preparing silica aerogel with excellent mechanical features has remained a big challenge.

Hu and co-workers [48] developed a green route for the synthesis of an ant inspired superelastic silicone aerogel. The robust network of this aerogel is carefully planned by regulating the hydrolytic condensation of the SiH_4 group. The properties of developed aerogel depend on precursor and reaction conditions [49,50]. The synthesis occurred by the sol-gel procedure in which methyltrimethoxysilane (MTMS) and dimethyldimethoxysilane (DMDMS) were used as a precursor with the presence of acetic acid in trace amount..Kwon et al. [51] prepared and characterized spherical polyimide aerogel microparticles where polyimide (PI) used as a precursor for the development of aerogel with high mechanical strength and chemical stability. Polyimide based aerogel widely studied as an organic aerogel because of tremendous electrical and thermal properties with good mechanical strength and wide application [52-55]. PI aerogel microparticles have a convincing application in insulation, medical implants and catalyst support over film-shaped aerogel. However, after further studies, Lee et al. [56] introduced a new swelling method based on a spherulitic formation mechanism to synthesize spherical polyimide aerogel with tunable pore size. The group focused on the development mechanism of thermally treated polyimide aerogel as a semi-crystalline spherical polymer. They studied the changes in the formation mechanism of polyimide aerogel and also the effect of the swelling method on the properties of the newly synthesized polyimide aerogel such as pore size and shape. The concentration of polyamic acid and interaction between the solvent and polymer is affected by the swelling agent and thus modified the lamellar arrangement.

Cellulose-based aerogel has become a hotspot in research in proposing great features such as sustainability, biodegradability, renewability, reactive surface, high surface area and porosity. Cellulose is an abundant natural resource and seems to be an excellent and novel oil adsorbent. Shi et al. [57] fabricated pomelo peel-based aerogel for usage as an

Materials Research Forum LLC
https://doi.org/10.21741/9781644900994-11

adsorbent for the extraction of lubricant and organic components. Different fibers and natural material act as a precursor to synthesized aerogel. The attractive feature of aerogel such as low density, three-dimensional network structure contributes majorly to the adsorption properties. The major motivation behind the intensive study on the development of cellulose aerogel because of eco- friendly, low cost, renewable properties make them potential alternative adsorbent [58]. To achieve the adsorption properties of aerogel, hydrophobic surface modification is a necessary step and can be done via esterification, nanocoating, carbonization, and salinization, etc. Pomelo peel used as a starting material which is pectin- rich fruit found in south-eastern Asia and it contains a rich amount of insoluble polysaccharide and lignin [59]. In this work, researchers prepared pomelo- peel based sponge aerogel via high-speed diffusion, along with the freeze-drying process, and silanization $CH_3Si(OCH_3)_3$ and surface modification is taking place to convert into hydrophobic sponge aerogel. This aerogel was synthesized by the simple method without any harm to the environment. The morphological study is done by scanning electron microscopy and energy dispersed X-ray microscopy and FTIR and energy dispersive X-ray spectroscopy (EDX) techniques used for the identification of the chemical composition of the aerogel surface. The adsorption capacity, reusability, thermal properties, and wettability were also studied. Ye et al. [60] synthesized reticulated SiC aerogel using a template method and studied mechanical and thermal properties. The SiC skeleton is a deposit in the 3-D network porous structure of carbon foam which functions as a template and silica aerogel deposit into the SiC network. The commercially available grade malemain foam (MF) acts as a precursor of carbon foam by pyrolysis at high temperature for 4 hours, lower density carbon foam template obtained with the retention in 3-D opening network with highly porous structure. The deposition of an ultra-thin layer of SiC takes place on the outer space of the carbon foam by the chemical deposition method. The density of starting material malemain foam is 7.73 Kg/m^3, and it is reduced to 6.8 Kg/m^3 of carbon foam skeleton. The silica aerogel is synthesized by using the sol-gel method. The Si/C aerogel composite is acquired by using the ambient pressure drying method. The carbon foam which is synthesized by pyrolysis of starting material MF is elastic and easily collapsed by shrinkage of silica aerogel. So, to prepare SiC skeleton MF used as a template that converts carbon foam into the SiC network and SiC coating is uniformly dispersed on the 3-D open skeleton of carbon foam. The morphology of SiC/ aerogel is investigated by SEM and X-ray diffraction (XRD). Thermal and mechanical properties study shown that the ultrathin net-like shape SiC skeleton remarkably improved compressive properties without affecting the heat insulation properties. The application of synthesized material may be increasing in the thermal insulation system. Wu and co-workers [61] developed fluorescent aerogel based on chemical bond formation between nanocellulose and carbon dots for the optic sensors.

Cellulose nanofiber offers various properties with low mechanical strength and hydrophilic nature which make aerogel structure collapse in water [62]. These drawbacks limited the application of nanocellulose based aerogel; so surface modification is one of the efficient approaches by using a crosslinking agent to react with the surface-active OH group of the nanocellulose to improve the poor mechanical properties of aerogel but most of the crosslinking agent are organic solvents and toxic which affects the nanotoxicology of the nanocellulose fiber (CNF) aerogel. In the following study novel and facile method is applied to synthesized nanocellulose based aerogel. The hazardous solvent-free fabrication path which depends on two natural bio-nanomaterial carboxylated cellulose nanofibers and carbon dots with an amino-modified surface was used to develop fluorescent aerogel. Cellulose nanofiber is an abundant natural material that is used as a precursor for the fabrication of aerogel and carbon dots are an appropriate fluorescent source to replace toxic metal-based quantum dots and organic dyes because they are biocompatible material and do not give photo-degradation reaction the presence of light [63-65]. There is covalent bonding that takes place between the carboxyl group of cellulose nanofiber and amino clusters of the carbon dots by a condensation reaction which gives structural stability to the aerogel. Carbon dots work as the fluorescent source and crosslinking agent which maintains the high porosity structure of the skeleton and improves the mechanical strength of aerogel by forming a three-dimensional network structure. Organic citric acid used as a source of carbon to produce carbon dots by hydrothermal treatment. The cellulose nanofiber aerogel was fabricated by the sol-gel process and subsequent freeze-drying process which is placed with amino-modified carbon dots aqueous suspension to form covalent bonding by a condensation reaction between the carboxyl group of carbon nano cellulose and the OH clusters of the carbon dots. The reaction started by the N-hydroxysuccinimideand N-(3-dimethylaminopropyl)-N'-ethylcarbodiimide hydrochloridecatalyst for one hour. Purification of aerogel takes place by washing with water 4-5 times to remove free carbon dots molecules and after the freeze-dried aerogel labels as fluorescent aerogel. The fabricated aerogel is highly sensitive to detect the NO_x and aldehyde group as an optic sensor with excellent characteristics such as a lightweight, high surface area with a highly porous structure. The morphology of cellulose nanofiber and carbon dots was observed by transmission electron microscopy. Fluorescent properties of cellulose nanofiber and carbon dots were detected on a UV spectrophotometer. When the pure cellulose nanofiber is placed in the UV radiation and day sunlight it does not display fluorescence. After the modification with carbon dots component, all the aerogels exhibit clear blue fluorescence, which proved that the fluorescence properties of aerogel are maintained during the formation of a covalent linkage between CD and CNF components. Lu et al. [66] synthesized resilient SiC nanowire aerogel with high-temperature stability. The researchers developed a novel method for preparing SiC nanowire aerogel (SiCNWAGs) in large amounts with the desired shape and low density. Recently few different ceramic aerogels developed from flexible nanostructure using various bottom-up methods, such as recoverable SiC nanowire fabricated by chemical vapor deposition method [67], boron nitride nanobelts aerogel developed by the freeze-drying method [68] and so on. But the complicated synthesis procedure and high-cost of raw material made their synthesis difficult at the commercial level. For the development of SiC nanowires, the microstructure of carbon fibers skeleton works as a three-dimensional space. Then the removal of carbon fiber skeleton is carried out using thermal-oxidation etching, which results in the formation of SiC nanowire aerogel. The aerogel possesses a high porosity, lightweight, good mechanical strength, high-temperature stability, and SiC nanowire are interconnected with each other. The thermal conductivity of bulk SiC material is 490 $Wm^{-1}K^{-1}$ in SiC nanowire aerogel.

4.   Future prospective of aerogel

Aerogel can be considered as a special form of solid that can be combined with a variety of components to produce a different type of composite aerogel such as silica, oxide, polymer, carbon, nanomaterial, and so on.   Because of their ultra-low thermal conductivity, acoustic properties, low refractive index and dielectric constant and low density among all solid, it is an excellent material with many applications in different fields and these properties are generated by choice of precursor; the method used for the fabrication and optimization techniques. The addition of new materials into the aerogel will enable them to be used in new technological developments like a new supercapacitor, biosensor, antimicrobial coating, oil spill absorbing pads much more. The drying process is very delicate for aerogel to remove the solvent from nanopores without disturbing its structure is achieved by different methods [69]. The supercritical drying is an expensive and complicated process so the ambient pressure drying techniques can be an alternative that will undoubtedly make industrial production of aerogel much cheaper. Some factors affect the structure of aerogel such as a starting material of preparation, catalyst, and conditions of preparation method. Aerogel structure, properties, and the application can be changed by changing these factors. Silicon dioxide aerogels have some physical and environmental advantages in comparison with other materials available in the market. The researchers continuously work on the design and development of next-generation aerogels with superior properties to overcome limitations of the application of these materials in various fields such as biomedical technology. To remove this gap recent development of scientific-technological networking, take an initiative to assemble various international experts and researchers to discuss the more effective and commercially available approach for the development of aerogel [70].

4.1 Thermal insulation

Aerogel can be deemed as the "future of thermal insulating material "of the new era. Ultra-low thermal conductivity with optical transparency is one of the exciting features of aerogel, which is 1-10% of a solid and allows its application in roofs, windowpanes, solar collector covers. As the porous media have a composition in both solid and gas phases, therefore the process of heat transfer could be subdivided into heat transfer through a gaseous phase and heat transfer through a solid medium. The conduction of heat in the gaseous phase is caused by different mediums, such as a collision between the gas molecules [13]. Silica aerogel considers as one of the promising materials for the insulation in the building. However, intensive work will reduce the manufacturing cost of the silica aerogel to make it commercially available. An aerogel blanket is a combination of silica aerogel with fibrous material to strengthening the aerogel and improves poor mechanical features into a durable, flexible and hydrophobic material, suitable for building envelopes, inside or outside and it manufactured and sold in various countries [71]. In recent years, some aerogel products used to remove indoor air pollutants and clean-up of the outdoor environment [72].

4.2 Drug delivery

Aerogel is a lightweight solid with high porous structure and surface area, biocompatible material that enhances the attention of researchers as a component of drug delivery vehicles [73-75]. Earlier, synthetic polymers such as polyethylene, polypropylene, and polydimethylsiloxane have been used for drug delivery systems; however, non-biocompatibility of these polymers is challenging. The polysaccharide is considered a natural polymer and considered for a drug delivery system.

4.3 Energy storage device

Increasing environmental pollution draws the attention of researchers to develop alternative renewable energy sources that provide clean energy. The supercapacitor batteries are a good source of energy storage with eco-friendly and renewable properties. Especially carbon material is frequently used as an electrochemical capacitor because of its lightweight structure, high electrical conductivity, and high surface area [76]. Carbon-based aerogel can be an attractive material because of its superior characteristics. Aerogel is a three-dimensional nanoporous skeleton that can store more electrical energy than a conventional capacitor because of a large specific surface area. Many aerogel composites are developed to be used in energy storage devices like in lithium-ion batteries; the high surface area of aerogel provides many reaction sites for lithium-ion batteries to improve energy storage capacity [77]. Graphene is an outstanding electrical and thermal conducting material with a 2-D structure, has great energy application but to improve its active sites, three-dimensional graphene aerogel is developed which becomes a promising material for the energy system because of its hierarchical structure [78].

There is a wide range of applications of aerogel like in tissue engineering, cosmetics, construction material, space application, separation technology, and sensing. It may increase in the coming years with the development of new methods and different aerogels.

Acknowledgments

Dinesh Kumar is thankful to DST, New Delhi, for the financial support offered to this work (sanctioned vide project Sanction Order F. No. DST/TM/WTI/WIC/2K17/124(C).

References

[1] I. Smirnova, P. Gurikov, Aerogel production: Current status, research directions, and future opportunities, J. Supercrit. Fluids. 134 (2018) 228–233. https://doi.org/10.1016/j.supflu.2017.12.037

[2] N. Hüsing, U. Schubert, Aerogels—Airy Materials: Chemistry, Structure, and Properties, Angew. Chemie Int. Ed. 37 (1998) 22–45. https://doi.org/10.1002/1521-3773(19980202)37:1/2<22::aid-anie22>3.3.co;2-9

[3] S. Gopi, P. Balakrishnan, V.G. Geethamma, A. Pius, S. Thomas, Applications of cellulose nanofibrils in drug delivery, Elsevier Inc., 2018. https://doi.org/10.1016/b978-0-12-813741-3.00004-2

[4] R.B. Malla, A. Maji, Engineering, construction, and operations in challenging environments Earth & space 2004: proceedings of the ninth bennal ASCE Aerospace Division international conference on engeneering, construction, and operations in challenging environments, March 7-10, 2004, League City, Houston, Texas, ASCE, 2004.

[5] J. Fricke, A. Emmerling, Aerogels - Recent Progress in Production Techniques and Novel Applications, J. Sol-Gel Sci. Technol. 13 (1998) 299–303. DOI:10.1023/A:1008663908431

[6] M. Schmidt, F. Schwertfeger, Applications for silica aerogel products, J. Non. Cryst. Solids. 225 (1998) 364–368. https://doi.org/10.1016/S0022-3093(98)00054-4

Materials Research Forum LLC
https://doi.org/10.21741/9781644900994-11

[7] G. Herrmann, R. Iden, M. Mielke, F. Teich, B. Ziegler, On the way to commercial production of silica aerogel, J. Non. Cryst. Solids. 186 (1995) 380–387. https://doi.org/10.1016/0022-3093(95)90076-4

[8] R.G. Jones, Compendium of polymer terminology and nomenclature: IUPAC recommendations, 2008, RSC Pub., Cambridge, 2009.

[9]   G.S. S, L.B. C, M. Engineering, S.N.D. Coe, A Review on Aerogel An Introduction, 0072 (2018) 4098–4101.

[10] S.S. Kistler, Coherent expanded aerogels and jellies [5], Nature. 127 (1931) 741. https://doi.org/10.1038/127741a0

[11] N. Bheekhun, A. Rahim, A. Talib, M.R. Hassan, JAIME AROCHAl Universidad Nacional de Colombia, 2013 (2013). https://doi.org/10.1155/2013/406065

[12] R.W. Pekala, Organic aerogels from the polycondensation of resorcinol with formaldehyde, J. Mater. Sci. 24 (1989) 3221–3227. https://doi.org/10.1007/BF01139044

[13] S. Montes, H. Maleki, Aerogels and their applications, Elsevier Inc., 2020. https://doi.org/10.1016/b978-0-12-813357-6.00015-2

[14] C. Tan, B.M. Fung, J.K. Newman, C. Vu, Organic Aerogels with Very High Impact Strength, Advanced Materials. 13 (2001) 644–646. doi:10.1002/1521-4095(200105)13:9<644::aid-adma644>3.0.co;2-#

[15] E. Cuce, P.M. Cuce, C.J. Wood, S.B. Riffat, Toward aerogel based thermal superinsulation in buildings: A comprehensive review, Renew. Sustain. Energy Rev. 34 (2014) 273–299. https://doi.org/10.1016/j.rser.2014.03.017

[16] T. Linhares, M.T. Pessoa De Amorim, L. Durães, Silica aerogel composites with embedded fibres: A review on their preparation, properties and applications, J. Mater. Chem. A. 7 (2019) 22768–22802. https://doi.org/10.1039/c9ta04811a

[17] U. Schubert, Part One Sol – Gel Chemistry and Methods, Sol-Gel Handb. Synth. Charact. Appl. (2015) 1–28. https://doi.org/10.1002/9783527670819.ch01

[18] A. Du, B. Zhou, Z. Zhang, J. Shen, A special material or a new state of matter: A review and reconsideration of the aerogel, Materials (Basel). 6 (2013) 941–968. https://doi.org/10.3390/ma6030941

[19] S. Araby, A. Qiu, R. Wang, Z. Zhao, C.H. Wang, J. Ma, Aerogels based on carbon nanomaterials, J. Mater. Sci. 51 (2016) 9157–9189. https://doi.org/10.1007/s10853-016-0141-z

[20] L.Wang, J. Feng, Y. Jiang, L. Li, J. Feng, Thermal conductivity of polyvinylpolymethylsiloxane aerogels with high specific surface area, RSC Adv. 9 (2019) 7833–7841. https://doi.org/10.1039/C8RA10493J

[21] G. Zu, T. Shimizu, K. Kanamori, Y. Zhu, A. Maeno, H. Kaji, J. Shen, K. Nakanishi, Transparent, Superflexible Doubly Cross-Linked Polyvinylpolymethylsiloxane Aerogel Superinsulators via Ambient Pressure Drying, ACS Nano. 12 (2018) 521– 532. https://doi.org/10.1021/acsnano.7b07117

[22] G. Zu, K. Kanamori, T. Shimizu, Y. Zhu, A. Maeno, H. Kaji, K. Nakanishi, J. Shen, Versatile Double-Cross-Linking Approach to Transparent, Machinable, Supercompressible, Highly Bendable Aerogel Thermal Superinsulators, Chem. Mater. 30 (2018) 2759–2770. https://doi.org/10.1021/acs.chemmater.8b00563

[23] G. Zu, K. Kanamori, A. Maeno, H. Kaji, K. Nakanishi, Superflexible Multifunctional Polyvinylpolydimethylsiloxane-Based Aerogels as Efficient Absorbents, Thermal Superinsulators, and Strain Sensors, Angew. Chemie - Int. Ed. 57 (2018) 9722–9727. https://doi.org/10.1002/anie.201804559

[24] M.A.B. Meador, E.J. Malow, R. Silva, S. Wright, D. Quade, S.L. Vivod, H. Guo, J. Guo, M. Cakmak, Mechanically strong, flexible polyimide aerogels cross-linked with aromatic triamine, ACS Appl. Mater. Interfaces. 4 (2012) 536–544. https://doi.org/10.1021/am2014635

[25] L. Li, B. Yalcin, B.N. Nguyen, M.A.B. Meador, M. Cakmak, Flexible nanofiber-reinforced aerogel (Xerogel) synthesis, manufacture, and characterization, ACS Appl. Mater. Interfaces. 1 (2009) 2491–2501. https://doi.org/10.1021/am900451x

[26] J.P. Randall, M.A.B. Meador, S.C. Jana, Tailoring mechanical properties of aerogels for aerospace applications, ACS Appl. Mater. Interfaces. 3 (2011) 613–626. https://doi.org/10.1021/am200007n

[27] K.E. Parmenter, F. Milstein, Mechanical properties of silica aerogels, J. Non. Cryst. Solids. 223 (1998) 179–189. https://doi.org/10.1016/S0022-3093(97)00430-4

[28] S. Motahari, A. Abolghasemi, Silica aerogel-glass fiber composites as fire shield for steel frame structures, J. Mater. Civ. Eng. 27 (2015) 1–7. https://doi.org/10.1061/(ASCE)MT.1943-5533.0001257

[29] Z. Deng, J. Wang, A. Wu, J. Shen, B. Zhou, High strength SiO2 aerogel insulation, J. Non. Cryst. Solids. 225 (1998) 101–104. https://doi.org/10.1016/S0022-3093(98)00106-9

Materials Research Forum LLC
https://doi.org/10.21741/9781644900994-11

[30] Y. Zhang, Y. Shen, D. Han, Z. Wang, J. Song, L. Niu, Reinforcement of silica with single-walled carbon nanotubes through covalent functionalization, J. Mater. Chem. 16 (2006) 4592–4597. https://doi.org/10.1039/b612317a

[31] M. Schwan, R. Tannert, L. Ratke, New soft and spongy resorcinol-formaldehyde aerogels, J. Supercrit. Fluids. 107 (2016) 201–208. https://doi.org/10.1016/j.supflu.2015.09.010

[32] A. Léonard, S. Blacher, M. Crine, W. Jomaa, Evolution of mechanical properties and final textural properties of resorcinol-formaldehyde xerogels during ambient air drying, J. Non. Cryst. Solids. 354 (2008) 831–838. https://doi.org/10.1016/j.jnoncrysol.2007.08.024

[33] L.A. Capadona, M.A.B. Meador, A. Alunni, E.F. Fabrizio, P. Vassilaras, N. Leventis, Flexible, low-density polymer crosslinked silica aerogels, Polymer (Guildf). 47 (2006) 5754–5761. https://doi.org/10.1016/j.polymer.2006.05.073

[34] B.N. Nguyen, M.A.B. Meador, A. Medoro, V. Arendt, J. Randall, L. McCorkle, B. Shonkwiler, Elastic behavior of methyltrimethoxysilane based aerogels reinforced with tri-isocyanate, ACS Appl. Mater. Interfaces. 2 (2010) 1430–1443. https://doi.org/10.1021/am100081a

[35] M. Aghabararpour, M. Mohsenpour, S. Motahari, A. Abolghasemi, Mechanical properties of isocyanate crosslinked resorcinol formaldehyde aerogels, J. Non. Cryst. Solids. 481 (2018) 548–555. https://doi.org/10.1016/j.jnoncrysol.2017.11.048

[36] A. Katti, N. Shimpi, S. Roy, H. Lu, E.F. Fabrizio, A. Dass, L.A. Capadona, N. Leventis, Chemical, physical, and mechanical characterization of isocyanate cross-Linked amine-modified silica aerogels, Chem. Mater. 18 (2006) 285–296. https://doi.org/10.1021/cm0513841

[37] N. Leventis, C. Sotiriou-Leventis, G. Zhang, A.M.M. Rawashdeh, Nanoengineering Strong Silica Aerogels, Nano Lett. 2 (2002) 957–960. https://doi.org/10.1021/nl025690e

[38] G. Zhang, A. Dass, A.M.M. Rawashdeh, J. Thomas, J.A. Counsil, C. Sotiriou-Leventis, E.F. Fabrizio, F. Ilhan, P. Vassilaras, D.A. Scheiman, L. McCorkle, A. Palczer, J.C. Johnston, M.A. Meador, N. Leventis, Isocyanate-crosslinked silica aerogel monoliths: Preparation and characterization, J. Non. Cryst. Solids. 350 (2004) 152–164. https://doi.org/10.1016/j.jnoncrysol.2004.06.041

[39] B.N. Nguyen, M.A.B. Meador, M.E. Tousley, B. Shonkwiler, L. McCorkle, D.A. Scheiman, A. Palczer, Tailoring elastic properties of silica aerogels cross-linked with

polystyrene, ACS Appl. Mater. Interfaces. 1 (2009) 621–630.
https://doi.org/10.1021/am8001617

[40] M.A.B. Meador, E.F. Fabrizio, F. Ilhan, A. Dass, G. Zhang, P. Vassilaras, J.C.
Johnston, N. Leventis, Cross-linking amine-modified silica aerogels with epoxies:
Mechanically strong lightweight porous materials, Chem. Mater. 17 (2005) 1085–
1098. https://doi.org/10.1021/cm048063u

[41] S. Shafi, R. Navik, X. Ding, Y. Zhao, Improved heat insulation and mechanical
properties of silica aerogel/glass fiber composite by impregnating silica gel, J. Non.
Cryst. Solids. 503–504 (2019) 78–83. https://doi.org/10.1016/j.jnoncrysol.2018.09.029

[42] C.Q. Hong, J.C. Han, X.H. Zhang, J.C. Du, Novel nanoporous silica aerogel
impregnated highly porous ceramics with low thermal conductivity and enhanced
mechanical properties, Scr. Mater. 68 (2013) 599–602.
https://doi.org/10.1016/j.scriptamat.2012.12.015

[43] D. Shi, Y. Sun, J. Feng, X. Yang, S. Han, C. Mi, Y. Jiang, H. Qi, Experimental
investigation on high temperature anisotropic compression properties of ceramic-fiber-
reinforced SiO2 aerogel, Mater. Sci. Eng. A. 585 (2013) 25–31.
https://doi.org/10.1016/j.msea.2013.07.029

[44] M. Aghabararpour, M. Mohsenpour, S. Motahari, A. Ghahreman, Mechanical and
thermal insulation properties of isocyanate crosslinked resorcinol formaldehyde
aerogel: Effect of isocyanate structure, J. Appl. Polym. Sci. 136 (2019).
https://doi.org/10.1002/app.48196

[45] Y. Pan, S. He, L. Gong, X. Cheng, C. Li, Z. Li, Z. Liu, H. Zhang, Low thermal-
conductivity and high thermal stable silica aerogel based on MTMS/Water-glass co-
precursor prepared by freeze drying, Mater. Des. 113 (2017) 246–253.
https://doi.org/10.1016/j.matdes.2016.09.083

[46] T. Zhou, X. Cheng, Y. Pan, C. Li, L. Gong, Mechanical performance and thermal
stability of polyvinyl alcohol–cellulose aerogels by freeze drying, Cellulose. 26 (2019)
1747–1755. https://doi.org/10.1007/s10570-018-2179-3

[47] A.C. Pierre, G.M. Pajonk, Chemistry of aerogels and their applications, Chem. Rev.
102 (2002) 4243–4265. https://doi.org/10.1021/cr0101306

[48] T. Hu, L. Li, J. Zhang, Green Synthesis of Ant Nest-Inspired Superelastic Silicone
Aerogels, ACS Sustain. Chem. Eng. 6 (2018) 11222–11227.
https://doi.org/10.1021/acssuschemeng.8b03141

[49] Š. Kadochová, J. Frouz, Thermoregulation strategies in ants in comparison to other social insects, with a focus on red wood ants (Formica rufa group), F1000Research. 2 (2014) 1–16. https://doi.org/10.12688/f1000research.2-280.v2

[50] K.W. Allen, Silane coupling agents, second edition, 1992. https://doi.org/10.1016/0143-7496(92)90011-j

[51] J. Kwon, J. Kim, T. Yoo, D. Park, H. Han, Preparation and characterization of spherical polyimide aerogel microparticles, Macromol. Mater. Eng. 299 (2014) 1081–1088. https://doi.org/10.1002/mame.201400010

[52] Y. Chen, G. Shao, Y. Kong, X. Shen, S. Cui, Facile preparation of cross-linked polyimide aerogels with carboxylic functionalization for CO 2 capture, Chem. Eng. J. 322 (2017) 1–9. https://doi.org/10.1016/j.cej.2017.04.003

[53] J. Feng, X. Wang, Y. Jiang, D. Du, J. Feng, Study on Thermal Conductivities of Aromatic Polyimide Aerogels, ACS Appl. Mater. Interfaces. 8 (2016) 12992–12996. https://doi.org/10.1021/acsami.6b02183

[54] J. Kim, J. Kwon, M. Kim, J. Do, D. Lee, H. Han, Low-dielectric-constant polyimide aerogel composite films with low water uptake, Polym. J. 48 (2016) 829–834. https://doi.org/10.1038/pj.2016.37

[55] C. Chidambareswarapattar, Z. Larimore, C. Sotiriou-Leventis, J.T. Mang, N. Leventis, One-step room-temperature synthesis of fibrous polyimide aerogels from anhydrides and isocyanates and conversion to isomorphic carbons, J. Mater. Chem. 20 (2010) 9666–9678. https://doi.org/10.1039/c0jm01844a

[56] D. Lee, J. Kim, S. Kim, G. Kim, J. Roh, S. Lee, H. Han, Tunable pore size and porosity of spherical polyimide aerogel by introducing swelling method based on spherulitic formation mechanism, Microporous Mesoporous Mater. 288 (2019) 109546. https://doi.org/10.1016/j.micromeso.2019.06.008

[57] G. Shi, Y. Qian, F. Tan, W. Cai, Y. Li, Y. Cao, Controllable synthesis of pomelo peel-based aerogel and its application in adsorption of oil/organic pollutants, R. Soc. Open Sci. 6 (2019). https://doi.org/10.1098/rsos.181823

[58] R. Lin, A. Li, T. Zheng, L. Lu, Y. Cao, Hydrophobic and flexible cellulose aerogel as an efficient, green and reusable oil sorbent, RSC Adv. 5 (2015) 82027–82033. https://doi.org/10.1039/c5ra15194e

[59] M.E. Argun, D. Güclü, M. Karatas, Adsorption of Reactive Blue 114 dye by using a new adsorbent: Pomelo peel, J. Ind. Eng. Chem. 20 (2014) 1079–1084. https://doi.org/10.1016/j.jiec.2013.06.045

[60] X. Ye, Z. Chen, S. Ai, J. Zhang, B. Hou, Q. Zhou, F. Wang, H. Liu, S. Cui, Mechanical and thermal properties of reticulated SiC aerogel composite prepared by template method, J. Compos. Mater. 53 (2019) 4117–4124. https://doi.org/10.1177/0021998319851190

[61] B. Wu, G. Zhu, A. Dufresne, N. Lin, Fluorescent Aerogels Based on Chemical Crosslinking between Nanocellulose and Carbon Dots for Optical Sensor, ACS Appl. Mater. Interfaces. 11 (2019) 16048–16058. https://doi.org/10.1021/acsami.9b02754

[62] K.J. De France, T. Hoare, E.D. Cranston, Review of Hydrogels and Aerogels Containing Nanocellulose, Chem. Mater. 29 (2017) 4609–4631. https://doi.org/10.1021/acs.chemmater.7b00531

[63] Z.L. Wu, Z.X. Liu, Y.H. Yuan, Carbon dots: Materials, synthesis, properties and approaches to long-wavelength and multicolor emission, J. Mater. Chem. B. 5 (2017) 3794–3809. https://doi.org/10.1039/c7tb00363c

[64] W. Liu, C. Li, Y. Ren, X. Sun, W. Pan, Y. Li, J. Wang, W. Wang, Carbon dots: Surface engineering and applications, J. Mater. Chem. B. 4 (2016) 5772–5788. https://doi.org/10.1039/c6tb00976j

[65] S.Y. Lim, W. Shen, Z. Gao, Carbon quantum dots and their applications, Chem. Soc. Rev. 44 (2015) 362–381. https://doi.org/10.1039/c4cs00269e

[66] D. Lu, L. Su, H. Wang, M. Niu, L. Xu, M. Ma, H. Gao, Z. Cai, X. Fan, Scalable Fabrication of Resilient SiC Nanowires Aerogels with Exceptional High-Temperature Stability, ACS Appl. Mater. Interfaces. (2019) acsami.9b16811. https://doi.org/10.1021/acsami.9b16811

[67] L. Su, H. Wang, M. Niu, X. Fan, M. Ma, Z. Shi, S.W. Guo, Ultralight, Recoverable, and High-Temperature-Resistant SiC Nanowire Aerogel, ACS Nano. 12 (2018) 3103–3111. https://doi.org/10.1021/acsnano.7b08577

[68] G. Li, M. Zhu, W. Gong, R. Du, A. Eychmüller, T. Li, W. Lv, X. Zhang, Boron Nitride Aerogels with Super-Flexibility Ranging from Liquid Nitrogen Temperature to 1000 °C, Adv. Funct. Mater. 29 (2019) 1–7. https://doi.org/10.1002/adfm.201900188

[69] T. Woignier, A. HafidiAlaoui, J. Primera, J. Phalippou, Mechanical Properties of Aerogels : Brittle or Plastic Solids?, Key Eng. Mater. 391 (2008) 27–44. https://doi.org/10.4028/www.scientific.net/kem.391.27

[70] C.A. García-González, T. Budtova, L. Durães, C. Erkey, P. Del Gaudio, P. Gurikov, M. Koebel, F. Liebner, M. Neagu, I. Smirnova, An opinion paper on aerogels for

biomedical and environmental applications, Molecules. 24 (2019) 1–15.
https://doi.org/10.3390/molecules24091815

[71] K. Kamiuto, T. Miyamoto, S. Saitoh, Thermal characteristics of a solar tank with aerogel surface insulation, Appl. Energy. 62 (1999) 113–123.
https://doi.org/10.1016/S0306-2619(99)00004-5

[72] E. Cuce, P.M. Cuce, C.J. Wood, S.B. Riffat, Toward aerogel based thermal superinsulation in buildings: A comprehensive review, Renew. Sustain. Energy Rev. 34 (2014) 273–299. https://doi.org/10.1016/j.rser.2014.03.017

[73] D.T. Fakult, G. Doktor-ingenieur, S. Suttiruengwong, Silica Aerogels and Hyperbranched Polymers as Drug Delivery Systems, Synthese. (2005).

[74] R. Media, Supercritical Fluids as Solvents and Reaction Media G. Brunner (editor) © 2004 Elsevier B.V. All rights reserved 39, Supercrit. Fluids as Solvents React. Media. (2004) 39–60.

[75] M.H.A. Alnaief, Process development for production of aerogels with controlled morphology as potential drug carrier systems, ProQuest Diss. Theses. (2011). https://doi.org/10.15480/882.1009

[76] Z. Ülker, D. Sanli, C. Erkey, Applications of Aerogels and Their Composites in Energy-Related Technologies, Supercrit. Fluid Technol. Energy Environ. Appl. (2014) 157–180. https://doi.org/10.1016/B978-0-444-62696-7.00008-3

[77] L. Zang, Energy Efficiency and Renewable Energy Through Nanotechnology, 2011. https://doi.org/10.1007/978-0-85729-638-2

[78] J. Mao, J. Iocozzia, J. Huang, K. Meng, Y. Lai, Z. Lin, Graphene aerogels for efficient energy storage and conversion, Energy Environ. Sci. 11 (2018) 772–799. https://doi.org/10.1039/c7ee03031b

Keyword Index

About the Editors

Dr. Inamuddin is currently working as Assistant Professor at the Department of Applied Chemistry, Aligarh Muslim University, Aligarh, India. He obtained Master of Science degree in Organic Chemistry from Chaudhary Charan Singh (CCS) University, Meerut, India, in 2002. He received his Master of Philosophy and Doctor of Philosophy degrees in Applied Chemistry from Aligarh Muslim University (AMU), India, in 2004 and 2007, respectively. He has extensive research experience in multidisciplinary fields of Analytical Chemistry, Materials Chemistry, and Electrochemistry and, more specifically, Renewable Energy and Environment. He has worked on different research projects as project fellow and senior research fellow funded by University Grants Commission (UGC), Government of India, and Council of Scientific and Industrial Research (CSIR), Government of India. He has received Fast Track Young Scientist Award from the Department of Science and Technology, India, to work in the area of bending actuators and artificial muscles. He has completed four major research projects sanctioned by University Grant Commission, Department of Science and Technology, Council of Scientific and Industrial Research, and Council of Science and Technology, India. He has published 171 research articles in international journals of repute and eighteen book chapters in knowledge-based book editions published by renowned international publishers. He has published 105 edited books with Springer (U.K.), Elsevier, Nova Science Publishers, Inc. (U.S.A.), CRC Press Taylor & Francis Asia Pacific, Trans Tech Publications Ltd. (Switzerland), IntechOpen Limited (U.K.), Wiley-Scrivener, (U.S.A.) and Materials Research Forum LLC (U.S.A). He is a member of various journals' editorial boards. He is also serving as Associate Editor for journals (Environmental Chemistry Letter, Applied Water Science and Euro-Mediterranean Journal for Environmental Integration, Springer-Nature), Frontiers Section Editor (Current Analytical Chemistry, Bentham Science Publishers), Editorial Board Member (Scientific Reports-Nature), Editor (Eurasian Journal of Analytical Chemistry), and Review Editor (Frontiers in Chemistry, Frontiers, U.K.) He is also guest-editing various special thematic special issues to the journals of Elsevier, Bentham Science Publishers, and John Wiley & Sons, Inc. He has attended as well as chaired sessions in various international and national conferences. He has worked as a Postdoctoral Fellow, leading a research team at the Creative Research Initiative Center for Bio-Artificial Muscle, Hanyang University, South Korea, in the field of renewable energy, especially biofuel cells. He has also worked as a Postdoctoral Fellow at the Center of Research Excellence in Renewable Energy, King Fahd University of Petroleum and Minerals, Saudi Arabia, in the field of polymer electrolyte membrane fuel cells and computational fluid dynamics of polymer electrolyte membrane fuel cells. He is a life member of the Journal of the Indian

Chemical Society. His research interest includes ion exchange materials, a sensor for heavy metal ions, biofuel cells, supercapacitors and bending actuators.

Dr. Tauseef Ahmad Rangreez is working as a postdoctoral fellow at National Institute of Technology, Srinagar, India. He completed his Ph.D in Applied Chemistry, from Aligarh Muslim University, Aligarh, India on the topic "Development of Nanostructure Organic-Inorganic Composite Materials based Sensors for Inorganic Pollutants". He worked as a Project Fellow under the UGC Funded Research Project entitled "Development of Nanostructured Conductive Organic Inorganic Composite Materials based sensors Functionalities for Organic and Inorganic Pollutants". He completed his Masters in Chemistry from Jamia Hamdard, New Delhi. He has published several research articles of international repute. He has edited books with Springer and Materials Research Forum LLC, U.S.A. His research interest includes ion exchange chromatography, development of nanocomposite sensors for heavy metals and biosensors.

Dr. Mohd Imran Ahamed received his Ph.D degree on the topic "Synthesis and characterization of inorganic-organic composite heavy metals selective cation-exchangers and their analytical applications", from Aligarh Muslim University, Aligarh, India in 2019. He has published several research and review articles in the journals of international recognition. He has also edited various books which are published by Springer, CRC Press Taylor & Francis Asia Pacific and Materials Research Forum LLC, U.S.A. He has completed his B.Sc. (Hons) Chemistry from Aligarh Muslim University, Aligarh, India, and M.Sc. (Organic Chemistry) from Dr. Bhimrao Ambedkar University, Agra, India. His research work includes ion-exchange chromatography, wastewater treatment, and analysis, bending actuator and electrospinning.

Dr. Rajender Boddula is currently working with Chinese Academy of Sciences-President's International Fellowship Initiative (CAS-PIFI) at National Center for Nanoscience and Technology (NCNST, Beijing). He obtained Master of Science in Organic Chemistry from Kakatiya University, Warangal, India, in 2008. He received his Doctor of Philosophy in Chemistry with the highest honours in 2014 for the work entitled "Synthesis and Characterization of Polyanilines for Supercapacitor and Catalytic Applications" at the CSIR-Indian Institute of Chemical Technology (CSIR-IICT) and Kakatiya University (India). Before joining National Center for Nanoscience and Technology (NCNST) as CAS-PIFI research fellow, China, worked as senior research associate and Postdoc at National Tsing-Hua University (NTHU, Taiwan) respectively in the fields of bio-fuel and CO_2 reduction applications. His academic honors include University Grants Commission National Fellowship and many merit scholarships, study-abroad fellowships from Australian Endeavour Research Fellowship, and CAS-PIFI. He has published many scientific articles in international peer-reviewed journals and has authored around twenty book chapters, and he is also serving as an editorial board member and a referee for reputed international peer-reviewed journals. He has published edited books with Springer (UK), Elsevier, Materials Research Forum LLC (USA), Wiley-Scrivener, (U.S.A.) and CRC Press Taylor & Francis group. His specialized areas of research are energy conversion and storage, which include sustainable nanomaterials, graphene, polymer composites, heterogeneous catalysis for organic transformations, environmental remediation technologies, photoelectrochemical water-splitting devices, biofuel cells, batteries and supercapacitors.

www.ingramcontent.com/pod-product-compliance
Lightning Source LLC
Chambersburg PA
CBHW071335210326

41597CB00015B/1456